Operator Theory
Advances and Applications
Vol. 92

Editor
I. Gohberg

Complete Second Order Linear Differential Equations in Hilbert Spaces

Alexander Ya. Shklyar

Birkhäuser Verlag
Basel · Boston · Berlin

Author's address:

Alexander Ya. Shklyar
Institute of Mathematics
Ukrainian Academy of Sciences
Tereshchenkivska str. 3
252 601 Kiev
Ukraine

1991 Mathematics Subject Classification 35G05, 35M99, 44-02, 47B99

A CIP catalogue record for this book is available from the Library of Congress, Washington D.C., USA

Deutsche Bibliothek Cataloging-in-Publication Data
Škljar, Aleksandr J.:
Complete second order linear differential equations in Hilbert spaces / Alexander Ya. Shklyar. - Basel ; Boston ; Berlin : Birkhäuser, 1997
(Operator theory ; Vol. 92)
ISBN-13:978-3-0348-9940-6 e-ISBN-13:978-3-0348-9187-5
DOI: 10.1007/978-3-0348-9187-5
NE: GT

Printed on acid-free paper produced from chlorine-free pulp. TCF ∞
Cover design: Heinz Hiltbrunner, Basel

ISBN 3-7643-5377-5
ISBN 0-8176-5377-5

9 8 7 6 5 4 3 2 1

To the memory
of my grandparents
Max Rabur and
Dora Vainerman

Contents

Chapter 4. Boundary-value problems on a finite segment

Part II. Initial data of solutions

Chapter 5. Boundary behaviour of an integral transform $R(t)$ as $t \to 0$ depending on the sub-integral measure

Chapter 6. Initial data of solutions

Part III. Extension, stability, and stabilization of weak solutions

Chapter 10. Stability and stabilization of weak solutions

Part IV. Boundary-value problems on a half-line

Chapter 11. The Dirichlet problem on a half-line

Chapter 12. The Neumann problem on a half-line

Preface

First order linear differential equations in Banach spaces

$$y'(t) + Ay(t) = 0 \tag{0.1}$$

and «incomplete» second order equations

$$y''(t) + By(t) = 0 \tag{0.2}$$

and related nonhomogeneous and nonlinear equations, as well as semigroups and cosine operator functions closely related to these equations, have been studied extensively during the second half of our century and have become to date a classical part of modern mathematical analysis (see, for example, the monographs [3, 57, 60, 92, 93, 131, 137, 150, 179, 211, 243, 268, 334, 375, 382] and the reviews [138, 180, 231, 352] and references therein).

The present monograph is an attempt to give a unified systematic theory of complete second order linear differential equations

$$y''(t) + Ay'(t) + By(t) = 0 \tag{0.3}$$

in Hilbert spaces, considering from a single point of view the following subjects: well-posedness of the Cauchy problem as well as the Dirichlet and Neumann problems; initial (boundary) conditions ensuring the solvability of initial (boundary) value problems; boundary behaviour and extension of solutions on a finite interval; stability and stabilization of solutions at infinity; boundary-value problems on a half-line. The theory is developed in a special but important case, which can be considered as model. We study equation (0.3) as an independent object without any references to equations (0.1) or (0.2).

Answers to all posed questions are given in the form of necessary and sufficient conditions. Particular emphasis is placed on new effects which do not take place for first order and incomplete second order equations.

To this end, we obtain some unexpected results relating to the spectral theory of pairs of operators and to the boundary behaviour of integral transforms (the latter results are considered as analogues of Tauberian theorems).

I hope that this monograph can also serve as an introductory course and a reference book in this subject. The presentation in the monograph is self-contained. The only prerequisite for its understanding is the knowledge of basic concepts and theorems on measure and integration, linear spaces, and linear operators in Banach and Hilbert spaces, according to the first course in functional analysis and operator theory. In particular, it is expected that the reader is familiar with the spectral theory for self-adjoint and, in general, normal operators in Hilbert space (recall that these operators do not need to be bounded or to have discrete spectrum). All necessary information may be found, for instance, in [287]; for an extended presentation see, for example, [2, 34, 75, 126, 141, 245, 355]. One of the new textbooks introducing all concepts, notations and results that are necessary for understanding this monograph is «Functional Analysis» (Vols. 1, 2) by Yu.M. Berezanskii, G.F. Us and Z.E. Sheftel just published in Birkhäuser's Operator Theory series.

A brief remark on notation. We will frequently use logical quantors; recall their meaning: «$\exists \ldots$» means «there exist(s)... such that»; «$\forall \ldots$» means «for an arbitrary (for all)...»; finally, «:» (used without a word before it) means «there holds the following condition:» («we have:»). The notation $\{x \in X | P(x)\}$ stands for the set of all those elements of X which satisfy the condition(s) $P(x)$.

I would like to emphasize a great impact on my investigations and the contents of the monograph made by Yu. M. Berezanskii and M. L. Gorbachuk.

My appreciation also to participants of seminars in the Institute of Mathematics of the National Academy of Sciences of Ukraine, especially to S. D. Eidelman, Yu. S. Samoilenko and G. F. Us, for useful discussions.

It is of great importance for me that S. G. Krein and J. A. Goldstein, authors of classical monographs on linear differential equations in Banach space, supported and encouraged this work.

My special thanks to M. I. Gekhtman for friendly help.

Kiev, February 1996 A. Ya. Shklyar

Introduction

Complete second order linear differential equations in Banach spaces

$$y''(t) + Ay'(t) + By(t) = 0 \tag{1}$$

have been extensively studied during the last thirty-five years. Nevertheless, as H.O. Fattorini remarked in the monograph «Second order linear differential equations in Banach spaces» (1985): «The theory of (1) is considerably more complicated than that of its incomplete counterpart $y''(t) + By(t) = 0$ and can hardly be said to be in definitive form» [93, p. 270].

In this work, we study systematically equation (1) in an important particular case. Namely, we consider equation (1) where A and B are arbitrary commuting normal operators in a separable Hilbert space H. We impose no other restrictions on operators A and B; in particular, we don't assume that one of operators A and B is subordinate to another or to its degree or that both operators A and B are polynomials of a certain operator.

The class of all equations (1) with commuting normal operators (c.n.o.) A and B which is of concern in the present work seems to be sufficiently wide. In particular, it covers naturally all equations of the form (1) where A and B are linear differential operators with constant complex-valued coefficients (l.d.o. with c.c.) in $H = L_2(R^m)$, for any $m \geq 1$. We believe that the class of all equations (1) with c.n.o. A and B can be considered as a model class of abstract second order linear operator differential equations.

A characteristic of this work is the following. For this important class of equations, we have obtained answers to all questions in the form of necessary and sufficient conditions. These conditions look quite simple and clear and can be applied immediately to some important specific situations. Some new effects arise which apparently have not been described earlier.

We were able to obtain all these results only by systematic application of the classical spectral theory of pairs of commuting normal operators in Hilbert space (see, for instance, [34]). Besides, in order to answer some questions connected with equation (1), we needed to obtain apparently new results on estimates for roots of second order polynomials (Chapter 1), on position in C^2 of the joint spectrum of commuting normal operators (Chapter 1), and also to introduce an apparently new integral transform of σ-finite measure $R(t)$ $(t \geq 0)$ in order to obtain some analogues of Tauberian theorems for it (Chapter 5).

Part I
Well-Posedness of Boundary-Value Problems

Introduction to Part I

Now let us explain the main purpose of the first part of the monograph. Earlier, many different sufficient conditions of unique solvability of the Cauchy problem for equation (1) have been obtained. In other words, one defined a certain space of initial data G, which depended on A and B (for instance: $G = (D(A) \cap D(B)) \times (D(A^2) \cap D(AB))$). Further, certain conditions on A and B were given. Then one showed that under these conditions on A and B, the Cauchy problem for (1) with initial conditions (i.c.) $y(0) = f_0$, $y'(0) = f_1$ is uniquely solvable for every $(f_0, f_1) \in G$.

Now, our main purpose is the following. We define a sufficiently wide space of boundary data G which depends on A and B. The question under study is: to find *all* equations (1) such that the Cauchy problem for (1) with boundary conditions (b.c.) $y(0) = f_0$, $y'(0) = f_1$ is uniquely solvable for every $(f_0, f_1) \in G$. In other words: to find a necessary and sufficient condition on A and B for equation (1) to have this property.

Such a question may be posed for other boundary-value problems as well. In the first part of the work, we answer this question for the Cauchy, Dirichlet, Neumann and the inverse Cauchy problems on a finite segment $[a, b]$, and for the Cauchy problem on a half-line $[a, +\infty)$ and the inverse Cauchy problem on a half-line $(-\infty, b]$ as well.

All these boundary-value problems are considered in the usual setting and in a «weak» setting as well.

Chapter 1
Joint spectrum of commuting normal operators and its position. Estimates for roots of second order polynomials. Definition of well-posedness of boundary-value problems

1 Introductory notes

For the sake of simplicity, we begin with an important particular case where A and B are arbitrary commuting self-adjoint operators (c.s.o.) in a separable Hilbert space H.

In some articles (see, for instance, [107–109, 133, 147–148, 169]) such an equation (1) was studied in the case of

$$A = P(C), \quad B = Q(C) \tag{1.1}$$

where C is a self-adjoint operator in H, P and Q are polynomials. In this special case, the c.s.o. A and B admit simultaneous spectral decomposition

$$A = \int\limits_{\sigma(C)} P(\xi) dE_C(\xi), \quad B = \int\limits_{\sigma(C)} Q(\xi) dE_C(\xi) \tag{1.2}$$

where E_C is the resolution of the identity of the operator C. The further analysis of (1) was based on (1.2) taking into account the behaviour of $(P(\xi), Q(\xi))$ for $\xi \in \sigma(C)$.

In the general case, c.s.o. A, B do not admit representations (1.1) and (1.2). We explain this with the example where A and B are self-adjoint l.d.o. with c.c. in $L_2(R^m)$, $m \geq 1$ (see [34]). For $m = 1$, i.e., if A and B are ordinary self-adjoint l.d.o. with c.c., (1.1) has general character. However for $m > 1$ condition (1.1) picks out a very small subclass from the class of all pairs of self-adjoint l.d.o. with c.c. in $L_2(R^m)$.

Example 1.1. Let $H = L_2(R^2)$, and let A and B be c.s.o. generated by the following differential expressions on $C_0^\infty(R^2)$:

$$\mathcal{A} = -a_1 \frac{\partial^2}{\partial x_1^2} - a_2 \frac{\partial^2}{\partial x_2^2} + a_3, \quad \mathcal{B} = -b_1 \frac{\partial^2}{\partial x_1^2} - b_2 \frac{\partial^2}{\partial x_2^2} + b_3, \tag{1.3}$$

where $a_i, b_i \in R^1$ $(i = 1, 2, 3)$ are constants. As will be proved after Example 1.2, A and B admit a representation (1.1) if and only if $a_1 b_2 = b_1 a_2$.

By von Neumann's theorem [291], arbitrary c.s.o. A and B admit a representation of the form (1.1) too. But in the general case P and Q are merely measurable functions. Let, for example,

$$A = -\frac{\partial^2}{\partial x_1^2}, \quad B = -\frac{\partial^2}{\partial x_2^2} \quad \text{in } L_2(R^2);$$

then the mapping $\sigma(C) \ni \xi \mapsto (P(\xi), Q(\xi))$ maps the ray $\{\xi | \xi \geq 0\}$ onto the right angle $\{(\lambda, \mu) | \lambda \geq 0,\ \mu \geq 0\} \subset R^2$ [291]. An attempt to analyze (1) taking into account the behaviour of such $P(\xi)$, $Q(\xi)$ on $\sigma(C)$ could apparently only complicate the problem.

That is why for general c.s.o. A and B we shall be based on another generalization of the theorem on the resolution of the identity of a self-adjoint operator.

2 Joint spectrum of operators

In order to answer the questions which were posed in «Introduction to Part I» for equation (1), the key concept for us will be the joint spectrum of operators A and B. Let us remind it. In what follows, $D(\cdot)$ denotes the domain of an operator, $\sigma(\cdot)$ denotes the spectrum of an operator, * denotes the adjoint operator. The notation $\mathcal{B}(P)$ stands for the set of all Borel subsets of P.

That self-adjoint operators A and B commute means, by definition, that their resolutions of the identity, E_A and E_B respectively, commute, i.e., for all Borel sets Δ_1, Δ_2 in R^1 : $E_A(\Delta_1)E_B(\Delta_2) = E_B(\Delta_2)E_A(\Delta_1)$. There exists a unique joint resolution of the identity of the operators A and B, that is a spectral measure E on Borel subsets of R^2 such that its values are orthogonal projectors on H, for all $f \in D(A) : Af = \int_{R^2} \lambda dE_{\lambda,\mu} f$, and for all $g \in D(B) : Bg = \int_{R^2} \mu dE_{\lambda,\mu} g$. For all Borel sets $\Delta_1, \Delta_2 \subseteq R^1 : E(\Delta_1 \times \Delta_2) = E_A(\Delta_1)E_B(\Delta_2)$. The support of this spectral measure $\operatorname{supp} E \subseteq R^2$ is called the joint spectrum of the operators A and B. It is denoted by $\sigma(A, B)$. See [34].

In what follows, we denote: $\square_k = \{(\lambda, \mu) |\ |\lambda| \leq k,\ |\mu| \leq k\}$; $E_k = E(\square_k)$.

By definition, if $\varphi(\lambda, \mu)$ is a continuous real-valued function on $\sigma(A, B)$ then $\varphi(A, B)$ is a self-adjoint operator in H with domain $D(\varphi(A, B)) = \{f \in H | \int_{\sigma(A,B)} |\varphi(\lambda, \mu)|^2 d(E_{\lambda,\mu} f, f) < +\infty\}$ such that for every $f \in D(\varphi(A, B))$:

$$\varphi(A, B) f \overset{\text{def}}{=} \int\limits_{\sigma(A,B)} \varphi(\lambda, \mu) dE_{\lambda,\mu} f = \lim_{k \to \infty} \int\limits_{\sigma(A,B) \cap \square_k} \varphi(\lambda, \mu) dE_{\lambda,\mu} f,$$

where the integrals converge in the space H. Furthermore, $\| \varphi(A, B) \| < +\infty$ if and only if

$$\sup_{(\lambda,\mu) \in \sigma(A,B)} |\varphi(\lambda, \mu)| < +\infty;$$

in this case

$$\| \varphi(A,B) \| = \sup_{(\lambda,\mu)\in\sigma(A,B)} |\varphi(\lambda,\mu)|.$$

Similarly to [34], one can prove the following theorem «on mapping of the joint spectrum».

Theorem 1.1. *Let f and g be real-valued continuous functions on R^m, let $T_1,\ldots,$ T_m be c.s.o. in H, and let $A = f(T_1,\ldots,T_m)$, $B = g(T_1,\ldots,T_m)$. Then $\sigma(A,B)$ is the closure of the set $\{(\lambda,\mu) \in R^2 | \lambda = f(t_1,\ldots,t_m),\ \mu = g(t_1,\ldots,t_m),\ (t_1,\ldots,$ $t_m) \in \sigma(T_1,\ldots,T_m)\}$.*

Corollary 1. *Let $A = f(T_1)$, $B = g(T_2)$. Then $\sigma(A,B)$ is the closure of the set $\{(\lambda,\mu) \in R^2 | \lambda = f(t_1),\ \mu = g(t_2),\ (t_1,t_2) \in \sigma(T_1,T_2)\}$.*

Corollary 2. *Let $A = f(T_1)$, $B = g(T_1)$. Then $\sigma(A,B)$ is the closure of the set $\{(\lambda,\mu) \in R^2 | \lambda = f(t),\ \mu = g(t),\ t \in \sigma(T_1)\}$.*

3 Position of the joint spectrum

It is obvious that $\sigma(A,B) \subseteq \sigma(A) \times \sigma(B)$. Moreover, the closure of $pr_{0\lambda}\sigma(A,B)$ is $\sigma(A)$; the closure of $pr_{0\mu}\sigma(A,B)$ is $\sigma(B)$.

If E_A, E_B were scalar measures then, what's more, $\operatorname{supp} E = \operatorname{supp}(E_A E_B)$ $= \operatorname{supp} E_A \times \operatorname{supp} E_B$. But in the present situation, when E_A, E_B are spectral measures, $\sigma(A,B)$ *may not coincide with* $\sigma(A) \times \sigma(B)$. For example, let $B = A$. Then for any $\Delta_1, \Delta_2 \in \mathcal{B}(R^1)$ such that $\Delta_1 \cap \Delta_2 = \emptyset$ we have: $E(\Delta_1 \times \Delta_2) =$ $E_A(\Delta_1)E_B(\Delta_2) = E_A(\Delta_1)E_A(\Delta_2) = 0$ by the definition of the resolution of the identity E_A. Thus, in this case $E(\cdot)$ is concentrated on the «diagonal» and $\sigma(A,B) = \{(\lambda,\mu) | \lambda \in \sigma(A),\ \mu = \lambda\}$. This was observed in [33, 34, 35].

Recall that an operator B is said to be subordinate to an operator A if $D(A) \subseteq$ $D(B)$ and for some $0 \le a, b < +\infty$:

$$\| Bx \|^2 \le a^2 \| Ax \|^2 + b^2 \| x \|^2 \qquad \text{for all } x \in D(A). \tag{1.4}$$

One says that B is strongly subordinate to the operator A if B is subordinate to A and for some $0 \le a < 1$, $0 \le b < +\infty$ (1.4) holds. The operators A and B are called equivalent if they are subordinate to each other.

Theorem 1.2. *For c.s.o. A and B the following conditions are equivalent:*

a) *B is subordinate to the operator A;*

b) $D(A) \subseteq D(B)$;

c) *for some* $0 < \mathrm{K},\ \mathrm{R} < +\infty$:

$$\sigma(A,B) \subseteq \{(\lambda,\mu) |\ |\lambda|^2 + |\mu|^2 \leq R^2\} \cup \{(\lambda,\mu) |\ |\mu| \leq \mathrm{K}|\lambda|\} \tag{1.5}$$

(see Fig. 1.1).

Moreover, $a_A(B) = \mathrm{K}_{\mathrm{inf}}$, *where* $a_A(B)$ *is equal to the infimum of a over all pairs* (a,b) *such that* $0 \leq a, b < +\infty$ *and* (1.4) *holds,* $\mathrm{K}_{\mathrm{inf}}$ *is equal to the infimum of* K *over all pairs* (K,R) *such that* $0 < \mathrm{K},\ \mathrm{R} < +\infty$ *and* (1.5) *holds.*

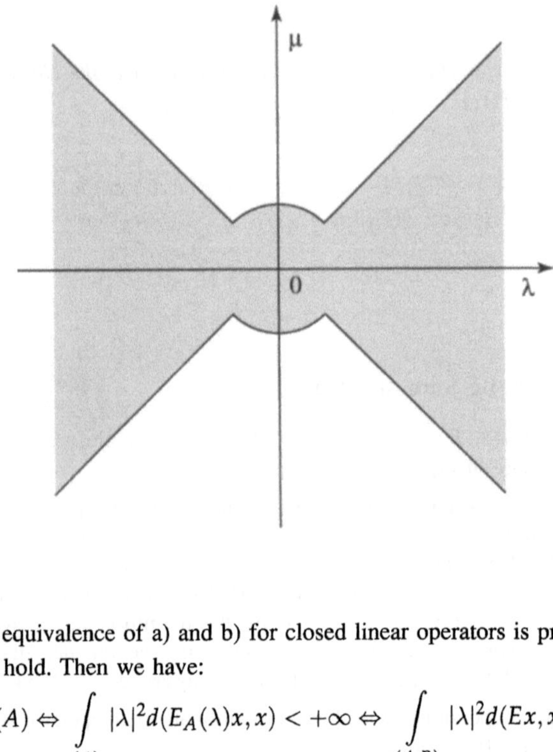

Fig. 1.1

Proof. The equivalence of a) and b) for closed linear operators is proved in [34].

Let c) hold. Then we have:

$$x \in D(A) \Leftrightarrow \int\limits_{\sigma(A)} |\lambda|^2 d(E_A(\lambda)x,x) < +\infty \Leftrightarrow \int\limits_{\sigma(A,B)} |\lambda|^2 d(Ex,x) < +\infty$$

$$\Rightarrow \int\limits_{\sigma(A,B)} |\mu|^2 d(Ex,x) \leq \int\limits_{\sigma(A,B)} (R^2 + K^2|\lambda|^2) d(Ex,x)$$

$$= R^2 \parallel x \parallel^2 + K^2 \parallel Ax \parallel^2 < +\infty$$

$$\Rightarrow x \in D(B) \text{ and } \parallel Bx \parallel^2 \leq R^2 \parallel x \parallel^2 + K^2 \parallel Ax \parallel^2 .$$

Thus, if c) holds then b) holds too and $a_A(B) \leq K$; so $a_A(B) \leq K_{\inf}$.

Conversely, assume that b) holds. Fix a $\gamma > 0$; we have: $H = H_1 \oplus H_2$, where $H_1 = E_A([-\gamma, \gamma])H$, $H_2 = E_A(R^1 \setminus [-\gamma, \gamma])H$, $H_1 \perp H_2$. For each of $i = 1, 2$, let $A \restriction H_i$ and $B \restriction H_i$ be the restrictions of A and B to H_i. Then

$$\sigma(A, B) = \cup_{i=1,2}\sigma(A \restriction H_i, B \restriction H_i).$$

The operator $A \restriction H_1$ is bounded, the operator $A \restriction H_2$ is invertible. For each of $i = 1, 2 : A \restriction H_i$ and $B \restriction H_i$ satisfy assumption b) on A and B (with H replaced by H_i).

We prove that b) implies c) for a bounded A with $\sigma(A) \subseteq [-\gamma, \gamma]$ and an invertible A with $\sigma(A) \subseteq R^1 \setminus (-\gamma, \gamma)$. Afterwards we substitute for A, B the operators $A \restriction H_1, B \restriction H_1$ and $A \restriction H_2$, $B \restriction H_2$, respectively.

If A is invertible, then $H \xrightarrow{A^{-1}} D(A) \xrightarrow{B} H$, since $D(A) \subseteq D(B)$. Thus, on all of H there is defined the linear operator BA^{-1}, which is closed as the composition of a bounded and a closed operators. By the Banach closed graph theorem, BA^{-1} is bounded. At the same time, this operator $BA^{-1} = \mu(A, B)\lambda^{-1}(A, B) = (\mu\lambda^{-1})(A, B)$ (see [34]); here $\varphi(\lambda, \mu) = \mu\lambda^{-1}$ is a continuous function on $\sigma(A, B)$. Thus, we have $\sup_{(\lambda,\mu)\in\sigma(A,B)} |\varphi(\lambda, \mu)| = \| \varphi(A, B) \| = \| BA^{-1} \| < +\infty$, i.e., $\exists K = \| BA^{-1} \| < +\infty \forall(\lambda, \mu) \in \sigma(A, B) : |\frac{\mu}{\lambda}| \leq K$. Moreover, if $\forall x \in D(A) : \| Bx \|^2 \leq a^2 \| Ax \|^2 + b^2 \| x \|^2$, then $\forall y \in H : \| BA^{-1}y \|^2 \leq a^2 \| y \|^2 + b^2 \| A^{-1}y \|^2 \leq (a^2 + b^2/\gamma^2) \| y \|^2$; consequently, as K^2 one can take $a^2 + b^2/\gamma^2$.

If A is bounded, then by the indicated Banach's theorem $D(B) \supseteq D(A)$ implies that B is bounded, and so $\sigma(A, B) \subseteq \sigma(A) \times \sigma(B)$ is bounded, i.e., $\exists R < +\infty : \sigma(A, B) \subseteq \{(\lambda, \mu) | \, |\lambda|^2 + |\mu|^2 \leq R^2\}$.

Returning to the original A and B satisfying b), we find that for them c) holds, where one can set $K \geq a$ arbitrarily close to a (namely: $K = \sqrt{a^2 + b^2/\gamma^2}$ for a sufficiently large γ and the corresponding R). That is, if b) holds then $K_{\inf} \leq a$; so $K_{\inf} \leq a_A(B)$. The theorem is proved.

Corollary 1. *Let A and B be c.s.o. The operator B is strongly subordinate to A if and only if for some $0 < K < 1$ and $0 < R < +\infty$ (1.5) holds.*

Corollary 2. *For c.s.o. A and B the following conditions are equivalent:*

a) *A is equivalent to B;*

b) *$D(A) = D(B)$;*

c) *for some $0 < R$, $K_1, K_2 < +\infty$:*
$\sigma(A, B) \subseteq \{(\lambda, \mu) | \, |\lambda|^2 + |\mu|^2 \leq R^2\} \cup \{(\lambda, \mu) | K_1|\lambda| \leq |\mu| \leq K_2|\lambda|\}$.

Let f, g be continuous real-valued functions on R^1. In Theorem 1.2 and its corollaries, one can replace A and (or) B by $f(A)$ and (or) $g(B)$ (in conditions a) and b)) and λ, μ by $f(\lambda)$, $g(\mu)$ (in the inequalities), correspondingly. This follows from Corollary 1 of Theorem 1.1.

Example 1.2. Let A, B be the c.s.o. from Example 1.1. Then $A = f(S,T)$, $B = g(S,T)$, where $f(s,t) = a_1 s + a_2 t + a_3$, $g(s,t) = b_1 s + b_2 t + b_3$, S, T are c.s.o. with $\sigma(S,T) = R_+ \times R_+$ (here $R_+ \stackrel{\text{def}}{=} [0, +\infty)$). It is seen from Theorem 1.1 («on mapping of the joint spectrum») that $\sigma(A,B)$ is the closure of the set $\{(a_1 s + a_2 t + a_3,\ b_1 s + b_2 t + b_3) | s,\ t \geq 0\}$. We restrict ourselves to the case of $a_3 = b_3 = 0$; passing to the general case, it is necessary to transfer $\sigma(A,B)$ by the vector (a_3, b_3) in the plane $R^2 = \{(\lambda, \mu)\}$. For $a_1 = a_2 = b_1 = b_2 = 0$ $\sigma(A,B)$ is the point $(0,0)$; otherwise it is a line or an angle less than π with vertex $(0,0)$ (which may degenerate into a ray). Up to interchange of a_1 by a_2 and, correspondingly, b_1 by b_2 the following cases are possible.

1. $a_1 = a_2 = 0$, $b_1^2 + b_2^2 > 0$. Then $\sigma(A,B) = \{0\} \times \sigma(B)$ lies on the axis 0μ and is a ray (directed upwards for $b_1 \geq 0$, $b_2 \geq 0$, downwards for $b_1 \leq 0$, $b_2 \leq 0$) or a line (for $b_1 b_2 < 0$).

2. $a_1 > 0$, $a_2 \geq 0$. Then $\sigma(A,B)$ is an angle lying in the right half-plane: for $a_2 \neq 0$ with sides $\{(\lambda,\mu) | \lambda \geq 0,\ \mu = (b_1/a_1)\lambda\}$ and $\{(\lambda,\mu) | \lambda \geq 0,\ \mu = (b_2/a_2)\lambda\}$, each of which is horizontal or sloping (in the special case of $b_1/a_1 = b_2/a_2$ we have: $B = (b_1/a_1)A$, and the angle degenerates into a ray); for $a_2 = 0$, $b_2 \neq 0$ one of the sides of the angle is $\{(\lambda,\mu) | \lambda \geq 0,\ \mu = (b_1/a_1)\lambda\}$, the second one is the upper (for $b_2 > 0$) or lower (for $b_2 < 0$) half-axis of 0μ; for $a_2 = 0$, $b_2 = 0$ the angle degenerates into the ray $\{(\lambda,\mu) | \lambda \geq 0,\ \mu = (b_1/a_1)\lambda\}$, here $B = (b_1/a_1)A$.

Notice that for $a_1 \geq 0$, $a_2 \geq 0$: $\sigma(A,B) = \sigma(A) \times \sigma(B)$ if and only if one of the following conditions holds:

1) $a_1 = a_2 = 0$;

2) $b_1 = b_2 = 0$;

3) $a_1 = b_2 = 0$;

4) $a_2 = b_1 = 0$.

In particular, $\sigma(A,B) = R_+ \times R^1$ is impossible for A, B from Example 1.1, although in the case of $a_1 > 0$, $a_2 \geq 0$ and $b_1 b_2 < 0$ one always has $\sigma(A) \times \sigma(B) = R_+ \times R^1$.

3. $a_1 < 0$, $a_2 \leq 0$. Then $\sigma(A,B)$ lies in the left half-plane. All the inferences of point 2 are preserved, with $\lambda \geq 0$ replaced by $\lambda \leq 0$.

4. $a_1 > 0$, $a_2 < 0$. If $b_1/a_1 = b_2/a_2$, then $\sigma(A,B) = \{(\lambda,\mu) | \lambda \in R^1,\ \mu = (b_1/a_1)\lambda\}$ is a horizontal or sloping line, and $B = (b_1/a_1)A$.

If $b_1/a_1 \neq b_2/a_2$, then $\sigma(A,B)$ is the smaller of the angles with sides $\{(\lambda,\mu) | \lambda \geq 0,\ \mu = (b_1/a_1)\lambda\}$ and $\{(\lambda,\mu) | \lambda \leq 0,\ \mu = (b_2/a_2)\lambda\}$.

Remark 1. These inferences about the properties of $\sigma(A,B)$ follow from the fact that $A = (a_1 s + a_2 t + a_3)(S,T)$, $B = (b_1 s + b_2 t + b_3)(S,T)$, where S, T are c.s.o., $\sigma(S,T) = R_+ \times R_+$, and are therefore preserved if realizing S and T as differential operators one puts R^{k+m} instead of R^2 and correspondingly replaces $\frac{\partial^2}{\partial x_1^2}$ in (1.3) by $\frac{\partial^2}{\partial x_1^2} + \cdots + \frac{\partial^2}{\partial x_k^2}$, $\frac{\partial^2}{\partial x_2^2}$ by $\frac{\partial^2}{\partial x_{k+1}^2} + \cdots + \frac{\partial^2}{\partial x_{k+m}^2}$.

Now we can prove that the operators A and B from Example 1.1 admit a representation (1.1) if and only if $a_1b_2 = b_1a_2$. The sufficiency is obvious: if $a_1b_2 = b_1a_2$ then one of the operators A and B is a first order polynomial of another. Conversely, let (1.1) hold; then, by Corollary 2 of Theorem 1.1, there exist polynomials P, Q such that $\sigma(A, B)$ is contained in the closure of the set $\{(\lambda, \mu) \in R^2 | \lambda = P(t),\ \mu = Q(t),\ t \in R^1\}$. In particular, there exists a finite number (namely, $p = \deg P$) of continuous real-valued functions $f_i(\lambda)$ on R^1 such that $\sigma(A, B) \subseteq \bigcup_{i=1}^{p} \{(\lambda, \mu) |\ \lambda \in R^1,\ \mu = f_i(\lambda)\}$. One can see from Example 1.2 that this is the case only if $a_1b_2 = b_1a_2$.

Remark 2. If we consider the general case where A, B are arbitrary commuting normal operators (c.n.o.) in a Hilbert space H, then all the definitions, notations and statements of Sections 2, 3 remain valid, with «self-adjoint» replaced by «normal»; R^1, R^2, R^m by C^1, C^2, C^m respectively; «real-valued» by «complex-valued». In particular, here $\sigma(A, B) \subseteq C^2$.

4 Estimates for roots of characteristic polynomials

Denote for each $(\lambda, \mu) \in C^2 : \lambda_1 = \operatorname{Re}\lambda,\ \lambda_2 = \operatorname{Im}\lambda,\ \mu_1 = \operatorname{Re}\mu,\ \mu_2 = \operatorname{Im}\mu$.

For each $(\lambda, \mu) \in C^2$, consider the characteristic polynomial $\omega^2 + \lambda\omega + \mu$ which corresponds to equation (1).

Denote by $\omega_{1,2}$ the roots of this characteristic polynomial; here

$$\omega_{1,2} = \omega_{1,2}(\lambda, \mu) = -\frac{\lambda}{2} \pm \sqrt{\frac{\lambda^2}{4} - \mu}.$$

We index ω_1, ω_2 so that $\operatorname{Re}\omega_1 \geq \operatorname{Re}\omega_2$, and if $\operatorname{Re}\omega_1 = \operatorname{Re}\omega_2$ then $\operatorname{Im}\omega_1 \geq \operatorname{Im}\omega_2$.

Lemma 1.1. *For each* $(\lambda, \mu) \in C^2$*:*

$$\frac{1}{2}\max(|\lambda|,\ \sqrt{|\mu|}) \leq \max(|\omega_1|,\ |\omega_2|) \leq 2\max(|\lambda|,\ \sqrt{|\mu|}),$$

$$\frac{1}{2}\min(|\lambda^{-1}\mu|,\ \sqrt{|\mu|}) \leq \min(|\omega_1|,\ |\omega_2|) \leq 2\min(|\lambda^{-1}\mu|,\ \sqrt{|\mu|}).$$

Proof. The second double inequality follows from the first one since $\max_{i=1,2} |\omega_i| \cdot \min_{i=1,2} |\omega_i| = |\omega_1| \cdot |\omega_2| = |\mu|$. In the first one, the left inequality is obvious:

$$|\lambda| = |\omega_1 + \omega_2| \leq 2\max_{i=1,2} |\omega_i|, \quad \sqrt{|\mu|} = \sqrt{|\omega_1\omega_2|} \leq \max_{i=1,2} |\omega_i|.$$

To prove the right one, we note that three cases are possible:

1) $\frac{1}{2}|\omega_2| \le |\omega_1| \le 2|\omega_2|$; then $\frac{1}{2}|\omega_1| \le |\omega_2| \le 2|\omega_1|$, $|\omega_1|^2 \le 2|\omega_1|\,|\omega_2| = 2|\mu|$, $|\omega_2|^2 \le 2|\mu|$;
2) $|\omega_1| \le \frac{1}{2}|\omega_2|$; then $|\lambda| = |\omega_1+\omega_2| \ge |\omega_2| - |\omega_1| \ge \frac{1}{2}|\omega_2| \Rightarrow |\omega_2| \le 2|\lambda|$, and hence $|\omega_1| \le 2|\lambda|$ as well;
3) $|\omega_1| \ge 2|\omega_2|$, i.e., $|\omega_2| \le \frac{1}{2}|\omega_1|$; this case is analogous to 2).

Lemma 1.2. *Let $\gamma \in R^1$, $(\lambda,\mu) \in R^2$. Then:*

a) $\max\limits_{i=1,2} \operatorname{Re}\omega_i(\lambda,\mu) < \gamma$ *if and only if* $\lambda > -2\gamma$, $\mu > -\gamma\lambda - \gamma^2$;
b) $\max\limits_{i=1,2} \operatorname{Re}\omega_i(\lambda,\mu) \ge -\gamma$ *if and only if* $(\lambda,\mu) \in R^2 \setminus \{(\lambda,\mu)|\lambda > 2\gamma,\ \mu > \gamma\lambda - \gamma^2\}$;
c) $\min\limits_{i=1,2} \operatorname{Re}\omega_i(\lambda,\mu) > -\gamma$ *if and only if* $\lambda < 2\gamma$, $\mu > \gamma\lambda - \gamma^2$;
d) $\min\limits_{i=1,2} \operatorname{Re}\omega_i(\lambda,\mu) \le \gamma$ *if and only if* $(\lambda,\mu) \in R^2 \setminus \{(\lambda,\mu)|\lambda < -2\gamma,\ \mu > -\gamma\lambda - \gamma^2\}$.

Proof. Let's prove a). We have: $\max\limits_{i=1,2} \operatorname{Re}\omega_i < \gamma \Leftrightarrow \operatorname{Re}\xi_i < 0$, $i = 1,2$, where ξ_i are the roots of the polynomial $P(\xi) = (\xi+\gamma)^2 + \lambda(\xi+\gamma) + \mu \equiv \xi^2 + (2\gamma + \lambda)\xi + (\gamma^2 + \gamma\lambda + \mu)$. Applying the Routh-Hurwitz criterion [118, 185] to $P(\xi)$, we obtain that $\operatorname{Re}\xi_i < 0$, $i = 1,2$, if and only if $\lambda + 2\gamma > 0$, $\mu + \gamma\lambda + \gamma^2 > 0$.

Further, $\min\limits_{i=1,2} \operatorname{Re}\omega_i > -\gamma \Leftrightarrow \max\limits_{i=1,2} \operatorname{Re}\omega_i' < \gamma$ where $\omega_i' = -\omega_i$, $i = 1,2$. Here ω_i' are the roots of the polynomial $(\omega_i')^2 - \lambda\omega' + \mu$. Applying a) to this polynomial, we get c).

Finally, b) follows from a), with γ replaced by $-\gamma$; d) follows from c) in the same way.

5 Definitions of well-posedness and weak well-posedness of boundary-value problems for equation (1)

We now return to equation (1).

Let $T \subseteq R^1$ be a finite or infinite closed, open, or half-open interval. Recall that $x(t) \in C^r(T,H)$ means that the vector-valued function $T \ni t \mapsto x(t) \in H$ is r times strongly continuously differentiable (i.e., it has r strong derivatives on T, the last of which is continuous on T in the strong sense). For $r = 0$, this means that the function $x(t)$ itself is continuous on T in the strong sense. The notation $C^r(T)$ stands for the space of all r times continuously differentiable on T complex-valued functions on T.

First of all, we give definitions of the usual well-posedness for boundary-value problems for equation (1) with c.n.o. A and B.

Definition 1.1. Let $T \subseteq R^1$ be an open or half-open interval or a segment on the real line R^1. A vector-valued function $y(t)$ $(t \in T)$ is said to be a solution of equation (1) on T if:

1) $y(t) \in C^2(T,H)$;
2) for each $t \in T$: $y(t) \in D(B)$, $y'(t) \in D(A)$, and moreover, $Ay'(t) \in C(T,H)$, $By(t) \in C(T,H)$;
3) $y''(t) + Ay'(t) + By(t) = 0$ for each $t \in T$.

Definition 1.1 follows S.G. Krein [179], H.O. Fattorini [93].

Definition 1.2. Let $-\infty < a < b < +\infty$. A solution $y(t)$ of equation (1) on $[a,b]$ is said to be:

- a solution of the Cauchy problem for (1) on $[a,b]$ with boundary (initial) data (f_0, f_1) if $y(a) = f_0$, $y'(a) = f_1$;
- a solution of the Dirichlet problem for (1) on $[a,b]$ with boundary data (f_0, f_1) if $y(a) = f_0$, $y(b) = f_1$;
- a solution of the Neumann problem for (1) on $[a,b]$ with boundary data (f_0, f_1) if $y'(a) = f_0$, $y'(b) = f_1$;
- a solution of the inverse Cauchy problem for (1) on $[a,b]$ with boundary (final) data (f_0, f_1) if $y(b) = f_0$, $y'(b) = f_1$.

Definition 1.3. Let $G \subseteq H \times H$. We say that a certain boundary-value problem for (1) on $[a,b]$ is *well-posed in the space* G if for an arbitrary $(f_0, f_1) \in G$ there exists a unique solution of this boundary-value problem for equation (1) on $[a,b]$ with boundary data (f_0, f_1).

We now go on to define *the weak well-posedness* for boundary-value problems for equation (1) with c.n.o. A and B.

Let Φ be a linear subspace of H which is dense in H, $\Phi \subseteq D(A) \cap D(B)$.

Definition 1.4. Let $T \subseteq R^1$ be an open or half-open interval or a segment on the real line R^1. A vector-valued function $y(t)$ $(t \in T)$ is said to be *a weak over* Φ *solution* of equation (1) on T if:

1) $y(t) \in C(T,H)$;
2) for each $g \in \Phi$: $(y(t),g) \in C^2(T)$, $(y(t),A^*g) \in C^1(T)$, $(y(t),B^*g) \in C(T)$;
3) for each $g \in \Phi$: $(y(t),g)'' + (y(t),A^*g)' + (y(t),B^*g) = 0$ for all $t \in T$.

Remark 1. All further results remain valid if we replace $[a,b]$ by (a,b) in points 2, 3 (but not in point 1) of the definition of a weak over Φ solution of equation (1) on $[a,b]$.

Denote by Φ' the space of all antilinear functionals on Φ. By virtue of Definition 1.4, if $y(t)$ is a weak over Φ solution of (1) on T then for each $\tau \in T$: $y(\tau) \in H$ and $(y(t),g)'|_{t=\tau}$ is an antilinear functional on $g \in \Phi$, i.e., an element of Φ'.

Definition 1.5. Let $-\infty < a < b < +\infty$. A weak over Φ solution of equation (1) on $[a, b]$ is said to be:

- a weak over Φ solution of the Cauchy problem for (1) on $[a, b]$ with boundary (initial) data $(f_0, f_1) \in H \times \Phi'$ if $y(a) = f_0$ and $\forall g \in \Phi : (y(t), g)'|_{t=a} = (f_1, g)$;
- a weak over Φ solution of the Dirichlet problem for (1) on $[a, b]$ with boundary data $(f_0, f_1) \in H \times H$ if $y(a) = f_0$, $y(b) = f_1$;
- a weak over Φ solution of the Neumann problem for (1) on $[a, b]$ with boundary data $(f_0, f_1) \in \Phi' \times \Phi'$ if $\forall g \in \Phi : (y(t), g)'|_{t=a} = (f_0, g)$, $(y(t), g)'|_{t=b} = (f_1, g)$;
- a weak over Φ solution of the inverse Cauchy problem for (1) on $[a, b]$ with boundary (final) data $(f_0, f_1) \in H \times \Phi'$ if $y(b) = f_0$ and $\forall g \in \Phi : (y(t), g)'|_{t=b} = (f_1, g)$.

It is obvious that a usual solution of a boundary-value problem for (1) on $[a, b]$ with boundary data (f_0, f_1) is a weak over Φ solution of this boundary-value problem for (1) on $[a, b]$ with the same boundary data.

Definition 1.6. Let $G \subseteq H \times \Phi'$. We say that a certain boundary-value problem for equation (1) on $[a, b]$ is *weakly over Φ well-posed in G* if for an arbitrary $(f_0, f_1) \in G$ there exists a unique weak over Φ solution (w.s.) of this boundary-value problem for (1) on $[a, b]$ with boundary data (b.d.) (f_0, f_1).

We preserve these definitions for the Cauchy problem in the case of $b = +\infty$ (in this case $[a, b] \stackrel{\text{def}}{=} [a, +\infty)$) and for the inverse Cauchy problem in the case of $a = -\infty$ (in this case $[a, b] \stackrel{\text{def}}{=} (-\infty, b]$).

For definiteness, in what follows we set

$$\Phi = D(A^\infty) \cap D(B^\infty) \stackrel{\text{def}}{=} \Big(\bigcap_{n=1}^{\infty} D(A^n)\Big) \cap \Big(\bigcap_{n=1}^{\infty} D(B^n)\Big) = \bigcap_{n=1}^{\infty} D((|A|^n + 1)\,(|B|^n + 1))$$

and omit the words «over Φ». In fact, all the results relating to the weak well-posedness of boundary-value problems for equation (1) which will be obtained in this part remain valid for an arbitrary Φ such that $\bigcup_{k=1}^{\infty} E_k H \subseteq \Phi \subseteq D(A^2) \cap D(B)$. For the other results of this monograph, analogous remarks hold.

Moreover, Theorem 2.1 on the weak well-posedness of the Cauchy problem for equation (1) in the case that A and B are commuting self-adjoint operators in H remains valid for $G = H \times H$ and $\Phi = D(A) \cap D(B)$ (see [312]). See also the remarks after Theorems 4.4 and 12.4, [313, 315].

6 Spaces of boundary data

In what follows, we study the weak well-posedness of boundary-value problems for equation (1) in $G = H \times H$ and in some wider spaces. Let us define these spaces following [31–33, 137].

Let C be a self-adjoint operator in H with $\inf\sigma(C) > 0$. Its domain $D(C)$ can be turned into a Hilbert space by introducing the following scalar product: for all $f,g \in D(C)$, $(f,g)_{H_+(C)} \stackrel{\text{def}}{=} (Cf, Cg)_H$. This Hilbert space $H_+(C)$ is dense and continuously imbedded in H.

Further, C^{-1} is a bounded operator on H. We introduce a new scalar product in H by the formula: $(f,g)_{H_-(C)} = (C^{-1}f, C^{-1}g)_H$ for all $f,g \in H$. Then we complete H in the norm generated by this scalar product, and denote the obtained Hilbert space by $H_-(C)$. H is dense and continuously imbedded in $H_-(C)$. C^{-1} is an isometric isomorphism of H onto $H_+(C)$ and an isometric isomorphism of $H_-(C)$ onto H.

$H_-(C)$ can also be considered as the dual space of $H_+(C)$. Namely, for any $f \in H_-(C)$, $g \in H_+(C)$ the action of f on g is defined by the formula: $(f,g) = (C^{-1}f, Cg)_H$. Note that $H_+(C)$ is said to be the positive space and $H_-(C)$ the negative space constructed by the operator C in H.

Suppose that a certain linear bounded operator on H can be extended up to a linear bounded operator from $H_-(C)$ into H which acts on all of $H_-(C)$. Then for the latter operator we retain the same notation as for the original operator, which acts on H.

This construction holds for an arbitrary self-adjoint operator C in H such that $\inf\sigma(C) > 0$. In particular, let $\varphi(\lambda,\mu)$ be a continuous function on $\mathbb{C}^2$ such that

$$\exists 0<K<+\infty, n_1\in\mathbb{N}, n_2\in\mathbb{N}\,\forall(\lambda,\mu)\in\mathbb{C}^2 : 1\le\varphi(\lambda,\mu)\le K(|\lambda|^{n_1}+|\mu|^{n_2}+1),$$

and let $C = \varphi(A,B)$. Then $H_+(\varphi(A,B))$ is $D(\varphi(A,B))$ with the scalar product $(f,g)_{H_+(\varphi(A,B))} = \int_{\sigma(A,B)} |\varphi(\lambda,\mu)|^2 d(E_{\lambda,\mu}f,g)$ for all $f,g \in D(\varphi(A,B))$. Obviously $\Phi = D(A^\infty) \cap D(B^\infty) = \bigcap_{n=1}^{\infty} D((|A|^n+1)(|B|^n+1)) \subseteq H_+(\varphi(A,B))$. Moreover, Φ is dense in $H_+(\varphi(A,B))$, since $\bigcup_{k=1}^{\infty} E_k H \subseteq \Phi$ is dense in $H_+(\varphi(A,B))$.

Further, $H_-(\varphi(A,B))$ is the completion of H in the norm

$$\| g \|_{H_-(\varphi(A,B))} = \| \varphi^{-1}(A,B)g \| = \left(\int_{\sigma(A,B)} \frac{1}{|\varphi(\lambda,\mu)|^2} d(E_{\lambda,\mu}g,g) \right)^{1/2}$$

On the other hand, $H_-(\varphi(A,B))$ can be interpreted as the space of all antilinear continuous functionals on $H_+(\varphi(A,B))$. For any $f \in H_-(\varphi(A,B))$ and any $g \in H_+(\varphi(A,B))$: $(f,g) \stackrel{\text{def}}{=} (\varphi^{-1}(A,B)f, \varphi(A,B)g)$. In particular, for any $f \in H_-(\varphi(A,B))$ and any $g \in \Phi = D(A^\infty)\cap D(B^\infty) \subseteq H_+(\varphi(A,B))$: $(f,g) = (\varphi^{-1}(A,B)f, \varphi(A,B)g)$. In this sense, $H_-(\varphi(A,B)) \subseteq \Phi'$.

In what follows, H_1 stands for $H_+(\varphi(A,B))$, H_{-1} for $H_-(\varphi(A,B))$ where $\varphi(\lambda,\mu) = |\lambda| + \sqrt{|\mu|} + 1$. It is clear that $H_1 = D(A) \cap D(|B|^{1/2})$ and $H \times H \subseteq H \times H_-(|A|+1) \subseteq H \times H_-((|A|^2+|B|+1)^{1/2}) = H \times H_{-1}$ as sets.

$H_1 = H_+((|A|^2 + |B| + 1)^{1/2}) = H_+((\max(|\lambda|, \sqrt{|\mu|}) + 1)(A,B)) = H_+((\max_{i=1,2} |\omega_i(\lambda,\mu)| + 1)(A,B))$ as topological spaces, by virtue of Lemma 1.1.

By H_2 we denote $H_+(\varphi(A,B))$ and by H_{-2} we denote $H_-(\varphi(A,B))$ where $\varphi(\lambda,\mu) = \min(|\lambda^{-1}\mu|, \sqrt{|\mu|}) + 1$. Here H_2 coincides with the linear span of $D(BA^{-1}) \cup D(|B|^{1/2})$ as a set.

Note that $H_2 = H_+((\min(|\lambda^{-1}\mu|, \sqrt{|\mu|}) + 1)(A,B)) = D((\min_{i=1,2} |\omega_i(\lambda,\mu)| + 1)(A,B))$ as topological spaces, by virtue of Lemma 1.1.

In the general case of c.n.o. A and B in H we have: $H_1 \subseteq H_2 \subseteq H$ and, correspondingly, $H \subseteq H_{-2} \subseteq H_{-1}$. Here $H_2 = H$ (and, correspondingly, $H_{-2} = H$) if and only if $D(A) \subseteq D(B)$. Further, $H_2 = H_1$ (and, correspondingly, $H_{-2} = H_{-1}$) if and only if $D(B) \subseteq D(A^2)$; in this case $H_1 = H_2 = D(|B|^{1/2})$, $H_{-2} = H_{-1} = H_-(|B|^{1/2} + 1)$. This follows from Theorem 1.2 for c.n.o. A and B, the remark to it, and Lemma 1.1.

In the general case of c.n.o. A and B in H, $\Phi = D(A^\infty) \cap D(B^\infty) \subseteq D(A^2) \cap D(B) \subseteq D(A) \cap D(|B|^{1/2}) = H_1 \subseteq H_2 \subseteq H \subseteq H_{-2} \subseteq H_{-1} \subseteq \Phi'$.

7 (Weak) well-posedness and uniform (weak) well-posedness

Let us now say some words on the relation between weak well-posedness and uniform weak well-posedness.

Definition 1.7. Let $G \subseteq H \times \Phi'$, and let G be a Hilbert space. A certain boundary-value problem for (1) on $[a,b]$ is said to be uniformly weakly well-posed in G if it is weakly well-posed in G in the sense of Definition 1.6, and moreover, for every finite segment T such that $T \subseteq [a,b]$ there exists $C_T < +\infty$ such that for each $(f_0, f_1) \in G$ and the weak solution $y(t)$ of this boundary-value problem for (1) on $[a,b]$ with boundary data (f_0, f_1) : $\| y(t) \| \leq C_T \| (f_0, f_1) \|_G$ for all $t \in T$.

Statement 1.1. In what follows, if we show that a certain boundary-value problem for (1) on $[a,b]$ is weakly well-posed in a Hilbert space $G \subseteq H \times \Phi'$, then it is uniformly weakly well-posed as well.

The similar results are valid for the «usual» well-posedness, with $\| y(t) \|$ in the definition replaced by $\max_{i=0,1,2} \| y^{(i)}(t) \|$. For G in the corresponding theorems to be Hilbert spaces, we consider $D(B)$ as a Hilbert space with such a norm that $\forall f \in D(B) : \| f \|^2_{D(B)} = \| Bf \|^2 + \| f \|^2$; $D(A) \cap D(|B|^{1/2})$ as a Hilbert space with such a norm that $\forall f \in D(A) \cap D(|B|^{1/2}) : \| f \|^2_{D(A) \cap D(|B|^{1/2})} = \| Af \|^2 + \| |B|^{1/2} f \|^2 + \| f \|^2$ (i.e., we consider $D(A) \cap D(|B|^{1/2})$ as $H_+((|A|^2 + |B| + 1)^{1/2})$); $D(A^2) \cap D(B)$ as a Hilbert space with such a norm that $\forall f \in D(A^2) \cap D(B) : \| f \|^2_{D(A^2) \cap D(B)} = \| A^2 f \|^2 + \| Bf \|^2 + \| f \|^2$ (i.e., we consider $D(A^2) \cap D(B)$ as $H_+((|A|^4 + |B|^2 + 1)^{1/2})$).

Chapter 2
Well-posedness of boundary-value problems for equation (1) in the case of commuting self-adjoint A and B

1 The Cauchy problem

Let $-\infty < a < b \leq +\infty$.

For each $(\lambda, \mu) \in C^2$ denote by $\psi_i(\lambda, \mu, t)$ $(t \in [a, b])$, $i = 0, 1$, solutions of the scalar ordinary differential equation (o.d.e.)

$$u''(t) + \lambda u'(t) + \mu u(t) = 0 \tag{2.1}$$

on $[a, b]$ such that $\psi_0(\lambda, \mu, a) = 1$, $\psi_0'(\lambda, \mu, a) = 0$, $\psi_1(\lambda, \mu, a) = 0$, $\psi_1'(\lambda, \mu, a) = 1$. Here for any $t \in [a, b]$:

$$\psi_0(\lambda, \mu, t) = \frac{\omega_1 e^{\omega_2(t-a)} - \omega_2 e^{\omega_1(t-a)}}{\omega_1 - \omega_2},$$

$$\psi_1(\lambda, \mu, t) = \frac{e^{\omega_1(t-a)} - e^{\omega_2(t-a)}}{\omega_1 - \omega_2}$$

for $(\lambda, \mu) \in C^2$ such that $\omega_1 \neq \omega_2$, and $\psi_i(\lambda, \mu, t)$, $i = 0, 1$, can be extended by continuity to the set $\{(\lambda, \mu) | \ \omega_1 = \omega_2\} = \{(\lambda, \mu) | \mu = \frac{\lambda^2}{4}\} \subset C^2$.

In what follows, $\psi_i^{(k)}(\lambda, \mu, t)$ stands for the k-th derivative of $\psi_i(\lambda, \mu, t)$ with respect to t at the point (λ, μ, t). Obviously $\psi_i^{(k)}(\lambda, \mu, t)$ is continuous on $C^2 \times [a, b]$, for each $k \geq 0$.

Theorem 2.1. *Let A and B be c.s.o. in H, and let $H \times H \subseteq G \subseteq H \times H_{-1}$. The Cauchy problem for equation* (1) *on* $[a, b]$ *is weakly well-posed in G if and only if one of the following equivalent conditions holds:*

1) $\exists 0 < \gamma < +\infty \ \forall (\lambda, \mu) \in \sigma(A, B) : \ \max_{i=1,2} \operatorname{Re} \omega_i(\lambda, \mu) \leq \gamma;$ (2.2)

2) $\exists 0 < \gamma < +\infty : \ \sigma(A, B) \subseteq \{(\lambda, \mu) | \lambda \geq -2\gamma, \ \mu \geq -\gamma\lambda - \gamma^2\}$ (Fig. 2.1); (2.3)

3) *A is semibounded below, $B_- \stackrel{\text{def}}{=} B E_B((-\infty, 0))$ (the negative part of B) is subordinate to A (i.e., $D(A) \subseteq D(B_-)$).*

In this case, for every $(f_0, f_1) \in H \times H_{-1}$ the weak solution of the Cauchy problem for (1) *on* $[a,b]$ *with boundary data (b.d.)* (f_0, f_1) *has the form*

$$y(t) = \sum_{i=0}^{1} \psi_i(A, B, t) f_i \quad (t \in [a,b]).$$

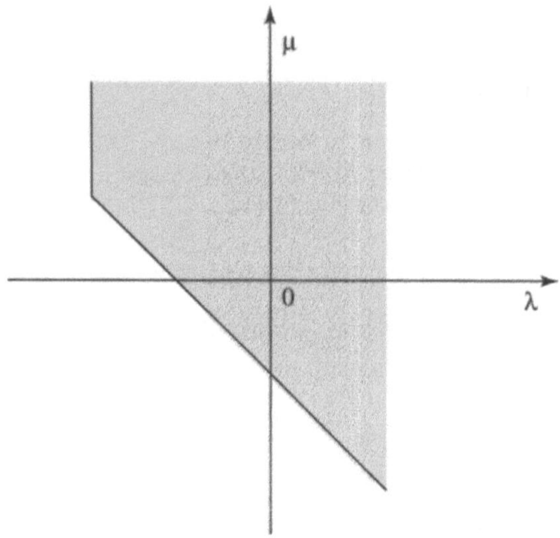

Fig. 2.1

Proof. The equivalence of conditions 1) and 2) follows from Lemma 1.2; the equivalence of conditions 2) and 3) follows from Theorem 1.2.

We now give the proof of Theorem 2.1 in the case that $[a,b] = [0, +\infty)$; for any other $[a,b]$ the proof is quite similar.

At first, we prove a lemma on uniqueness of a weak solution. The proof is based on the Holmgren principle and analogous to the proofs of Lemma 6.2 in [33, Chapter 2] and Lemma 5.1 in [137, Chapter 3].

Lemma 2.1. *If there exists a weak solution (w.s.) of equation* (1) *on* $R_+ = [0, +\infty)$ *with initial data (i.d.)* $(f_0, f_1) \in H \times H_{-1}$, *then it is unique and has the form*

$$y(t) = \lim_{k \to \infty} \sum_{i=0}^{1} \psi_i(A, B, t) E_k f_i \quad (t \in R_+).$$

Proof. Denote this w.s. by $y(t)$. It suffices to prove that $\forall k \geq 1 : E_k y(t) = \sum_{i=0}^{1} \psi_i(A, B, t) E_k f_i \ (t \in R_+)$. But $E_k y(t)$ is a w.s. on R_+ of the equation in $E_k H$ $y'' + Ay' + By = 0$, with i.d. $\{E_k f_i\}_{i=0}^{1}$. Here, $A \restriction E_k H$, $B \restriction E_k H$ are bounded c.s.o. on $E_k H$, therefore $\sum_{i=0}^{1} \psi_i(A, B, t) E_k f_i \ (t \in R_+)$ is a solution (and so a weak solution) of this equation on R_+ with i.d. $\{E_k f_i\}_{i=0}^{1}$. It remains to show that this weak solution is unique. Thus, it suffices to prove Lemma 2.1 for bounded A, B and $f_0 = f_1 = 0$.

We have to prove that for bounded c.s.o. A and B on H: $y(t) \in C(R_+, H)$, $y(0) = 0$, and $\forall g \in H : (y(t), g)'' \in C(R_+)$, $(y(t), g)'' + (y(t), Ag)' + (y(t), Bg) = 0$ $(t \in R_+)$, $(y'(0), g) \overset{\text{def}}{=} (y(t), g)'|_{t=0} = 0$ imply that $y(t) \equiv 0 \ (t \in R_+)$. Note that here $y'(t) \in H$, $y''(t) \in H \ (t \in R_+)$ as weak limits of antilinear continuous functionals on H (by virtue of a corollary from the Banach-Steinhaus theorem).

Denote by $\eta_1(\lambda, \mu, t)$ a solution of the scalar ordinary differential equation $\eta_1'' - \lambda \eta_1' + \mu \eta_1 = 0$ on R^1 such that $\eta_1(\lambda, \mu, 0) = 0$, $\eta_1'(\lambda, \mu, 0) = 1$. Fix arbitrary $g \in H$ and $s > 0$. Then

$$(\eta_1(A, B, t - s)g)'' - A(\eta_1(A, B, t - s)g)' + B\eta_1(A, B, t - s)g = 0, \quad 0 \leq t \leq s, \tag{2.4}$$

with $\eta_1(A, B, t - s)g|_{t=s} = 0$, $(\eta_1(A, B, t - s)g)'|_{t=s} = g$.

Now we multiply the left-hand side of (2.4) by $y(t)$ and integrate this scalar product over t from 0 to s:

$$\int_0^s (y(t), (\eta_1(A, B, t - s)g)'')dt - \int_0^s (y(t), (\eta_1(A, B, t - s)Ag)')dt + \int_0^s (y(t), \eta_1(A, B, t - s)Bg)dt = 0.$$

Let's integrate by parts. We have the right to do this: if $v(t)$, $u(t)$ are vector-valued functions with values in H, $v(t)$ is over H weakly continuously differentiable on $[0, s]$, and $u(t)$ is strongly continuously differentiable on $[0, s]$, then there exists $(v, u)' = (v', u) + (v, u')$ at each $t \in [0, s]$, with (v, u'), (v', u) being continuous on $[0, s]$ (this follows from the Banach-Steinhaus theorem). Now y, y' play the role of $v(t)$; $(\eta_1(A, B, t - s)h)^{(j)}$, $j = 0, 1$, $h \in \{g, Ag\}$, play the role of $u(t)$.

After the integration by parts, we obtain (η_1 stands for $\eta_1(A, B, t-s)$):

$$\int_0^s (y(t), (\eta_1 g)'' - A(\eta_1 g)' + B\eta_1 g)dt$$

$$= \int_0^s (y''(t) + Ay'(t) + By(t), \eta_1 g)dt + [(y(t), (\eta_1 g)')$$

$$- (y'(t), \eta_1 g) - (y(t), A\eta_1 g)]|_0^s = 0.$$

Let's analyze the last equality. By the definition of a weak solution, $(y''(t) + Ay'(t) + By(t), h) = 0$, $\forall h \in H$, $t \geq 0$. By the assumption, $y(0) = y'(0) = 0$. By the definition of $\eta_1(\lambda, \mu, t)$, $\eta_1 g|_{t=s} = \eta_1(A, B, 0)g = 0$, $(\eta_1 g)'|_{t=s} = (\eta_1(A, B, t)g)'|_{t=0} = g$. Thus, the last equality can be rewritten as $(y(s), g) = 0$. Since $s > 0$ and $g \in H$ are arbitrary, we conclude that $y(t) \equiv 0$ $(t \geq 0)$. The proof of Lemma 2.1 is over.

We now prove the *sufficiency* in Theorem 2.1. It remains to show that if (2.2) holds then for each $(f_0, f_1) \in H \times H_{-1} : y(t) = \psi_0(A, B, t)f_0 + \psi_1(A, B, t)f_1$ $(t \in R_+)$ is a weak solution of (1) on R_+ with i.d. (f_0, f_1). To show this, we give the following chain of lemmas.

Lemma 2.2. *Let* (2.2) *hold. Then* $\forall 0 < T < +\infty \exists C_T < +\infty$:

$$|\psi_0(\lambda, \mu, t)| \leq C_T, \; |\psi_1(\lambda, \mu, t)| \leq \frac{C_T}{|\lambda| + \sqrt{|\mu|} + 1}, \qquad \forall (\lambda, \mu) \in \sigma(A, B), \quad t \in [0, T]. \tag{2.5}$$

Proof. At first, we estimate $|\frac{e^{(\omega_2-\omega_1)t}-1}{(\omega_2-\omega_1)t}|$. For each $(\lambda, \mu) \in \sigma(A, B)$: $\operatorname{Re}\omega_2 \leq \operatorname{Re}\omega_1$; so, for any $t \geq 0$ such that $|\omega_2 - \omega_1|t \leq 1$:

$$\left|\frac{e^{(\omega_2-\omega_1)t} - 1}{(\omega_2 - \omega_1)t}\right| \leq \max_{z \in C^1, |z| \leq 1} \left|\frac{e^z - 1}{z}\right| = \max_{z \in C^1, |z| \leq 1} \left|\sum_{k=1}^{\infty} \frac{z^{k-1}}{k!}\right|$$

$$\leq \max_{z \in C^1, |z| \leq 1} \sum_{k=1}^{\infty} \frac{|z|^{k-1}}{k!} = e - 1 < 2;$$

for any $t \geq 0$ such that $|\omega_2 - \omega_1|t > 1$:

$$\left|\frac{e^{(\omega_2-\omega_1)t} - 1}{(\omega_2 - \omega_1)t}\right| \leq |e^{(\omega_2-\omega_1)t} - 1| \leq 2.$$

Summarizing,

$$\left|\frac{e^{(\omega_2-\omega_1)t}-1}{(\omega_2-\omega_1)t}\right| \le 2, \quad \forall(\lambda,\mu)\in\sigma(A,B), \quad t\ge 0.$$

Fix an arbitrary $0<T<+\infty$. For any $(\lambda,\mu)\in\sigma(A,B)$, $t\in[0,T]$:

$$|\psi_1(\lambda,\mu,t)| = t|e^{\omega_1 t}|\left|\frac{e^{(\omega_2-\omega_1)t}-1}{(\omega_2-\omega_1)t}\right| \le 2Te^{\gamma T}.$$

Further, $\sigma(A,B)\subseteq R^2$. Therefore, for each $(\lambda,\mu)\in\sigma(A,B)$ one of the following conditions 1), 2) holds:

1) $\omega_1\in R^1$ and $\omega_2\in R^1$, or: $\omega_1=\bar{\omega}_2$ and $|\frac{\operatorname{Im}\omega_1}{\operatorname{Re}\omega_1}|\le 1$; then $\forall t\in[0,T]$:

$$\left|\frac{\omega_1(e^{\omega_2 t}-e^{\omega_1 t})}{\omega_1-\omega_2}\right| = |\omega_1 t e^{\omega_1 t}|\cdot\left|\frac{e^{(\omega_2-\omega_1)t}-1}{(\omega_2-\omega_1)t}\right| \le$$

$$\sqrt{2}|\operatorname{Re}\omega_1 t e^{\operatorname{Re}\omega_1 t}|\cdot\left|\frac{e^{(\omega_2-\omega_1)t}-1}{(\omega_2-\omega_1)t}\right| \le 2\sqrt{2}(e^{-1}+\gamma T e^{\gamma T});$$

further,

$$\left|\frac{\omega_2(e^{\omega_2 t}-e^{\omega_1 t})}{\omega_1-\omega_2}\right| = \left|\frac{\omega_1(e^{\omega_2 t}-e^{\omega_1 t})}{\omega_1-\omega_2}-(e^{\omega_2 t}-e^{\omega_1 t})\right| \le 2\sqrt{2}(e^{-1}+\gamma T e^{\gamma T}+e^{\gamma T});$$

2) $\omega_1=\bar{\omega}_2$ and $|\frac{\operatorname{Im}\omega_1}{\operatorname{Re}\omega_1}|>1$; then

$$|\operatorname{Im}\omega_1| = \frac{1}{2}|\omega_1-\omega_2|, \quad \left|\frac{\omega_1}{\omega_1-\omega_2}\right| = \frac{\sqrt{|\operatorname{Re}\omega_1|^2+|\operatorname{Im}\omega_1|^2}}{2|\operatorname{Im}\omega_1|} \le \frac{\sqrt{2}}{2},$$

and

$$\left|\frac{\omega_2}{\omega_1-\omega_2}\right|\cdot|e^{\omega_1 t}-e^{\omega_2 t}| = \left|\frac{\omega_1}{\omega_1-\omega_2}\right|\cdot|e^{\omega_1 t}-e^{\omega_2 t}| \le \sqrt{2}e^{\gamma T}.$$

Taking into account that $\forall(\lambda,\mu)\in\sigma(A,B)$, $t\in[0,T]$:

$$|\psi_0(\lambda,\mu,t)| = \left|e^{\omega_1 t}+\frac{\omega_1(e^{\omega_2 t}-e^{\omega_1 t})}{\omega_1-\omega_2}\right| \le e^{\gamma T}+\left|\frac{\omega_1(e^{\omega_2}-e^{\omega_1 t})}{\omega_1-\omega_2}\right|,$$

$$(|\lambda|+\sqrt{|\mu|}+1)|\psi_1(\lambda,\mu,t)| \le 4\max(|\omega_1|,|\omega_2|)\left|\frac{e^{\omega_1 t}-e^{\omega_2 t}}{\omega_1-\omega_2}\right| + |\psi_1(\lambda,\mu,t)|$$

(by Lemma 1.1), we obtain that there exists $C_T<+\infty$ such that (2.5) holds.

Lemma 2.3. *Fix an arbitrary* $0 < T < +\infty$. *Suppose that there exists* $C_T < +\infty$ *such that* (2.5) *holds. Then* $\forall(\lambda,\mu) \in \sigma(A,B)$, $t \in [0,T]$:

$$|\psi_0^{(j)}(\lambda,\mu,t)|\,(j=0,1,2), |\lambda\psi_0'(\lambda,\mu,t)|, |\mu\psi_0(\lambda,\mu,t)| \le 2C_T(|\mu|+1),$$

$$|\psi_1^{(j)}(\lambda,\mu,t)|\,(j=0,1,2), |\lambda\psi_1'(\lambda,\mu,t)|, |\mu\psi_1(\lambda,\mu,t)| \le 2C_T(|\lambda|+\sqrt{|\mu|}+1).$$

Proof. Indeed, $\forall(\lambda,\mu,t) \in \sigma(A,B) \times [0,T]$:

$$|\mu\psi_0| \le C_T|\mu|; \quad |\psi_0'| = |\mu\psi_1| \le C_T|\mu|; \quad |\lambda\psi_0'| = |\mu\lambda\psi_1| \le C_T|\mu|;$$

$$|\mu\psi_1| \le C_T\sqrt{|\mu|}; \quad \psi_1' = -\lambda\psi_1 + \psi_0; \quad |\psi_1'| \le |\lambda|\,|\psi_1| + |\psi_0| \le 2C_T;$$

$$|\lambda\psi_1'| \le 2C_T|\lambda|; \quad |\psi_0''| = |\mu\psi_1'| \le 2C_T|\mu|;$$

$$|\psi_1''| \le |\lambda|\,|\psi_1'| + |\psi_0'| = |\lambda|\,|\psi_1'| + |\mu|\,|\psi_1| \le 2C_T(|\lambda| + \sqrt{|\mu|}).$$

Lemma 2.4. *Let* $\forall 0 < T < +\infty \exists C_T < +\infty$ *such that* (2.5) *is valid. Then:*

1) *for any* $g \in D(B)$, $y(t) = \psi_0(A,B,t)g$ *is a solution of* (1) *on* R_+ *such that* $y(0) = g$, $y'(0) = 0$, *with* $y^{(j)}(t) = \psi_0^{(j)}(A,B,t)g$, $j = 0,1,2$, *for all* $t \in R_+$; *if, moreover,* $g \in D(A)$ *then* $(\psi_0(A,B,t)Ag)' = A(\psi_0(A,B,t)g)' \in C(R_+,H)$;

2) *for any* $g \in D(A) \cap D(\sqrt{|B|})$, $y(t) = \psi_1(A,B,t)g$ *is a solution of* (1) *on* R_+ *such that* $y(0) = 0$, $y'(0) = g$, *with* $y^{(j)}(t) = \psi_1^{(j)}(A,B,t)g$, $j = 0,1,2$, *for all* $t \in R_+$; *moreover,* $(\psi_1(A,B,t)Ag)' = A(\psi_1(A,B,t)g)' \in C(R_+,H)$.

Proof. To prove Lemma 2.4, we check immediately the fulfilment of the definitions. It follows from the fact that $\psi_i(A,B,t)g \stackrel{\text{def}}{=} \int_{\sigma(A,B)} \psi_i(\lambda,\mu,t)dE_{\lambda,\mu}g$ $(t \in R_+)$, $i = 0,1$, and the Lebesgue theorem, taking into account:

1) the estimates of $|\psi_i^{(j)}(\lambda,\mu,t)|$ $(j = 0,1,2)$, $|\lambda\psi_i'(\lambda,\mu,t)|$, $|\mu\psi_i(\lambda,\mu,t)|$, $i = 0,1$, by majorizing functions on $\sigma(A,B)$, obtained in Lemma 2.3; for an arbitrary $0 < T < +\infty$, these estimates hold uniformly in $t \in [0,T]$;

2) the integrability of these majorizing functions with respect to the measure $d(Eg,g)$:
for any $g \in D(B)$, $\int_{\sigma(A,B)}(|\mu|+1)^2 d(Eg,g) < +\infty$;
for any $g \in D(A) \cap D(\sqrt{|B|})$, $\int_{\sigma(A,B)}(|\lambda| + \sqrt{|\mu|} + 1)^2 d(Eg,g) < +\infty$;

3) the continuity of $\psi_i^{(j)}(\lambda,\mu,t)$ $(j = 0,1,2,\ i = 0,1)$ on R_+ for each $(\lambda,\mu) \in \sigma(A,B)$.

That $y(t) = \psi_k(A,B,t)g$, $k = 0,1$, satisfy the initial conditions now follows from the fact that $\forall(\lambda,\mu) \in \sigma(A,B) : \psi_0(\lambda,\mu,0) = 1$, $\psi_0'(\lambda,\mu,0) = 0$, $\psi_1(\lambda,\mu,0) = 0$, $\psi_1'(\lambda,\mu,0) = 1$.

Lemma 2.5. *Let* $\forall 0 < T < +\infty \exists C_T < +\infty$ *such that* (2.5) *is valid. Then for any* $(f_0, f_1) \in H \times H_{-1} : y(t) = \psi_0(A, B, t)f_0 + \psi_1(A, B, t)f_1$ *is a weak solution of the Cauchy problem for* (1) *on* R_+ *with i.d.* (f_0, f_1).

Remark. Sometimes we shall omit the words «the Cauchy problem for».

*Proof.*Let us show that for any $f_0 \in H : y(t) = \psi_0(A, B, t)f_0$ is a weak solution of (1) on R_+ with i.d. $(f_0, 0)$. The fact that

$$\psi_0(A, B, t)f_0 \overset{\text{def}}{=} \int\limits_{\sigma(A,B)} \psi_0(\lambda, \mu, t)dE_{\lambda,\mu}f_0 \in C(R_+, H)$$

follows from the Lebesgue theorem, since for any $0 < T < +\infty$:

1) $\exists C_T < +\infty : |\psi_0(\lambda, \mu, t)| \leq C_T$ for all $(\lambda, \mu) \in \sigma(A, B)$, $0 \leq t \leq T$;
2) $f_0 \in H \Rightarrow \int_{\sigma(A,B)} C_T^2 d(Ef_0, f_0) < +\infty$;
3) $\psi_0(\lambda, \mu, t)$ is continuous on $[0, T]$, for each $(\lambda, \mu) \in \sigma(A, B)$.

Consider any $g \in \Phi = D(A^\infty) \cap D(B^\infty) \subseteq D(A^2) \cap D(B)$. Taking into account Lemmas 2.3, 2.4, we have:

$$\begin{aligned}(y(t), g)'' = (\psi_0(A, B, t)f_0, g)'' &= (f_0, \psi_0(A, B, t)g)'' \\ &= (f_0, (\psi_0(A, B, t)g)'') \in C(R_+),\end{aligned}$$

similarly

$$(y(t), Ag)' = (f_0, \psi_0(A, B, t)Ag)' \in C(R_+),$$

$$(y(t), Bg) = (f_0, B\psi_0(A, B, t)g) \in C(R_+);$$

$$\begin{aligned}&(y(t), g)'' + (y(t), Ag)' + (y(t), Bg) \\ &\quad = (f_0, (\psi_0(A, B, t)g)'' + A(\psi_0(A, B, t)g)' + B\psi_0(A, B, t)g) \equiv 0;\end{aligned}$$

$$y(0) = \psi_0(A, B, 0)f_0 = f_0; \quad (y(t), g)'|_{t=0} = (f_0, (\psi_0(A, B, t)g)'|_{t=0}) = 0.$$

Since this is true for an arbitrary $g \in \Phi$, $y(t) = \psi_0(A, B, t)f_0$ is a weak solution of (1) on R_+ with i.d. $(f_0, 0)$.

We can prove analogously that for any $f_1 \in H_{-1} : y(t) = \psi_1(A, B, t)f_1$ is a weak solution of (1) on R_+ with i.d. $(0, f_1)$. To do this, we notice that for any $f_1 \in H_{-1} : f_1' = (|A| + \sqrt{|B|} + 1)^{-1}f_1 \in H$; for any $g \in D(A^2) \cap D(B)$: $g_1 = (|A| + \sqrt{|B|} + 1)g \in D(A) \cap D(\sqrt{|B|})$. Now, just as previously, we can show that

$$y(t) = \psi_1(A, B, t)f_1 = (\psi_1(\lambda, \mu, t)\,(|\lambda| + \sqrt{|\mu|} + 1))\,(A, B)f_1' \in C(R_+, H);$$

$(y(t), g)'' = (f_1', \psi_1(A, B, t)g_1)'' \in C(R_+)$ and so on.

The proof of Lemma 2.5 is completed, and with it the proof of the sufficiency in Theorem 2.1.

We now go on to prove the *necessity* in Theorem 2.1. We must prove that if the Cauchy problem for equation (1) on R_+ is weakly well-posed in $G = H \times H$ then (2.2) holds, i.e., $\exists 0 < \gamma < +\infty \forall (\lambda, \mu) \in \sigma(A, B) : \operatorname{Re} \omega_1 \leq \gamma$.

Lemma 2.6. *Let the Cauchy problem for* (1) *on* R_+ *be weakly well-posed in* $G = H \times H$. *Then* $\forall 0 < T < +\infty \exists C_T < +\infty$ *such that* $|\psi_0(\lambda, \mu, t)| \leq C_T$, $|\psi_1(\lambda, \mu, t)| \leq C_T$, $\forall (\lambda, \mu) \in \sigma(A, B)$, $t \in [0, T]$.

Proof. For an arbitrary $0 < T < +\infty$ we prove that

$$\sup_{(\lambda,\mu)\in\sigma(A,B),t\in[0,T]} |\psi_0(\lambda, \mu, t)| < +\infty;$$

one can prove that

$$\sup_{(\lambda,\mu)\in\sigma(A,B),t\in[0,T]} |\psi_1(\lambda, \mu, t)| < +\infty$$

similarly.

Since for every $f_0 \in H$ there exists a unique weak solution of (1) on R_+ with i.d. $(f_0, 0)$, and it has the form $y(t) = \lim_{k\to\infty} \psi_0(A, B, t)E_k f_0 = \psi_0(A, B, t) f_0$ $(t \in R_+)$ by Lemma 2.1, we conclude that a normal and consequently closed operator $\psi_0(A, B, t)$ is defined on the whole H, for any $t \geq 0$. By the Banach closed graph theorem, $\psi_0(A, B, t)$ is a linear bounded operator on H, for any $t \geq 0$.

Now fix an arbitrary $0 < T < +\infty$. For every $f_0 \in H : y(t) = \psi_0(A, B, t) f_0$ $\in C(R_+, H)$, since it is a weak solution of (1) on R_+. $\psi_0(A, B, t) f_0 \in C([0, T], H)$ implies that there exists $C_{T,f_0} < +\infty$ such that $\| \psi_0(A, B, t) f_0 \| \leq C_{T,f_0}$ for all $0 \leq t \leq T$. Since this is valid for any $f_0 \in H$, we deduce from the Banach-Steinhaus theorem that there exists $C_T < +\infty$ such that $\| \psi_0(A, B, t) \| \leq C_T$, $\forall 0 \leq t \leq T$. But for any $0 \leq t \leq T : \| \psi_0(A, B, t) \| = \sup_{(\lambda,\mu)\in\sigma(A,B)} |\psi_0(\lambda, \mu, t)|$.

We obtain finally:

$$\sup_{0\leq t\leq T} \sup_{(\lambda,\mu)\in\sigma(A,B)} |\psi_0(\lambda, \mu, t)| \leq C_T,$$

i.e., $|\psi_0(\lambda, \mu, t)| \leq C_T$ for all $(\lambda, \mu) \in \sigma(A, B)$, $t \in [0, T]$.

We now complete the proof of the necessity in Theorem 2.1. Let the Cauchy problem for (1) on R_+ be weakly well-posed in $G = H \times H$. For any $f \in D(\omega_1(A, B)) : f \in H$, $\omega_1(A, B) f \in H$; thus, (1) has a w.s. on R_+ with i.d. $(f, \omega_1(A, B) f)$. By Lemma 2.1, this w.s. has the form: $y(t) = \lim_{k\to\infty} e^{\omega_1 t}(A, B) E_k f$ $(t \in R_+)$. So, $\forall f' \in H : (e^{\omega_1 t}(|\omega_1|^2 + 1)^{-1/2})(A, B) f'$ is a weak solution of (1) on R_+. Hence, analogously to Lemma 2.6, we conclude that

$$\forall 0 < T < +\infty \exists C'_T < +\infty : |e^{\omega_1 t}(|\omega_1|^2 + 1)^{-1/2}| \leq C'_T,$$
$$\forall (\lambda, \mu) \in \sigma(A, B), \quad t \in [0, T].$$

Taking into account this fact and Lemma 2.6, we obtain:

$$\forall 0 < T_1 < T_2 < +\infty \exists C = C(T_1, T_2) < +\infty : \int_{T_1}^{T_2} |\psi_0(\lambda, \mu, t)|^2 dt \le C,$$

$$\forall (\lambda, \mu) \in \sigma(A, B); \tag{2.6}$$

$$\forall 0 < T < +\infty \exists 0 < C(T) < +\infty : \frac{e^{2\operatorname{Re}\omega_1 T}}{(\operatorname{Re}\omega_1)^2 + (\operatorname{Im}\omega_1)^2 + 1} \le C(T) \Leftrightarrow$$

$$\Leftrightarrow \frac{(\operatorname{Re}\omega_1)^2 + (\operatorname{Im}\omega_1)^2 + 1}{e^{2\operatorname{Re}\omega_1 T}} \ge \frac{1}{C(T)}, \quad \forall (\lambda, \mu) \in \sigma(A, B). \tag{2.7}$$

Using (2.6) and (2.7) with $T_1 = 1$, $T_2 = 2$, $T = 1$, we show that $\exists 0 < \gamma < +\infty \forall (\lambda, \mu) \in \sigma(A, B) : \operatorname{Re}\omega_1 \le \gamma$. Indeed, (2.7) $\Rightarrow \exists 0 < \gamma_1 < +\infty$, $C_0 = \frac{1}{\sqrt{3C(1)}} > 0 : \sigma(A, B) = \Delta_1 \cup \Delta_2$, where $\Delta_1 = \{(\lambda, \mu) \in \sigma(A, B) | \operatorname{Re}\omega_1 \le \gamma_1\}$, $\Delta_2 = \{(\lambda, \mu) \in \sigma(A, B) | \operatorname{Re}\omega_1 > \gamma_1, \, |\operatorname{Im}\omega_1| \ge C_0 e^{\operatorname{Re}\omega_1}\}$. It remains to show that $\exists 0 < \gamma_3 < +\infty \forall (\lambda, \mu) \in \Delta_2 : \operatorname{Re}\omega_1 \le \gamma_3$.

Recall that $\forall (\lambda, \mu) \in R^2 : \omega_i \in R^1$ $(i = 1, 2)$ or $\omega_2 = \bar{\omega}_1$. Since $\omega_i \in R^1$ $(i = 1, 2)$ is impossible for $(\lambda, \mu) \in \Delta_2$, we conclude that $\forall (\lambda, \mu) \in \Delta_2 : \omega_2 = \bar{\omega}_1$. Therefore, $\forall (\lambda, \mu) \in \Delta_2 \exists a \ge \gamma_1 > 0$, $b \ge C_0 e^a > 0 : \omega_1 = a + ib$, $\omega_2 = a - ib$, whence $\forall t \in [1, 2] : \psi_0(\lambda, \mu, t) = e^{at}(\cos bt - \frac{a}{b} \sin bt)$. Now (2.6) implies that

$$C = C(1, 2) \ge \int_1^2 e^{2at}(\cos bt - \frac{a}{b} \sin bt)^2 dt \ge e^{2a} \int_1^2 (\cos bt - \frac{a}{b} \sin bt)^2 dt.$$

Here

$$\begin{aligned} I(a, b) &\stackrel{\text{def}}{=} \int_1^2 (\cos bt - \frac{a}{b} \sin bt)^2 dt \\ &= \frac{1}{2} + \frac{1}{4b}(\sin 4b - \sin 2b) + \frac{a}{2b^2}(\cos 4b - \cos 2b) \\ &\quad + \frac{a^2}{b^2}(\frac{1}{2} - \frac{1}{4b}(\sin 4b - \sin 2b)). \end{aligned}$$

Since $\forall (\lambda, \mu) \in \Delta_2 : a > 0$, $b \ge C_0 e^a$, it is easy to see that $\exists 0 < \gamma_2 < +\infty \forall (\lambda, \mu) \in \Delta_2$: if $a \ge \gamma_2$, then $I(a, b) \ge \frac{1}{4}$. Thus, for $(\lambda, \mu) \in \Delta_2 : a \ge \gamma_2 \Rightarrow a \le \ln(2\sqrt{C}) < +\infty$. But $a = \operatorname{Re}\omega_1$. As a result, $\forall (\lambda, \mu) \in \Delta_2 : \operatorname{Re}\omega_1 \le \gamma_3 = \max(\gamma_2, \ln(2\sqrt{C})) < +\infty$.

The theorem is proved.

Theorem 2.2. *Let A and B be c.s.o. in H, $G = D(B) \times (D(A) \cap D(|B|^{1/2})) = D(B) \times H_1$. The Cauchy problem for equation* (1) *on* $[a, b]$ *is well-posed in G if and only if one of the equivalent conditions* 1), 2), 3) *of Theorem* 2.1 *holds.*

Proof. We give the proof for $[a, b] = R_+$; otherwise the proof is quite similar. In fact, the sufficiency is proved (see Lemmas 2.1, 2.5). It remains to prove the necessity of (2.2) for the Cauchy problem for (1) on R_+ to be well-posed in $G = D(B) \times (D(A) \cap D(|B|^{1/2}))$. It follows from Theorem 3.2 below.

Example 2.1. Let A and B be the operators from Example 1.1. $\sigma(A, B)$ is found in Example 1.2. Now Theorems 2.1, 2.2 imply that the Cauchy problem for equation (1) on $[a, b]$ is weakly well-posed in $G = H \times H$ (weakly well-posed in $G = H \times H_{-1}$, well-posed in $G = D(B) \times H_1$) if and only if for each of $i = 1, 2$ there holds one of the following conditions:

a) $a_i > 0$;

b) $a_i = 0$, $b_i \geq 0$.

Finally, let's consider the (weak) well-posedness of the Cauchy problem on $[a, b]$ for equation (1) with c.s.o. A, B in particular cases which are closely related to «incomplete» equations $y''(t) + By(t) = 0$ and $y''(t) + Ay'(t) = 0$ respectively.

Corollary 1. *Let A and B be c.s.o. in H and, moreover, let A be bounded; let $H \times H \subseteq G \subseteq H \times H_{-1}$. The Cauchy problem for* (1) *on* $[a, b]$ *is weakly well-posed in G (well-posed in $D(B) \times H_1$) if and only if B is semibounded below.*

Proof. Indeed, let A be bounded; in this case (2.3) means that $\inf \sigma(B) > -\infty$.

Remark 1. Notice that in this case $H \times H_{-1} = H \times H_{-}(|B|^{1/2} + 1)$; $D(B) \times H_1 = D(B) \times D(|B|^{1/2})$.

Corollary 2. *Let A and B be c.s.o. in H and, moreover, let $D(A) \subseteq D(B)$; let $H \times H \subseteq G \subseteq H \times H_{-1}$. The Cauchy problem for* (1) *on* $[a, b]$ *is weakly well-posed in G (well-posed in $D(B) \times H_1$) if and only if A is semibounded below.*

Proof. Indeed, let $D(A) \subseteq D(B)$; by Theorem 1.2, this means that $\exists 0 < K$, $R < +\infty : \sigma(A, B) \subseteq \{(\lambda, \mu) \in R^2 | \lambda^2 + \mu^2 \leq R^2\} \cup \{(\lambda, \mu) \in R^2 |\ |\mu| \leq K|\lambda|\}$. In this case, (2.3) means that $\inf \sigma(A) > -\infty$.

Remark 2. Notice that in this case $H \times H_{-1} = H \times H_{-}(|A| + 1)$; $D(B) \times H_1 = D(B) \times D(A)$.

2 The Dirichlet problem

Let $-\infty < a < b < +\infty$.

For each $(\lambda,\mu) \in C^2$, except those (λ,μ) where $\omega_1 - \omega_2 = \frac{2\pi i l}{b-a}$, $l \in \mathbb{Z}\setminus\{0\}$ (i.e., where $\mu = \frac{\lambda^2}{4} + \frac{\pi^2 l^2}{(b-a)^2}$, $l \in \mathbb{N}$), denote by $\varphi_i(\lambda,\mu,t)$ $(t \in [a,b])$, $i = 0,1$, solutions of the scalar o.d.e. (2.1) on $[a,b]$ such that $\varphi_0(\lambda,\mu,a) = 1$, $\varphi_0(\lambda,\mu,b) = 0$, $\varphi_1(\lambda,\mu,a) = 0$, $\varphi_1(\lambda,\mu,b) = 1$. Here for any $t \in [a,b]$:

$$\varphi_0(\lambda,\mu,t) = \frac{e^{\omega_2(t-a)} - e^{\omega_2(b-a)}e^{\omega_1(t-b)}}{1 - e^{(\omega_2-\omega_1)(b-a)}},$$

$$\varphi_1(\lambda,\mu,t) = \frac{e^{\omega_1(t-b)} - e^{\omega_1(a-b)}e^{\omega_2(t-a)}}{1 - e^{(\omega_2-\omega_1)(b-a)}}$$

for $(\lambda,\mu) \in C^2 \setminus \{(\lambda,\mu)|\omega_1 - \omega_2 = \frac{2\pi i l}{b-a},\ l \in \mathbb{Z}\} = C^2 \setminus \{(\lambda,\mu)|\mu = \frac{\lambda^2}{4} + \frac{\pi^2 l^2}{(b-a)^2},\ l \in \mathbb{Z}\}$, and $\varphi_0(\lambda,\mu,t)$, $\varphi_1(\lambda,\mu,t)$ can be extended by continuity to the set $\{(\lambda,\mu)|\omega_1 = \omega_2\} = \{(\lambda,\mu)|\mu = \frac{\lambda^2}{4}\} \subset C^2$.

Theorem 2.3. *Let A and B be c.s.o. in H. The Dirichlet problem for equation* (1) *on $[a,b]$ is weakly well-posed in $G = H \times H$ if and only if one of the following equivalent conditions* 1), 2) *holds:*

1) *a)* $\exists 0 < \gamma < +\infty \forall (\lambda,\mu) \in \sigma(A,B)$:

$$\max_{i=1,2} \operatorname{Re}\omega_i(\lambda,\mu) \geq -\gamma, \quad \min_{i=1,2} \operatorname{Re}\omega_i(\lambda,\mu) \leq \gamma \tag{2.8}$$

b) $\exists 0 < \varepsilon < +\infty \forall (\lambda,\mu) \in \sigma(A,B) : |(\omega_1 - \omega_2) - \frac{2\pi i l}{b-a}| \geq \varepsilon$ *for all* $l \in \mathbb{Z}\setminus\{0\}$; (2.9)

2) $\exists 0 < \gamma,\ \varepsilon < +\infty : \sigma(A,B) \subseteq (\{(\lambda,\mu)|\ |\lambda| \leq 2\gamma\} \cup \{(\lambda,\mu)|\ |\lambda| > 2\gamma,\ \mu \leq \gamma|\lambda| - \gamma^2\}) \setminus \bigcup_{l=1}^{\infty}\{(\lambda,\mu)|\mu - \frac{\lambda^2}{4} \in (\frac{\pi^2 l^2}{(b-a)^2} - \varepsilon l,\ \frac{\pi^2 l^2}{(b-a)^2} + \varepsilon l)\}$ (Fig. 2.2).

In this case, for every $(f_0, f_1) \in H \times H$ the w.s. of the Dirichlet problem for (1) *on $[a,b]$ with b.d. (f_0, f_1) has the form*

$$y(t) = \sum_{i=0}^{1} \varphi_i(A,B,t)f_i, \quad t \in [a,b].$$

The equivalence of conditions 1) and 2) follows from Lemma 1.2. The further proof see in Theorem 4.1.

Theorem 2.4. *Let A and B be c.s.o. in H. The Dirichlet problem for equation* (1) *on $[a,b]$ is well-posed in $G = (D(B) \cap D(A^2)) \times (D(B) \cap D(A^2))$ if and only if one of the equivalent conditions* 1), 2) *of Theorem* 2.3 *holds.* (See Theorem 4.2.)

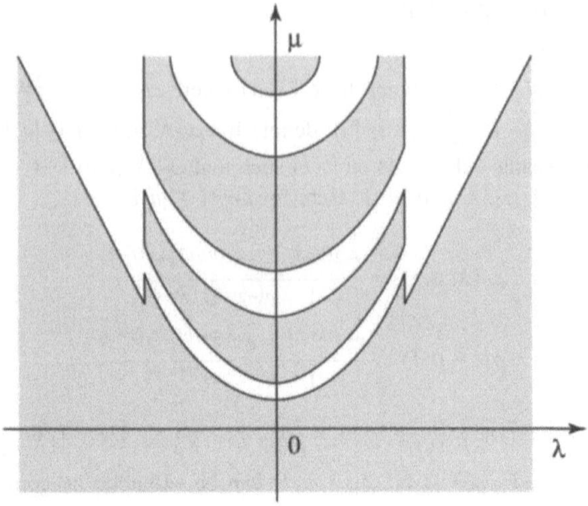

Fig. 2.2

Example 2.2. Let A and B be the operators from Example 1.1. $\sigma(A,B)$ is found in Example 1.2. Theorems 2.3, 2.4 imply that, with the exception of the trivial case of $a_1 = a_2 = b_1 = b_2 = 0$ and $b_3 - \frac{a_3^2}{4} \neq \frac{\pi^2 l^2}{(b-a)^2}$ for all $l \in \mathbb{N}$, the Dirichlet problem for equation (1) on $[a,b]$ is weakly well-posed in $G = H \times H$ (well-posed in $G = (D(B) \cap D(A^2)) \times (D(B) \cap D(A^2))$) if and only if simultaneously:

1) $k_3 = \frac{a_3^2}{4} + \frac{\pi^2}{(b-a)^2} - b_3 > 0$;
2) for each of $i = 1, 2$ there holds one of the following conditions:
   a) $a_i > 0$, $\frac{b_i}{a_i} < \frac{a_3}{2} + \sqrt{k_3}$,
   b) $a_i = 0$, $b_i \leq 0$,
   c) $a_i < 0$ $\frac{b_i}{a_i} > \frac{a_3}{2} - \sqrt{k_3}$;
3) a) if $a_1 > 0$ and $a_2 < 0$ then $\frac{b_1}{a_1} \leq \frac{b_2}{a_2}$;
   b) if $a_1 < 0$ and $a_2 > 0$ then $\frac{b_2}{a_2} \leq \frac{b_1}{a_1}$.

Here condition 1) means that the vertex of the angle $\sigma(A,B)$ lies in $R^2 = \{(\lambda, \mu)\}$ below the parabola $\mu = \frac{\lambda^2}{4} + \frac{\pi^2}{(b-a)^2}$. Condition 2) means that both of the sides of this angle lie below this parabola too. Condition 3) means that if the projection of the angle $\sigma(A,B)$ on the axis 0λ is the whole $R^1 = (-\infty, +\infty)$ then the interior angle lies below its sides or $\sigma(A,B)$ degenerates into a line.

Finally, let's consider the (weak) well-posedness of the Dirichlet problem on $[a,b]$ for equation (1) with c.s.o. A, B in some particular cases.

Corollary 1. *Let A and B be c.s.o. in H, and moreover, let A be bounded. Then* (2.8) *holds. The Dirichlet problem for* (1) *on* $[a,b]$ *is weakly well-posed in* $G = H \times H$ *(well-posed in* $G = D(B) \times D(B) = (D(B) \cap D(A^2)) \times (D(B) \cap D(A^2))$*) if and only if* (2.9) *is valid, i.e., for some* $0 < \varepsilon < +\infty$:

$$\sigma(A,B) \subseteq R^2 \setminus \bigcup_{l=1}^{\infty} \left\{ (\lambda,\mu) | \mu - \frac{\lambda^2}{4} \in \left(\frac{\pi^2 l^2}{(b-a)^2} - \varepsilon l, \frac{\pi^2 l^2}{(b-a)^2} + \varepsilon l \right) \right\}. \quad (2.10)$$

This follows from Lemma 1.2.

Corollary 2. *Let A and B be c.s.o. in H and, moreover, let* $B_+ \overset{\text{def}}{=} BE_B([0,+\infty))$ *(the nonnegative part of B) be subordinate to A (i.e.,* $D(A) \subseteq D(B_+)$*). Then* (2.8) *holds. The Dirichlet problem for* (1) *on* $[a,b]$ *is weakly well-posed in* $G = H \times H$ *(well-posed in* $G = (D(B) \cap D(A^2)) \times (D(B) \cap D(A^2))$*) if and only if* (2.9) *(i.e.,* (2.10)*) is valid.*

This follows from Theorem 1.2 and Lemma 1.2.

Corollary 3. *Let A and B be c.s.o. in H such that* $\sup \sigma(B) < \frac{\pi^2}{(b-a)^2}$*. Then the Dirichlet problem for* (1) *on* $[a,b]$ *is weakly well-posed in* $G = H \times H$ *(well-posed in* $G = (D(B) \cap D(A^2)) \times (D(B) \cap D(A^2))$*).*

Indeed, in this case condition 2) of Theorem 2.3 holds.

3 The Neumann problem

Let $-\infty < a < b < +\infty$.

For each $(\lambda,\mu) \in C^2$, except those (λ,μ) where $\omega_1 - \omega_2 = \frac{2\pi i l}{b-a}$, $l \in \mathbb{Z} \setminus \{0\}$ (i.e., $\mu = \frac{\lambda^2}{4} + \frac{\pi^2 l^2}{(b-a)^2}$, $l \in \mathbb{N}$) and those (λ,μ) where $\min_{i=1,2} |\omega_i| = 0$ (i.e., $\mu = 0$), denote by $\chi_i(\lambda,\mu,t)$ ($t \in [a,b]$), $i = 0,1$, solutions of the scalar o.d.e. (2.1) on $[a,b]$ such that $\chi_0'(\lambda,\mu,a) = 1$, $\chi_0'(\lambda,\mu,b) = 0$, $\chi_1'(\lambda,\mu,a) = 0$, $\chi_1'(\lambda,\mu,b) = 1$. Here for any $t \in [a,b]$:

$$\chi_0(\lambda,\mu,t) = \frac{\omega_2^{-1} e^{\omega_2(t-a)} - \omega_1^{-1} e^{\omega_2(b-a)} e^{\omega_1(t-b)}}{1 - e^{(\omega_2-\omega_1)(b-a)}},$$

$$\chi_1(\lambda,\mu,t) = \frac{\omega_1^{-1} e^{\omega_1(t-b)} - \omega_2^{-1} e^{\omega_1(a-b)} e^{\omega_2(t-a)}}{1 - e^{(\omega_2-\omega_1)(b-a)}}$$

for $(\lambda,\mu) \in C^2 \setminus (\{(\lambda,\mu) | \omega_1 - \omega_2 = \frac{2\pi i l}{b-a}, l \in \mathbb{Z}\} \cup \{(\lambda,\mu) | \min_{i=1,2} |\omega_i| = 0\}) = C^2 \setminus (\{(\lambda,\mu) | \mu = \frac{\lambda^2}{4} + \frac{\pi^2 l^2}{(b-a)^2}, l \in \mathbb{Z}\} \cup \{(\lambda,\mu) | \mu = 0\})$, and $\chi_i(\lambda,\mu,t)$, $i = 0,1$, can be extended by continuity to the set $\{(\lambda,\mu) | \omega_1 = \omega_2, \omega_1 \neq 0\} = \{(\lambda,\mu) | \mu = \frac{\lambda^2}{4}, \mu \neq 0\} \subset C^2$.

Theorem 2.5. *Let A and B be c.s.o. in H. The Neumann problem for equation* (1) *on* $[a,b]$ *is weakly well-posed in* $G = H_{-2} \times H_{-2}$ *if and only if one of the following equivalent conditions* 1), 2), 3) *holds:*

1) (2.8), (2.9) *are valid, and moreover,*

$$\exists 0 < \varepsilon_1 < +\infty \forall (\lambda, \mu) \in \sigma(A,B) : \min_{i=1,2} |\omega_i| \geq \varepsilon_1; \qquad (2.11)$$

2) $\exists 0 < \gamma, \varepsilon < +\infty : \sigma(A,B) \subseteq (\{(\lambda,\mu) |\, |\lambda| \leq 2\gamma\} \cup \{(\lambda,\mu) |\, |\lambda| > 2\gamma,\ \mu \leq \gamma|\lambda| - \gamma^2\}) \setminus (\bigcup_{l=1}^{\infty} \{(\lambda,\mu) | \mu - \frac{\lambda^2}{4} \in (\frac{\pi^2 l^2}{(b-a)^2} - \varepsilon l, \frac{\pi^2 l^2}{(b-a)^2} + \varepsilon l)\} \cup \{(\lambda,\mu) |\, |\lambda| \leq 2\gamma,\ |\mu| < \varepsilon\} \cup \{(\lambda,\mu) |\, |\lambda| > 2\gamma,\ |\mu| < \varepsilon|\lambda|\})$ (Fig. 2.3a);

3) *the Dirichlet problem for equation* (1) *on* $[a,b]$ *is weakly well-posed in* $G = H \times H$*, and moreover,* B^{-1} *is bounded (i.e.,* $0 \notin \sigma(B)$*) and* A *is subordinate to* B *(i.e.,* $D(B) \subseteq D(A)$*).*

In this case, for every $(f_0, f_1) \in H_{-2} \times H_{-2}$ *the w.s. of the Neumann problem for* (1) *on* $[a,b]$ *with b.d.* (f_0, f_1) *has the form*

$$y(t) = \sum_{i=0}^{1} \chi_i(A,B,t) f_i, \quad t \in [a,b].$$

See Theorem 4.4.

Theorem 2.6. *Let A and B be c.s.o. in H. The Neumann problem for equation* (1) *on* $[a,b]$ *is well-posed in* $G = (D(A) \cap D(|B|^{1/2})) \times (D(A) \cap D(|B|^{1/2})) = H_1 \times H_1$ *if and only if one of the following equivalent conditions* 1), 2), 3) *holds:*

1) (2.8), (2.9) *are valid, and moreover,*

$$\exists 0 < \varepsilon_1 < +\infty \forall (\lambda, \mu) \in \sigma(A,B) : |\omega_1 \omega_2| = |\mu| \geq \varepsilon_1; \qquad (2.12)$$

2) $\exists 0 < \gamma, \varepsilon < +\infty : \sigma(A,B) \subseteq (\{(\lambda,\mu) |\, |\lambda| \leq 2\gamma\} \cup \{(\lambda,\mu) |\, |\lambda| > 2\gamma,\ |\mu| \leq \gamma|\lambda| - \gamma^2\}) \setminus (\bigcup_{l=1}^{\infty} \{(\lambda,\mu) | \mu - \frac{\lambda^2}{4} \in (\frac{\pi^2 l^2}{(b-a)^2} - \varepsilon l, \frac{\pi^2 l^2}{(b-a)^2} + \varepsilon l)\} \cup \{(\lambda,\mu) |\, |\mu| \leq \varepsilon\})$ (Fig. 2.3b);

3) *the Dirichlet problem for* (1) *on* $[a,b]$ *is well-posed in* $G = (D(B) \cap D(A^2)) \times (D(B) \cap D(A^2))$*, and moreover,* B^{-1} *is bounded (i.e.,* $0 \notin \sigma(B)$*).*

In this case, for every $(f_0, f_1) \in H_1 \times H_1$ *the solution of the Neumann problem for* (1) *on* $[a,b]$ *with b.d.* (f_0, f_1) *is the following:*

$$y(t) = \sum_{i=0}^{1} \chi_i(A,B,t) f_i, \quad t \in [a,b].$$

(See Theorem 4.5.)

Notice that the condition

$$\exists \varepsilon_1 > 0 \forall (\lambda, \mu) \in \sigma(A, B) : \min_{i=1,2} |\omega_i| \geq \varepsilon_1,$$

or, what is the same in view of Lemma 1.1,

$$\exists \varepsilon_1 > 0 \forall (\lambda, \mu) \in \sigma(A, B) : |\mu| \geq \varepsilon_1, \ |\lambda^{-1}\mu| \geq \varepsilon_1,$$

is stronger than

$$\exists \varepsilon_1 > 0 \forall (\lambda, \mu) \in \sigma(A, B) : |\omega_1 \omega_2| = |\mu| \geq \varepsilon_1.$$

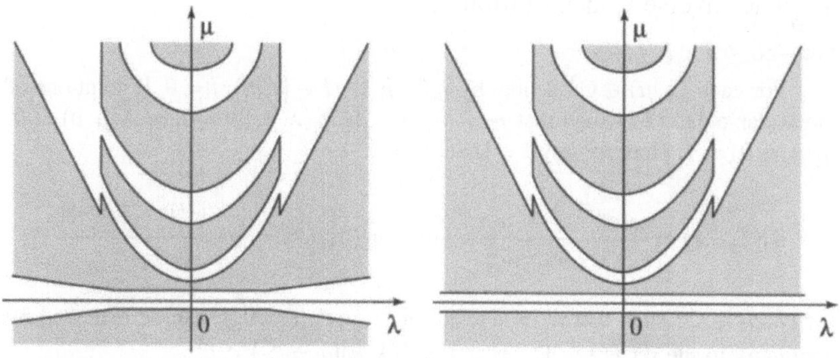

Fig. 2.3a, b

Example 2.3. Let A and B be from Example 1.1. $\sigma(A, B)$ is found in Example 1.2. Theorem 2.5 implies that, with the exception of the trivial case of $a_1 = a_2 = b_1 = b_2 = 0$, $b_3 \neq 0$, and $b_3 - \frac{a_3^2}{4} \neq \frac{\pi^2 l^2}{(b-a)^2}$ for all $l \in \mathbb{N}$, the Neumann problem for (1) on $[a, b]$ is weakly well-posed in $G = H_{-2} \times H_{-2}$ if and only if all conditions 1), 2), 3) of Example 2.2 hold, and moreover, one of the following cases holds:

1) $b_3 > 0$, $b_1 \geq 0$, $b_2 \geq 0$, and for each of $i = 1, 2$:
   a) if $b_i = 0$ then $a_i = 0$,
   b) if $b_i > 0$ then $a_i > 0$;
2) $b_3 > 0$, $b_1 \geq 0$, $b_2 \geq 0$, and for each of $i = 1, 2$:
   a) if $b_i = 0$ then $a_i = 0$,
   b) if $b_i > 0$ then $a_i < 0$;
3) $b_3 < 0$, $b_1 < 0$, $b_2 < 0$.

Theorem 2.6 implies that, with the indicated exception, the Neumann problem for (1) on $[a,b]$ is well-posed in $G = H_1 \times H_1$ if and only if all conditions 1), 2), 3) of Example 2.2 hold, and moreover, one of the following cases holds:

1) $b_3 > 0$, $b_1 \geq 0$, $b_2 \geq 0$, and for each of $i = 1, 2$:
   a) if $b_i = 0$ then $a_i \geq 0$,
   b) if $b_i > 0$ then $a_i > 0$;
2) $b_3 > 0$, $b_1 \geq 0$, $b_2 \geq 0$, and for each of $i = 1, 2$:
   a) if $b_i = 0$ then $a_i \leq 0$,
   b) if $b_i > 0$ then $a_i < 0$;
3) $b_3 < 0$, $b_1 \leq 0$, $b_2 \leq 0$.

4 The inverse Cauchy problem

Let $-\infty \leq a < b < +\infty$.

For each $(\lambda, \mu) \in \mathbf{C}^2$ denote by $\eta_i(\lambda, \mu, t)$ $(t \in [a,b])$, $i = 0, 1$, solutions of the scalar o.d.e. (2.1) such that $\eta_0(\lambda, \mu, b) = 1$, $\eta_0'(\lambda, \mu, b) = 0$, $\eta_1(\lambda, \mu, b) = 0$, $\eta_1'(\lambda, \mu, b) = 1$. Here for any $t \in [a,b]$:

$$\eta_0(\lambda, \mu, t) = \frac{\omega_1 e^{\omega_2(t-b)} - \omega_2 e^{\omega_1(t-b)}}{\omega_1 - \omega_2}, \quad \eta_1(\lambda, \mu, t) = \frac{e^{\omega_1(t-b)} - e^{\omega_2(t-b)}}{\omega_1 - \omega_2}$$

for $(\lambda, \mu) \in \mathbf{C}^2$ such that $\omega_1 \neq \omega_2$, and $\eta_i(\lambda, \mu, t)$, $i = 0, 1$, can be extended by continuity to the set $\{(\lambda, \mu) | \omega_1 = \omega_2\} = \{(\lambda, \mu) | \mu = \frac{\lambda^2}{4}\} \subset \mathbf{C}^2$.

Theorem 2.7. *Let A and B be c.s.o. in H, and let $H \times H \subseteq G \subseteq H \times H_{-1}$. The inverse Cauchy problem for equation* (1) *on* $[a,b]$ *is weakly well-posed in G if and only if one of the following equivalent conditions* 1), 2), 3) *holds:*

1) $$\exists 0 < \gamma < +\infty \forall (\lambda, \mu) \in \sigma(A, B) : \min_{i=1,2} \operatorname{Re} \omega_i(\lambda, \mu) \geq -\gamma; \tag{2.13}$$
2) $\exists 0 < \gamma < +\infty : \sigma(A, B) \subseteq \{(\lambda, \mu) | \lambda \leq 2\gamma,\ \mu \geq \gamma\lambda - \gamma^2\}$ (Fig. 2.4);
3) *A is semibounded above, $B_- \stackrel{\text{def}}{=} BE_B((-\infty, 0))$ (the negative part of B) is subordinate to A $(D(A) \subseteq D(B_-))$.*

In this case, for every $(f_0, f_1) \in H \times H_{-1}$ the w.s. of the inverse Cauchy problem for (1) *on* $[a,b]$ *with b.d.* (f_0, f_1) *has the form*

$$y(t) = \sum_{i=0}^{1} \eta_i(A, B, t) f_i, \quad t \in [a,b].$$

Proof. It is easy to see that $y(t)$ is a weak solution of the inverse Cauchy problem for equation (1) on $[a,b]$ with b.d. (f_0, f_1) if and only if $u(t) = y(b-t)$ is a weak solution of the equation

$$u''(t) - Au'(t) + Bu(t) = 0 \tag{2.14}$$

on $[0, b-a]$ with i.d. (f_0, f_1). Thus, the inverse Cauchy problem for equation (1) on $[a,b]$ is weakly well-posed in G if and only if the Cauchy problem for equation (2.14) on $[0, b-a]$ is weakly well-posed in G. It remains to apply Theorem 2.1 to equation (2.14) and to notice that $\sigma(-A, B) = \{(-\lambda, \mu) | (\lambda, \mu) \in \sigma(A, B)\}$ and for any $(\lambda, \mu) \in \mathbb{C}^2$ the roots of the polynomial $\omega^2 - \lambda\omega + \mu$ are $(-\omega_1)$, $(-\omega_2)$.

Theorem 2.8. *Let A and B be c.s.o. in H, $G = D(B) \times (D(A) \cap D(|B|^{1/2})) = D(B) \times H_1$. The inverse Cauchy problem for equation* (1) *on* $[a,b]$ *is well-posed in G if and only if one of the equivalent conditions* 1), 2), 3) *of Theorem* 2.7 *holds.*

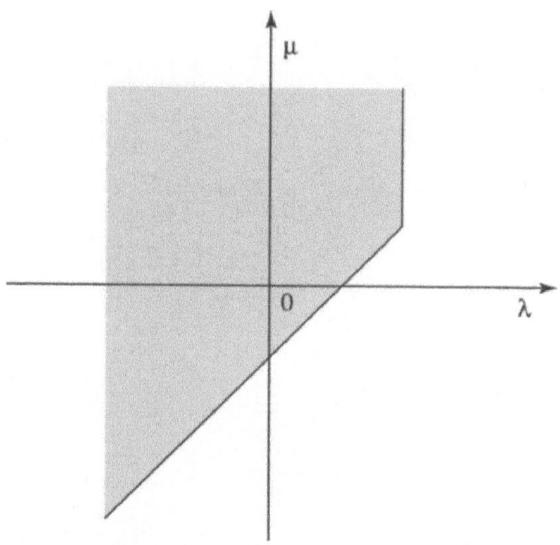

Fig. 2.4

Example 2.4. Let A and B be the operators from Example 1.1. $\sigma(A, B)$ is found in Example 1.2. Theorems 2.7, 2.8 imply that the inverse Cauchy problem for (1) on $[a,b]$ is weakly well-posed in $G = H \times H$ (weakly well-posed in $G = H \times H_{-1}$, well-posed in $G = D(B) \times H_1$) if and only if for each of $i = 1, 2$ there holds one of the following cases:

a) $a_i < 0$; b) $a_i = 0$, $b_i \geq 0$.

Corollaries 1, 2 to Theorems 2.1, 2.2 and the remarks to these corollaries remain valid if one replaces «the Cauchy problem» by «the inverse Cauchy problem» and replaces in Corollary 2 «A is semibounded below» by «A is semibounded above».

Chapter 3
The Cauchy problem

We now go on to consider the Cauchy problem for equation (1) in a more general case where A and B are arbitrary commuting *normal* operators (c.n.o.) in H. We consider the Cauchy problem for (1) on $R_+ = [0, +\infty)$; for any other $[a, b]$ such that $-\infty < a < b \leq +\infty$, all the results remain valid.

1 Distinction of the general case of commuting normal operators A and B

Theorem 2.1 shows that for commuting *self-adjoint* A and B, (2.2) is a necessary and sufficient condition for the weak well-posedness of the Cauchy problem for (1) in $G = H \times H$.

Statement 3.1. In the general case where A and B are commuting normal operators (c.n.o.) in a Hilbert space H, condition (2.2) is neither necessary nor sufficient for the weak well-posedness of the Cauchy problem for equation (1) on R_+ in $G = H \times H$.

To show this, we need the following lemma.

Lemma 3.1. *Let A and B be c.n.o. in H. For the Cauchy problem for equation (1) on R_+ to be weakly well-posed in $G = H \times H$, it is necessary and sufficient that $\forall 0 < T < +\infty \exists C_T < +\infty$ such that*

$$|\psi_0(\lambda, \mu, t)| \leq C_T,\ |\psi_1(\lambda, \mu, t)| \leq C_T,\ \forall (\lambda, \mu) \in \sigma(A, B),\ t \in [0, T]. \quad (3.1)$$

One can prove it similarly to Theorem 2.1, with Ag, Bg replaced by A^*g, B^*g and so on; note that $\Phi = D(A^\infty) \cap D(B^\infty) = D((A^*)^\infty) \cap D((B^*)^\infty)$, $|\psi_i^{(k)}(\bar{\lambda}, \bar{\mu}, t)| = |\overline{\psi_i^{(k)}(\lambda, \mu, t)}| = |\psi_i^{(k)}(\lambda, \mu, t)|$.

Examples 3.1, 3.2 now complete the proof of Statement 3.1.

Example 3.1. Let $B = 0$, $\sigma(\operatorname{Re} A) = (-\infty, 0]$, $\operatorname{Im} A = \exp[(\operatorname{Re} A)^2]$ (in other words, $B = 0$, $\sigma(A) = \{\lambda_1 + i\lambda_2 | \lambda_1 \leq 0,\ \lambda_2 = \exp(\lambda_1^2)\}$). Then

$$\sup_{(\lambda,\mu)\in\sigma(A,B),\, i=1,2} \operatorname{Re} \omega_i(\lambda, \mu) = \sup_{(\lambda,\mu)\in\sigma(A,B)} (-\lambda_1) = +\infty,$$

i.e., (2.2) is not valid. Nevertheless, the Cauchy problem for equation (1) on R_+ is weakly well-posed in $H \times H$, by Lemma 3.1. Indeed, for any $0 < T < +\infty$ we have: $\psi_0(\lambda, \mu, t) = 1$, $|\psi_1(\lambda, \mu, t)| = |\frac{e^{-\lambda t}-1}{-\lambda}| \leq \frac{e^{-\lambda_1 t}+1}{\exp(\lambda_1^2)} \leq 1 + \exp(\frac{T^2}{4})$ for all $(\lambda, \mu) \in \sigma(A, B)$, $t \in [0, T]$.

Example 3.2. Let $\operatorname{Re} A = 0$, $\sigma(\operatorname{Im} A) = (-\infty, +\infty)$, $B = \frac{A^2}{4}$ (for instance: $A = \frac{d}{dx}$, $B = \frac{1}{4}\frac{d^2}{dx^2}$ in $H = L_2(R^1)$, defined as usually). Here

$$\forall(\lambda,\mu) \in \sigma(A,B) : \omega_1(\lambda,\mu) = \omega_2(\lambda,\mu) = -\frac{1}{2}i\lambda_2 \Rightarrow \operatorname{Re}\omega_i(\lambda,\mu) = 0,\ i = 1,2,$$

so condition (2.2) is satisfied. Nevertheless, the Cauchy problem for equation (1) on R_+ is not weakly well-posed in $H \times H$, by Lemma 3.1. Indeed, fix an arbitrary $0 < t < +\infty$; then $\forall(\lambda,\mu) \in \sigma(A,B)$:

$$|\psi_0(\lambda,\mu,t)| = \left|\left(\frac{1}{2}i\lambda_2 t - 1\right)\exp\left(-\frac{1}{2}i\lambda_2 t\right)\right| \geq |\lambda| - 1;$$

so in view of $\sup\limits_{(\lambda,\mu)\in\sigma(A,B)} |\lambda_2| = +\infty$ we have: $\sup\limits_{(\lambda,\mu)\in\sigma(A,B)} |\psi_0(\lambda,\mu,t)| = +\infty$.

The proof of Statement 3.1 is completed.

The following theorem shows that if one takes the wider space $G = H \times H_-(|A|+1)$ in place of $G = H \times H$, then condition (2.2) becomes necessary for the weak well-posedness of the Cauchy problem for (1) on R_+ in the space G (in the general case of c.n.o. A and B as in the case of c.s.o. A and B).

2 A criterion for the weak well-posedness

Theorem 3.1. *Let A and B be c.n.o. in H, and let $H \times H_-(|A|+1) \subseteq G \subseteq H \times H_-(|A| + |B|^{1/2} + 1) = H \times H_{-1}$. The Cauchy problem for equation* (1) *on R_+ is weakly well-posed in G if and only if one of the following equivalent conditions* 1), 2), 3) *holds:*

1) *a)* (2.2) *is valid;*
   b) $\exists 0 < \varepsilon < +\infty : \sigma(A,B) \subseteq \{(\lambda,\mu) |\ |\omega_i| \leq \frac{1}{\varepsilon}\ (i = 1,2)\} \cup \{(\lambda,\mu) |$ $\left|\frac{\omega_1-\omega_2}{\omega_1+\omega_2}\right| \geq \varepsilon\} \cup \{(\lambda,\mu) |\ |\operatorname{Im}\omega_i| \leq \frac{1}{\varepsilon}|\operatorname{Re}\omega_i| (i = 1,2)\}$; (3.2)
2) *a)* $\exists 0 < \gamma < +\infty : \sigma(A,B) \subseteq \{(\lambda,\mu) | \lambda_1 > -2\gamma,$ $\mu_1 > \frac{1}{(\lambda_1+2\gamma)^2}(\mu_2^2 - \lambda_1\lambda_2 \cdot \mu_2 - \gamma(\lambda_1+\gamma)(\lambda_2^2 + (\lambda_1+2\gamma)^2))\}$, (3.3)
   b) $\exists 0 < \varepsilon < +\infty : \sigma(A,B) \subseteq \{(\lambda,\mu) |\ |\lambda| \leq \frac{1}{\varepsilon}\} \cup \{(\lambda,\mu) |\ |\lambda_2| \leq \frac{1}{\varepsilon}|\lambda_1|\} \cup$ $\{(\lambda,\mu) |\ |\frac{4\mu}{\lambda^2} - 1| \geq \varepsilon\}$; (3.4)
3) *there exist $0 < \gamma, \varepsilon < +\infty$ such that for each $(\lambda,\mu) \in \sigma(A,B)$ one of the following conditions is satisfied:*
   a) $\lambda_1 > -2\gamma$, $|\lambda_2| \leq \frac{1}{\varepsilon}(1 + |\lambda_1|)$, *$(\mu_1,\mu_2)$ lies in R^2 to the right of the parabola*

$$\mu_1 - \left(-\frac{\lambda_2^2}{4} - \lambda_1\gamma - \gamma^2\right) = \frac{1}{(\lambda_1+2\gamma)^2}\left(\mu_2 - \frac{\lambda_1\lambda_2}{2}\right)^2, \tag{3.5}$$

b) $\lambda_1 > -2\gamma$, $|\lambda_2| > \frac{1}{\varepsilon}(1+|\lambda_1|)$, (μ_1, μ_2) *lies in* R^2 *to the right of the parabola* (3.5) *and outside the circle of radius* $\varepsilon|\lambda|^2$ *the center of which is situated on the axis of the parabola* (3.5) *to the right of its vertex by* $(\frac{\lambda_1}{2}+\gamma)^2$.

In this case, for every $(f_0, f_1) \in H \times H_{-1}$ *the weak solution of the Cauchy problem on* R_+ *for equation* (1) *with initial conditions* $y(0) = f_0$, $y'(0) = f_1$ *(i.e., with b.d.* (f_0, f_1)*) can be written in the following form:*

$$\begin{aligned} y(t) &= \psi_0(A, B, t) f_0 + \psi_1(A, B, t) f_1 \\ &= \psi_0(A, B, t) f_0 + \lim_{k \to \infty} \psi_1(A, B, t) E_k f_1 \ (t \in R_+). \end{aligned}$$

Corollary 1. *Let* A *and* B *be c.n.o. in* H*, and moreover, let* A *be bounded. Then for an arbitrary* G *such that* $H \times H \subseteq G \subseteq H \times H_-(|B|^{1/2}+1) = H \times H_{-1}$, *condition* (2.2) *is necessary and sufficient for the Cauchy problem for equation* (1) *on* R_+ *to be weakly well-posed in* G.

In particular, this is the case for every second order «incomplete» equation $y''(t) + By(t) = 0$ where B is normal.

So, condition (2.2) may not be necessary for the weak well-posedness of the Cauchy problem for (1) with c.n.o. A and B in $G = H \times H$ only in the case of an unbounded A.

3 Proof of Theorem 3.1

The proof of Theorem 3.1 consists of four parts.

I We prove the equivalence of conditions 1), 2), 3).

II We show that for every $0 < T < +\infty$ there exists $C_T < +\infty$ such that

$$|\psi_0(\lambda, \mu, t)| \le C_T, \quad |\psi_1(\lambda, \mu, t)| \le \frac{C_T}{|\lambda| + \sqrt{|\mu|} + 1}, \qquad \forall (\lambda, \mu) \in \sigma(A, B), \quad t \in [0, T] \tag{3.6}$$

if and only if condition 1) holds.

III We show that for a positive T there exists $C_T < +\infty$ such that (3.6) holds if and only if for this T there exists $C'_T < +\infty$ such that

$$|\psi_0(\lambda, \mu, t)| \le C'_T, \quad (|\lambda|+1)|\psi_1(\lambda, \mu, t)| \le C'_T, \qquad \forall (\lambda, \mu) \in \sigma(A, B), \quad t \in [0, T]. \tag{3.7}$$

IV We show that the Cauchy problem for equation (1) on R_+ is weakly well-posed in $G = H \times H_{-1} = H \times H_-(|A| + |B|^{1/2} + 1)$ if and only if for every $0 < T < +\infty$ there exists $C_T < +\infty$ such that (3.6) holds.

Similarly, we show that the Cauchy problem for (1) on R_+ is weakly well-posed in $G = H \times H_-(|A|+1)$ if and only if for every $0 < T < +\infty$ there exists $C'_T < +\infty$ such that (3.7) holds.

Here I, II, III can be proved even if $\sigma(A, B)$ is replaced by an arbitrary $\Omega \subseteq C^2$.

We now give the proof of Theorem 3.1 in more detail.

I We prove the equivalence of conditions 1), 2), 3).

It is obvious that condition 3) is the reformulated 2). We show that condition 2) is equivalent to 1), and moreover, (2.2) is equivalent to (3.3), while (3.2) is equivalent to (3.4).

Lemma 3.2. *Let $\Omega \subseteq C^2$, $\gamma \in R^1$. The following conditions are equivalent:*

1) $\forall(\lambda, \mu) \in \Omega : \operatorname{Re}\omega_i(\lambda, \mu) < \gamma \ (i = 1, 2)$;
2) $\forall(\lambda, \mu) \in \Omega : \lambda_1 > -2\gamma$,

$$\mu_1 > \frac{1}{(\lambda_1 + 2\gamma)^2}(\mu_2^2 - \lambda_1\lambda_2 \cdot \mu_2 - \gamma(\lambda_1 + \gamma)(\lambda_2^2 + (\lambda_1 + 2\gamma)^2)).$$

Proof. Fix an arbitrary $(\lambda, \mu) \in \Omega$. $\max_{i=1,2} \operatorname{Re}\omega_i(\lambda, \mu) < \gamma \Leftrightarrow \operatorname{Re}\xi_1 < 0$, $\operatorname{Re}\xi_2 < 0$, where ξ_1, ξ_2 are the roots of the polynomial $P(\xi) = (\xi+\gamma)^2 + \lambda(\xi+\gamma) + \mu \equiv \xi^2 + (2\gamma+\lambda)\xi + (\gamma^2 + \lambda\gamma + \mu)$. Applying the Routh-Hurwitz criterion for polynomials with complex-valued coefficients [118, pp. 500–501] to the polynomial $P(\xi)$, we obtain: $\operatorname{Re}\xi_1 < 0$, $\operatorname{Re}\xi_2 < 0 \Leftrightarrow \lambda_1 > -2\gamma$,

$$\begin{aligned}\mu_1 &> \frac{1}{(\lambda_1 + 2\gamma)^2}(\mu_2 + \gamma\lambda_2)(\mu_2 - \gamma\lambda_2 - \lambda_1\lambda_2) - \lambda_1\gamma - \gamma^2 \\ &= \frac{1}{(\lambda_1 + 2\gamma)^2}(\mu_2^2 - \lambda_1\lambda_2 \cdot \mu_2 - \gamma(\lambda_1 + \gamma)(\lambda_2^2 + (\lambda_1 + 2\gamma)^2)).\end{aligned}$$

Lemma 3.2 is proved.

Lemma 3.3. *Let $\Omega \subseteq C^2$. The following conditions are equivalent:*

1) *there exists $\varepsilon > 0$ such that for each $(\lambda, \mu) \in \Omega : |\omega_i| \le \frac{1}{\varepsilon}$ $(i = 1, 2)$, or $|\frac{\omega_1 - \omega_2}{\omega_1 + \omega_2}| \ge \varepsilon$, or $|\operatorname{Im}\omega_i| \le \frac{1}{\varepsilon}|\operatorname{Re}\omega_i|$ $(i = 1, 2)$;*
2) *there exists $\varepsilon > 0$ such that for every $(\lambda, \mu) \in \Omega : |\lambda| \le \frac{1}{\varepsilon}$, or $|\operatorname{Im}\lambda| \le \frac{1}{\varepsilon}|\operatorname{Re}\lambda|$, or $|\frac{4\mu}{\lambda^2} - 1| \ge \varepsilon$.*

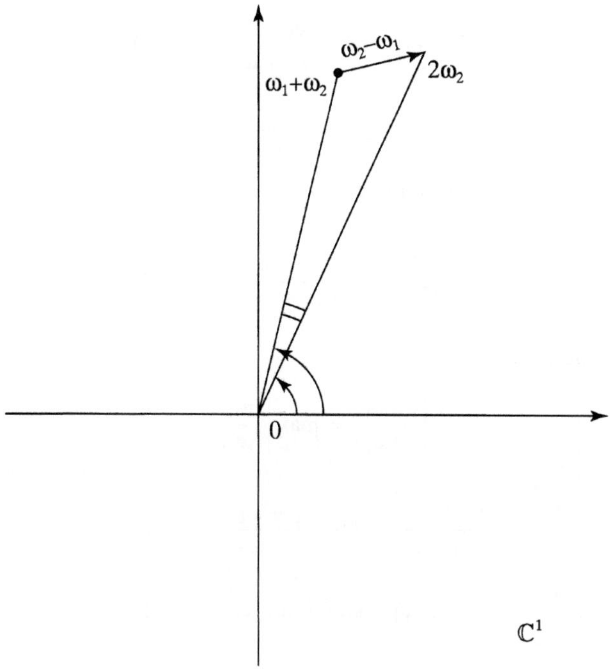

Fig. 3.1

Proof. Assume that 2) does not hold, i.e., for every $k \in \mathbb{N}$ there exists $(\lambda_k, \mu_k) \in \Omega$ such that $|\lambda_k| > k$, $|\operatorname{Im}\lambda_k| > k|\operatorname{Re}\lambda_k|$, $|\frac{\omega_1-\omega_2}{\omega_1+\omega_2}| < \frac{1}{k}$. Then for the sequence $\{(\lambda_k, \mu_k)\}_{k=1}^{\infty}$: $\max\limits_{i=1,2}|\omega_i| \to +\infty$, $|\frac{\omega_1-\omega_2}{\omega_1+\omega_2}| \to 0$ as $k \to +\infty$. Further, $|\frac{\omega_1-\omega_2}{\omega_1+\omega_2}| \xrightarrow[k\to+\infty]{} 0$, $|\omega_1+\omega_2| \geq 1$, and $2\omega_2 = (\omega_1+\omega_2) - (\omega_1-\omega_2)$ imply that $\arg(2\omega_2) - \arg(\omega_1+\omega_2) \to 0$ as $k \to +\infty$ (here $-\pi < \arg z \leq \pi$, $\forall z \in C^1$) (see Fig. 3.1). But $\arg(2\omega_2) = \arg\omega_2$; $|\arg(\omega_1+\omega_2)| \xrightarrow[k\to+\infty]{} \frac{\pi}{2}$ since $|\frac{\operatorname{Im}(\omega_1+\omega_2)}{\operatorname{Re}(\omega_1+\omega_2)}| = |\frac{\operatorname{Im}\lambda_k}{\operatorname{Re}\lambda_k}| > k$ for (λ_k, μ_k); so, $|\arg\omega_2| \xrightarrow[k\to+\infty]{} \frac{\pi}{2}$, i.e., $|\frac{\operatorname{Im}\omega_2}{\operatorname{Re}\omega_2}| \xrightarrow[k\to+\infty]{} +\infty$. Similarly, $|\frac{\operatorname{Im}\omega_1}{\operatorname{Re}\omega_1}| \xrightarrow[k\to+\infty]{} +\infty$. Thus, 1) is not satisfied.

Conversely, assume that 1) does not hold, i.e., for every $k \in \mathbb{N}$ there exists $(\lambda_k, \mu_k) \in \Omega$ such that

$$\max_{i=1,2}|\omega_i| > k, \quad \left|\frac{\omega_1-\omega_2}{\omega_1+\omega_2}\right| < \frac{1}{k}, \text{ and } \max_{i=1,2}\left|\frac{\operatorname{Im}\omega_i}{\operatorname{Re}\omega_i}\right| > k.$$

Then, setting for each (λ_k, μ_k):

$$|\omega_1| = \max_{i=1,2}|\omega_i|,$$

we have:

$$\left|\frac{2\omega_1}{\omega_1+\omega_2}\right| = \left|1+\frac{\omega_1-\omega_2}{\omega_1+\omega_2}\right| \xrightarrow[k\to+\infty]{} 1,$$

whence

$$|\lambda_k| = |\omega_1+\omega_2| \xrightarrow[k\to+\infty]{} +\infty$$

in view of $|\omega_1| \xrightarrow[k\to+\infty]{} +\infty$. Further,

$$\left|\frac{\lambda_k^2-4\mu_k}{\lambda_k^2}\right| = \left|\frac{\omega_1-\omega_2}{\omega_1+\omega_2}\right|^2 \xrightarrow[k\to+\infty]{} 0.$$

Finally, setting for each $k \geq 1$:

$$\left|\frac{\operatorname{Im}\omega_1}{\operatorname{Re}\omega_1}\right| = \max_{i=1,2}\left|\frac{\operatorname{Im}\omega_i}{\operatorname{Re}\omega_i}\right|,$$

we have:

$$\frac{2\omega_1}{\omega_1+\omega_2} = 1+\frac{\omega_1-\omega_2}{\omega_1+\omega_2} \xrightarrow[k\to+\infty]{} 1,$$

whence

$$\arg(2\omega_1)-\arg(\omega_1+\omega_2) \xrightarrow[k\to+\infty]{} 0;$$

but

$$|\arg(2\omega_1)| = |\arg\omega_1| \xrightarrow[k\to+\infty]{} \frac{\pi}{2}$$

since

$$\left|\frac{\operatorname{Im}\omega_1}{\operatorname{Re}\omega_1}\right| \xrightarrow[k\to+\infty]{} +\infty;$$

so,

$$|\arg(\omega_1+\omega_2)| \xrightarrow[k\to+\infty]{} \frac{\pi}{2},$$

i.e.,

$$\left|\frac{\operatorname{Im}\lambda_k}{\operatorname{Re}\lambda_k}\right| = \left|\frac{\operatorname{Im}(\omega_1+\omega_2)}{\operatorname{Re}(\omega_1+\omega_2)}\right| \xrightarrow[k\to+\infty]{} +\infty.$$

Thus, 2) is not satisfied.

Lemma 3.3 is proved.

II We show that for every $0 < T < +\infty$ there exists $C_T < +\infty$ such that

$$|\psi_0(\lambda,\mu,t)| \leq C_T,\ |\psi_1(\lambda,\mu,t)| \leq \frac{C_T}{|\lambda|+\sqrt{|\mu|}+1}, \qquad (3.6)$$
$$\forall(\lambda,\mu)\in\sigma(A,B),\ t\in[0,T]$$

if and only if condition 1) of Theorem 3.1 is satisfied.

Lemma 3.4. *Let condition* 1) *of Theorem* 3.1 *be satisfied. Then for every* $0 < T < +\infty$ *there exists* $C_T < +\infty$ *such that* (3.6) *is valid.*

Proof. Fix an arbitrary $0 < T < +\infty$. Let's show that there exists $C_T < +\infty$ such that $|\psi_0(\lambda, \mu, t)| \leq C_T$, $\forall(\lambda, \mu) \in \sigma(A, B)$, $t \in [0, T]$. Indeed, $\sigma(A, B)$ satisfies condition 1) of Theorem 3.1; therefore, there exists $0 < \varepsilon < \frac{1}{3}$ such that for each $(\lambda, \mu) \in \sigma(A, B)$ one of the following cases holds.

1. $\max\limits_{i=1,2} |\omega_i| \leq \frac{1}{\varepsilon}$. Then $|\lambda| \leq \frac{2}{\varepsilon}$, $|\mu| \leq \frac{1}{\varepsilon^2}$ by the Vieta formulas; so

$$\forall t \in [0, T] : |\psi_0(\lambda, \mu, t)| \leq C_{T,\varepsilon} \stackrel{\text{def}}{=} \sup_{\{(\lambda,\mu,t) \mid |\lambda| \leq \frac{2}{\varepsilon}, |\mu| \leq \frac{1}{\varepsilon^2}, 0 \leq t \leq T\}} |\psi_0(\lambda, \mu, t)|.$$

Here $C_{T,\varepsilon} < +\infty$, since $\psi_0(\lambda, \mu, t)$ is continuous on the compact set

$$\{(\lambda, \mu, t) \mid |\lambda| \leq \frac{2}{\varepsilon},\ |\mu| \leq \frac{1}{\varepsilon^2},\ 0 \leq t \leq T\} \subset C^2 \times R_+.$$

2. $|\frac{\omega_1 - \omega_2}{\omega_1 + \omega_2}| \geq \varepsilon$; then $|\frac{2\omega_1}{\omega_1 - \omega_2}| = |1 + \frac{\omega_1 + \omega_2}{\omega_1 - \omega_2}| \leq \frac{1}{\varepsilon} + 1 \leq \frac{4}{3\varepsilon} \Rightarrow |\frac{\omega_1 - \omega_2}{\omega_1}| \geq \frac{3}{2}\varepsilon \geq \varepsilon$; similarly, $|\frac{\omega_1 - \omega_2}{\omega_2}| \geq \varepsilon$; so, $\forall t \in [0, T]$:

$$|\psi_0(\lambda, \mu, t)| \leq \left|\frac{\omega_1}{\omega_1 - \omega_2}\right| |e^{\omega_2 t}| + \left|\frac{\omega_2}{\omega_1 - \omega_2}\right| |e^{\omega_1 t}| \leq \frac{2}{\varepsilon} e^{\gamma t} \leq \frac{2}{\varepsilon} e^{\gamma T}.$$

3. $|\operatorname{Im} \omega_1| \leq \varepsilon^{-1} |\operatorname{Re} \omega_1|$, $|\operatorname{Im} \omega_2| \leq \varepsilon^{-1} |\operatorname{Re} \omega_2|$. Note that $|\psi_0(\lambda, \mu, t)| = |e^{\omega_1 t} + \omega_1 t e^{\omega_1 t} \cdot \frac{1 - e^{(\omega_2 - \omega_1)t}}{(\omega_2 - \omega_1)t}|$. Here $\forall t \in [0, T] : |e^{\omega_1 t}| \leq e^{\gamma t} \leq e^{\gamma T}$. Further, $|\frac{1 - e^{(\omega_2 - \omega_1)t}}{(\omega_2 - \omega_1)t}| \leq 2$ (see the proof of Lemma 2.2). Finally,

$$|\omega_1 t e^{\omega_1 t}| = \sqrt{|\operatorname{Re} \omega_1|^2 + |\operatorname{Im} \omega_1|^2} t e^{\operatorname{Re} \omega_1 t} \leq \sqrt{1 + \varepsilon^{-2}} |\operatorname{Re} \omega_1 t \cdot e^{\operatorname{Re} \omega_1 t}|$$
$$\leq \sqrt{1 + \varepsilon^{-2}} \max_{\{x \in R^1 \mid x \leq \gamma t\}} |x e^x|.$$

Here

$$\max_{\{x \in R^1 \mid x \leq \gamma t\}} |x e^x| \leq e^{-1} + \max_{0 \leq x \leq \gamma T} (x e^x) = e^{-1} + \gamma T e^{\gamma T}.$$

We obtain that $|\psi_0(\lambda, \mu, t)|$ is uniformly bounded on $\sigma(A, B) \times [0, T]$.

In view of Lemma 1.1, the existence of $C_T < +\infty$ such that $|\psi_1(\lambda, \mu, t)| \leq \frac{C_T}{|\lambda| + \sqrt{|\mu|} + 1}$ for all $(\lambda, \mu, t) \in \sigma(A, B) \times [0, T]$ is equivalent to the uniform boundedness of $(\max(|\omega_1|, |\omega_2|) + 1) \cdot |\psi_1(\lambda, \mu, t)|$ on $\sigma(A, B) \times [0, T]$.

For the set $\{(\lambda,\mu) \in \sigma(A,B) | \max(|\omega_1|,|\omega_2|) \le \varepsilon\} \times [0,T]$, this uniform boundedness is the fact similarly to the case 1 above. For the set $\{(\lambda,\mu) \in \sigma(A,B) | \max(|\omega_1|,|\omega_2|) \ge \varepsilon\} \times [0,T]$, we have:

$$|\omega_1\psi_1(\lambda,\mu,t)| = |\psi_0(\lambda,\mu,t) - e^{\omega_1 t}| \le |\psi_0(\lambda,\mu,t)| + e^{\gamma T};$$

similarly,

$$|\omega_2\psi_1(\lambda,\mu,t)| \le |\psi_0(\lambda,\mu,t)| + e^{\gamma T};$$

finally,

$$\begin{aligned}|\psi_1(\lambda,\mu,t)| &\le \varepsilon^{-1}\max(|\omega_1|,|\omega_2|)\cdot|\psi_1(\lambda,\mu,t)|\\ &= \varepsilon^{-1}\max(|\omega_1\psi_1(\lambda,\mu,t)|,|\omega_2\psi_1(\lambda,\mu,t)|).\end{aligned}$$

Now the uniform boundedness of $(\max(|\omega_1|,|\omega_2|)+1)\cdot|\psi_1(\lambda,\mu,t)|$ on this set follows from the previous estimates for $|\psi_0(\lambda,\mu,t)|$.

Lemma 3.5. *Let for a certain* $0 < T < +\infty$ *there exist* $C_T < +\infty$ *such that* (3.6) *is valid. Then* (2.2) *and* (3.2) *hold, i.e., condition* 1) *of Theorem* 3.1 *is satisfied.*

Proof. In view of Lemma 1.1, for any $(\lambda,\mu) \in \sigma(A,B)$, $t \in [0,T]$:

$$\max(|\omega_1|,|\omega_2|)\cdot|\psi_1(\lambda,\mu,t)| \le 2(|\lambda| + \sqrt{|\mu|} + 1)\cdot|\psi_1(\lambda,\mu,t)| \le 2C_T.$$

Further,

$$\begin{aligned}&\forall(\lambda,\mu) \in \sigma(A,B),\ t \in [0,T]:\ \psi_0(\lambda,\mu,t) = e^{\omega_1 t} - \omega_1\frac{e^{\omega_1 t} - e^{\omega_2 t}}{\omega_1 - \omega_2}\\ &\Rightarrow e^{\operatorname{Re}\omega_1 t} = |e^{\omega_1 t}| \le |\psi_0(\lambda,\mu,t)| + |\omega_1|\cdot|\psi_1(\lambda,\mu,t)| \le 3C_T;\end{aligned}$$

so,

$$\sup_{(\lambda,\mu)\in\sigma(A,B)} e^{\operatorname{Re}\omega_1 T} \le 3C_T < +\infty \Rightarrow \sup_{(\lambda,\mu)\in\sigma(A,B)} \operatorname{Re}\omega_1 < +\infty.$$

Since $\forall(\lambda,\mu) \in \sigma(A,B) : \operatorname{Re}\omega_2 \le \operatorname{Re}\omega_1$, (2.2) holds.

Further, $\forall(\lambda,\mu,t) \in \sigma(A,B) \times [0,T]$:

$$\begin{aligned}C_T \ge |(\omega_1+\omega_2)\psi_1(\lambda,\mu,t)| &= \left|\frac{\omega_1+\omega_2}{\omega_1-\omega_2}(e^{\omega_1 t} - e^{\omega_2 t})\right|\\ &= \left|\frac{\frac{\omega_1+\omega_2}{2}}{\frac{\omega_1-\omega_2}{2}}\right| \cdot e^{\operatorname{Re}\frac{\omega_1+\omega_2}{2}t} \cdot \left|e^{\frac{\omega_1-\omega_2}{2}t} - e^{-\frac{\omega_1-\omega_2}{2}t}\right|.\end{aligned}$$

Denote for each $(\lambda,\mu) \in C^2$: $a_1 = -\operatorname{Re}\frac{\omega_1+\omega_2}{2}$, $a_2 = \operatorname{Im}\frac{\omega_1+\omega_2}{2}$, $b = \frac{\omega_1-\omega_2}{2}$; we have obtained that there exists $C = C_T^2 < +\infty$ such that $\forall(\lambda,\mu) \in \sigma(A,B)$, $\forall t \in [0,T]$:

$$\frac{a_1^2 + a_2^2}{|b|^2}e^{-2a_1 t}|\sinh^2 bt| \le C. \tag{3.8}$$

Consider the continuous function $w(z) = \frac{\sinh z}{z}$ $(z \in C^1)$; here $w(0) = 1$. Choose a $0 < \delta < +\infty$ such that $|w(z)| \geq \frac{1}{2}$ for all $z \in C^1$ having $|z| \leq \delta$; set $\tau = \min(\delta, T)$.

For each $(\lambda, \mu) \in \sigma(A, B)$ one of the following cases holds.

1. $a_1 \leq 0$ and $\min(\tau, \frac{T}{|b|}) = \frac{T}{|b|}$, or: $a_1 > 0$ and $\min(\tau, \frac{T}{a_1}, \frac{T}{|b|}) = \frac{T}{|b|}$. Then $a_1 \leq 0$ or: $a_1 > 0$ and $|b| \geq a_1$, and for $t = \frac{\tau}{|b|} \leq T$ we have: $t|b| = \tau \leq \delta \Rightarrow |\sinh^2 bt| \geq \frac{1}{4}\tau^2$; $e^{-2a_1 t} \geq e^{-2\tau}$; so, (3.8) $\Rightarrow \frac{a_1^2+a_2^2}{|b|^2} \leq \frac{4Ce^{2\tau}}{\tau^2}$.

2. $a_1 > 0$ and $\min(\tau, \frac{T}{a_1}, \frac{T}{|b|}) = \frac{T}{a_1}$. Then $a_1 \geq |b|$, and for $t = \frac{\tau}{a_1} \leq \tau$ we have: $|bt| = \frac{|b|}{a_1}\tau \leq \tau \leq \delta \Rightarrow |\frac{\sinh bt}{b}| \geq \frac{t}{2}$; so, (3.8) $\Rightarrow \frac{1}{4} \cdot \frac{a_1^2+a_2^2}{a_1^2} \cdot a_1^2 t^2 e^{-2a_1 t} = \frac{1}{4}(1 + \frac{a_2^2}{a_1^2})\tau^2 e^{-2\tau} \leq C \Rightarrow 1 + \frac{a_2^2}{a_1^2} \leq \frac{4Ce^{2\tau}}{\tau^2}$.

3. $a_1 \leq 0$ and $\min(\tau, \frac{T}{|b|}) = \tau$, or: $a_1 > 0$ and $\min(\tau, \frac{T}{a_1}, \frac{T}{|b|}) = \tau$. Then $a_1 \leq 1$, $|b| \leq 1$, and for $t = \tau$ we have: $|b|t = |b|\tau \leq \tau \Rightarrow |\frac{\sinh bt}{b}| \geq \frac{1}{2}t$; $e^{-a_1 t} \geq e^{-\tau}$; so, (3.8) $\Rightarrow a_1^2 + a_2^2 \leq \frac{4Ce^{2\tau}}{\tau^2}$.

Summarizing, there exists $C' = \frac{4Ce^{2\tau}}{\tau^2} < +\infty$ such that for each $(\lambda, \mu) \in \sigma(A, B)$:

$$\frac{a_1^2 + a_2^2}{|b|^2} = \left|\frac{\omega_1 + \omega_2}{\omega_1 - \omega_2}\right|^2 = \left|\frac{\lambda^2}{\lambda^2 - 4\mu}\right| \leq C' \Leftrightarrow \left|\frac{4\mu}{\lambda^2} - 1\right| \geq \frac{1}{C'} > 0,$$

or

$$\frac{a_2^2}{a_1^2} = \frac{|\operatorname{Im} \lambda|^2}{|\operatorname{Re} \lambda|^2} \leq C' \Leftrightarrow |\lambda_2| \leq \sqrt{C'}|\lambda_1|,$$

or

$$a_1^2 + a_2^2 = \frac{1}{4}|\lambda|^2 \leq C' \Leftrightarrow |\lambda| \leq 2\sqrt{C'}.$$

This means that (3.4) is satisfied. Since (3.2) is equivalent to (3.4) by Lemma 3.3, (3.2) is valid as well.

III We show that for a positive T there exists $C_T < +\infty$ such that (3.6) holds if and only if for this T there exists $C'_T < +\infty$ such that (3.7) holds.

To do this, it suffices to prove the following lemma.

Lemma 3.6. *Assume that* $0 < T < +\infty$ *and there exists* $C'_T < +\infty$ *such that* (3.7) *holds. Then there exists a* $C < +\infty$ *such that*

$$\sqrt{|\mu|}|\psi_1(\lambda, \mu, t)| \leq C \textit{ for all } (\lambda, \mu) \in \sigma(A, B), \quad t \in [0, T].$$

Proof. Since for any $(\lambda,\mu)\in C^2: \sqrt{|\mu|}\le 2\max(|\omega_1|,|\omega_2|)$ (by Lemma 1.1), it is sufficient to prove that

$$\exists C<+\infty:\ \max(|\omega_1|,|\omega_2|)\cdot|\psi_1(\lambda,\mu,t)|\le C,\ \forall(\lambda,\mu)\in\sigma(A,B),\ t\in[0,T].$$

Take arbitrary $0<\delta,\ K<+\infty$. Then $\sigma(A,B)=\Delta_1\cup\Delta_2$, where $\Delta_1=\{(\lambda,\mu)\in\sigma(A,B)|\ |\frac{\omega_2}{\omega_1}+1|\ge\delta\}\cup\{(\lambda,\mu)\in\sigma(A,B)|\max(|\omega_1|,|\omega_2|)\le K\}$, $\Delta_2=\{(\lambda,\mu)\in\sigma(A,B)|\ |\frac{\omega_2}{\omega_1}+1|<\delta,\max(|\omega_1|,|\omega_2|)>K\}$. For each $(\lambda,\mu)\in\Delta_1:\max(|\omega_1|,|\omega_2|)\le K$ or: $|\frac{\omega_1}{\omega_1+\omega_2}|\le\frac{1}{\delta}$, $|\frac{\omega_2}{\omega_1+\omega_2}|=|1-\frac{\omega_1}{\omega_1+\omega_2}|\le 1+\frac{1}{\delta}\Rightarrow \max(|\omega_1|,|\omega_2|)\le(1+\delta^{-1})|\omega_1+\omega_2|=(1+\delta^{-1})|\lambda|$. Thus,

$$\begin{aligned}\max(|\omega_1|,|\omega_2|)\cdot|\psi_1(\lambda,\mu,t)|&\le(K+\delta^{-1}+1)\,(|\lambda|+1)|\psi_1(\lambda,\mu,t)|\\&\le(K+\delta^{-1}+1)C'_T,\ \forall(\lambda,\mu)\in\Delta_1,\ t\in[0,T].\end{aligned}$$

If, moreover, $\exists 0<\gamma<+\infty\forall(\lambda,\mu)\in\Delta_2:\operatorname{Re}\omega_1=\max\limits_{i=1,2}\operatorname{Re}\omega_i\le\gamma$, then Lemma 3.6 is valid. Indeed, in this case for any $(\lambda,\mu)\in\Delta_2,\ t\in[0,T]$:

$$\psi_0(\lambda,\mu,t)=-\omega_1\psi_1(\lambda,\mu,t)+e^{\omega_1 t}\Rightarrow$$

$$|\omega_1\psi_1(\lambda,\mu,t)|\le|\psi_0(\lambda,\mu,t)|+|e^{\omega_1 t}|\le C'_T+e^{\gamma t},$$

similarly $|\omega_2\psi_1(\lambda,\mu,t)|\le C'_T+e^{\gamma t}$, and thus $\max(|\omega_1|,|\omega_2|)\cdot|\psi_1(\lambda,\mu,t)|\le C'_T+e^{\gamma T},\ \forall(\lambda,\mu)\in\Delta_2,\ t\in[0,T]$.

So, we shall have the proof completed if we find $0<\delta,\ K<+\infty$ such that $\exists 0<\gamma<+\infty\forall(\lambda,\mu)\in\Delta_2:\operatorname{Re}\omega_1\le\gamma$. For such a γ to exist, it is sufficient that $\forall(\lambda,\mu)\in\Delta_2\exists\frac{T}{2}\le t_0\le T$ such that $|e^{(\omega_2-\omega_1)t_0}-(-1)|\ge 2\delta$. Indeed, in this case we have for each $(\lambda,\mu)\in\Delta_2$ such that $\operatorname{Re}\omega_1>0$:

$$\left|\frac{\omega_1}{\omega_1-\omega_2}\right|=\left|1-\frac{\omega_2}{\omega_1}\right|^{-1}=\left|2-\left(1+\frac{\omega_2}{\omega_1}\right)\right|^{-1}\ge(2+\delta)^{-1},$$

$$\left|e^{(\omega_2-\omega_1)t_0}-\frac{\omega_2}{\omega_1}\right|\ge\left|e^{(\omega_2-\omega_1)t_0}-(-1)\right|-\left|\frac{\omega_2}{\omega_1}-(-1)\right|\ge 2\delta-\delta=\delta,$$

thus

$$\begin{aligned}(2+\delta)^{-1}\cdot e^{\operatorname{Re}\omega_1\cdot\frac{T}{2}}\cdot\delta&\le\left|\left(\frac{\omega_1}{\omega_1-\omega_2}\right)\cdot e^{\omega_1 t_0}\cdot\left(e^{(\omega_2-\omega_1)t_0}-\frac{\omega_2}{\omega_1}\right)\right|\\&=|\psi_0(\lambda,\mu,t_0)|\le C'_T\Rightarrow\operatorname{Re}\omega_1\le\frac{2}{T}\ln\left(\frac{2+\delta}{\delta}\cdot C'_T\right).\end{aligned}$$

Now we set $0<\delta<\frac{1}{2}$, $K>\frac{2\pi+|\ln(1-2\delta)|}{T}$ and show that in this case

$$\forall(\lambda,\mu)\in\Delta_2\exists\frac{T}{2}\le t_0\le T\ \text{such that}\ |e^{(\omega_2-\omega_1)t_0}-(-1)|\ge 2\delta.$$

If this does not hold, then $\exists(\lambda,\mu)\in\Delta_2$ such that $\forall\frac{T}{2}\leq t\leq T: |e^{(\omega_2-\omega_1)t}-(-1)|<2\delta$. In other words, for each $\frac{T}{2}\leq t\leq T$ the number $e^{(\omega_2-\omega_1)t}$ lies in the complex plane C^1 inside the circle of radius 2δ centered at -1. In particular, for each $\frac{T}{2}\leq t\leq T$ the number $e^{(\omega_2-\omega_1)t}$ is situated in the half-plane $\{z\in C^1|\operatorname{Re}z<0\}$; hence $|\operatorname{Im}((\omega_2-\omega_1)T)-\operatorname{Im}((\omega_2-\omega_1)\frac{T}{2})|<\pi\Rightarrow|\operatorname{Im}(\omega_2-\omega_1)|<\frac{2\pi}{T}$. Furthermore, for $t=T: |e^{(\omega_2-\omega_1)t}|\geq 1-2\delta>0$, therefore $\operatorname{Re}(\omega_2-\omega_1)T\geq \ln(1-2\delta)\Rightarrow|\operatorname{Re}(\omega_2-\omega_1)|\leq\frac{|\ln(1-2\delta)|}{T}$. Thus, for this $(\lambda,\mu): |\omega_1-\omega_2|\leq |\operatorname{Re}(\omega_1-\omega_2)|+|\operatorname{Im}(\omega_1-\omega_2)|\leq\frac{2\pi+|\ln(1-2\delta)|}{T}<K$.

On the other hand, by virtue of the definition of Δ_2, we have for this (λ,μ) : $|\frac{\omega_2}{\omega_1}|<1+\delta$; hence $|\omega_1|>(1+\delta)^{-1}\max(|\omega_1|,|\omega_2|)$; finally, $|\omega_1-\omega_2|=|\omega_1|\cdot|2-(1+\frac{\omega_2}{\omega_1})|>\frac{2-\delta}{1+\delta}\max(|\omega_1|,|\omega_2|)>\frac{2-\delta}{1+\delta}K>K$. We arrive at a contradiction.

The proposition is proved, and with it Lemma 3.6.

IV We show that the Cauchy problem for equation (1) on R_+ is weakly well-posed in $G=H\times H_{-1}=H\times H_-(|A|+|B|^{1/2}+1)$ if and only if for every $0<T<+\infty$ there exists $C_T<+\infty$ such that (3.6) holds. Similarly, we show that the Cauchy problem for (1) on R_+ is weakly well-posed in $G=H\times H_-(|A|+1)$ if and only if for every $0<T<+\infty$ there exists $C'_T<+\infty$ such that (3.7) holds.

The *proof* is omitted here; it is quite similar to the proof of Theorem 2.1 (see Lemmas 2.1, 2.3–2.6), with Ag, Bg replaced by A^*g, B^*g and so on (note that $\Phi=D(A^\infty)\cap D(B^\infty)=D((A^*)^\infty)\cap D((B^*)^\infty)$; $|\psi_i^{(k)}(\bar\lambda,\bar\mu,t)|=|\overline{\psi_i^{(k)}(\lambda,\mu,t)}|=|\psi_i^{(k)}(\lambda,\mu,t)|$).

4 A criterion for the well-posedness

Up to this point, we considered the weak well-posedness of the Cauchy problem for equation (1) on R_+. Now we go on to study the «usual» well-posedness.

Theorem 3.2. *The Cauchy problem on R_+ for equation* (1) *with c.n.o. A and B is well-posed in $G=D(B)\times(D(A)\cap D(|B|^{1/2}))=D(B)\times H_1$ if and only if one of the equivalent conditions* 1), 2), 3) *of Theorem* 3.1 *holds. If this is the case, then for every $(f_0,f_1)\in D(B)\times(D(A)\cap D(|B|^{1/2}))$ the solution of the Cauchy problem for* (1) *on R_+ with initial conditions $y(0)=f_0$, $y'(0)=f_1$ (i.e., with initial data (f_0,f_1)) has the form $y(t)=\psi_0(A,B,t)f_0+\psi_1(A,B,t)f_1$ $(t\in R_+)$.*

Proof. Let $f_0,f_1\in H$. If a usual solution of the Cauchy problem for (1) on R_+ with i.d. (f_0,f_1) exists then, by the definitions, it is also a weak solution of the Cauchy problem for (1) on R_+ with these i.d.; similarly to Lemma 2.1, such a weak solution is unique and can be written in the following form: $y(t)=\psi_0(A,B,t)f_0+\psi_1(A,B,t)f_1$ $(t\in R_+)$.

Let condition 1) of Theorem 3.1 hold. Then, by Lemma 3.4, $\forall 0<T<+\infty\exists C_T<+\infty$ such that (3.6) holds. Now one can prove, just as in Lemma 2.4, that $\forall f_0 \in D(B)$, $f_1 \in D(A) \cap D(|B|^{1/2}) : y(t) = \psi_0(A,B,t)f_0 + \psi_1(A,B,t)f_1$ is a solution of (1) on R_+ such that $y(0) = f_0$, $y'(0) = f_1$. The sufficiency is proved.

To prove the necessity of condition 1) of Theorem 3.1, it suffices to show that $\forall 0 < T < +\infty \exists C_T < +\infty$ such that (3.6) is valid; if this is the case, then the application of Lemma 3.5 completes the proof of the necessity.

Let the Cauchy problem for equation (1) on R_+ be well-posed in $D(B) \times H_1$. For any $f_0 \in D(B)$ there exists a unique solution $y(t)$ of (1) on R_+ such that $y(0) = f_0$, $y'(0) = 0$; it was shown above that $y(t) = \psi_0(A,B,t)f_0$ $(t \in R_+)$; so, by the definition of a solution of equation (1) on $R_+ : \psi_0(A,B,t)f_0 \in C(R_+,H)$, $B\psi_0(A,B,t)f_0 \in C(R_+,H)$. For any $f_0' \in H : (|B|+1)^{-1}f_0' \in D(B)$; thus, $\psi_0(A,B,t)\,(|B|+1)^{-1}f_0' \in C(R_+,H)$, $B\psi_0(A,B,t)\,(|B|+1)^{-1}f_0' \in C(R_+,H)$, $\forall f_0' \in H$.

For any $f_1 \in H_1$ there exists a unique solution $y(t)$ of (1) on R_+ such that $y(0) = 0$, $y'(0) = f_1$; it was shown above that $y(t) = \psi_1(A,B,t)f_1$ $(t \in R_+)$. For each $t \in R_+ : (\psi_1(A,B,t)f_1)' = \psi_1'(A,B,t)f_1$; indeed, for every $k \geq 1 :$ $E_k(\psi_1(A,B,t)f_1)' = (\psi_1(A,B,t)E_kf_1)' = \psi_1'(A,B,t)E_kf_1$, thus $\exists \psi_1'(A,B,t)f_1 = \lim_{k\to\infty} \psi_1'(A,B,t)E_kf_1 = \lim_{k\to\infty} E_k(\psi_1(A,B,t)f_1)' = (\psi_1(A,B,t)f_1)' \in H$.

Now we have by the definition of a solution of equation (1) on R_+: $\psi_1(A,B,t)f_1$, $B\psi_1(A,B,t)f_1$, $A\psi_1'(A,B,t)f_1 \in C(R_+,H)$. For every $f_1' \in H :$ $(|A| + |B|^{1/2} + 1)^{-1}f_1' \in H_1 = D(|A| + |B|^{1/2} + 1)$; thus, $\forall f_1' \in H$:

$$\psi_1(A,B,t)\,(|A| + |B|^{1/2} + 1)^{-1}f_1', \quad B\psi_1(A,B,t)\,(|A| + |B|^{1/2} + 1)^{-1}f_1',$$
$$A\psi_1'(A,B,t)\,(|A| + |B|^{1/2} + 1)^{-1}f_1' \in C(R_+,H).$$

Fix an arbitrary $0 < T < +\infty$. Just as in the proof of Lemma 2.6, we deduce from the previous conclusions that there exists $C < +\infty$ such that $\forall(\lambda,\mu) \in \sigma(A,B)$, $t \in [0,T]$:

$$\frac{|\mu|+1}{|\mu|+1}|\psi_0(\lambda,\mu,t)| = |\psi_0(\lambda,\mu,t)| \leq C,$$
$$\frac{|\mu|+1}{|\lambda| + \sqrt{|\mu|} + 1}|\psi_1(\lambda,\mu,t)| \leq C, \quad \frac{|\lambda|}{|\lambda| + \sqrt{|\mu|} + 1}|\psi_1'(\lambda,\mu,t)| \leq C.$$

For each $(\lambda,\mu) \in \sigma(A,B)$ one of the following cases holds:
1) $\max(|\lambda|, \sqrt{|\mu|}, 1) = |\lambda|$; then for any $t \in [0,T]$:

$$\frac{1}{3}|\psi_1'(\lambda,\mu,t)| \leq \frac{|\lambda|}{|\lambda| + \sqrt{|\mu|} + 1}|\psi_1'(\lambda,\mu,t)| \leq C \Rightarrow |\psi_1'| \leq 3C \Rightarrow$$
$$|\lambda|\,|\psi_1| = |\psi_0 - \psi_1'| \leq |\psi_0| + |\psi_1'| \leq 4C$$
$$\Rightarrow (|\lambda| + \sqrt{|\mu|} + 1)|\psi_1(\lambda,\mu,t)| \leq 12C;$$

2) $\max(|\lambda|, \sqrt{|\mu|}, 1) = \sqrt{|\mu|}$; then for any $t \in [0, T]$:

$$\frac{1}{3}\sqrt{|\mu|}|\psi_1(\lambda,\mu,t)| \le \frac{\sqrt{|\mu|}}{|\lambda| + \sqrt{|\mu|} + 1}\sqrt{|\mu|}|\psi_1| = \frac{|\mu|}{|\lambda| + \sqrt{|\mu|} + 1}|\psi_1| \le C$$

$$\Rightarrow \sqrt{|\mu|}|\psi_1| \le 3C \Rightarrow (|\lambda| + \sqrt{|\mu|} + 1)|\psi_1(\lambda,\mu,t)| \le 9C;$$

3) $\max(|\lambda|, \sqrt{|\mu|}, 1) = 1$; then for any $t \in [0, T]$:

$$\frac{1}{3}|\psi_1(\lambda,\mu,t)| \le \frac{1}{|\lambda| + \sqrt{|\mu|} + 1}|\psi_1| \le C$$

$$\Rightarrow |\psi_1| \le 3C \Rightarrow (|\lambda| + \sqrt{|\mu|} + 1)|\psi_1(\lambda,\mu,t)| \le 9C.$$

Summarizing, we have for all $(\lambda, \mu) \in \sigma(A, B)$, $t \in [0, T] : |\psi_0(\lambda, \mu, t)| \le C$, $(|\lambda|+\sqrt{|\mu|}+1)|\psi_1(\lambda,\mu,t)| \le 12C$; i.e., $\exists C_T = 12C < +\infty$ such that (3.6) holds. The theorem is proved.

5 The (weak) well-posedness in particular cases

Let us analyze condition 1) of Theorem 3.1, which is necessary and sufficient for the weak well-posedness of the Cauchy problem for equation (1) on R_+ in $G = H \times H_{-1}$. In addition to the traditional condition (2.2) and the equivalent one (3.3), which can also be rewritten in the following form:

$$\sup_{(\lambda,\mu)\in\sigma(A,B),\, i=1,2} \operatorname{Re}\omega_i(\lambda,\mu) < +\infty,$$

there arise a new condition (3.2) and the equivalent one (3.4). This condition (3.4) can also be rewritten in the following form:

$$\exists C < +\infty \forall (\lambda, \mu) \in \sigma(A, B) : \ |\omega_1 + \omega_2| \le C(1 + |\operatorname{Re}(\omega_1 + \omega_2)| + |\omega_1 - \omega_2|), \tag{3.9}$$

or, what is the same:

$$\exists C < +\infty \forall (\lambda, \mu) \in \sigma(A, B) : \ |\lambda| \le C(1 + |\operatorname{Re}\lambda| + \sqrt{|4\mu - \lambda^2|}).$$

In a series of important particular cases, this condition (3.2) is satisfied automatically, and so (2.2) (i.e., (3.3)) becomes a necessary and sufficient condition for the weak well-posedness.

Corollary 2. *Let A and B be c.n.o. in H, and moreover, let $D(\operatorname{Re}A) \subseteq D(\operatorname{Im}A)$, i.e., $\operatorname{Im}A$ be subordinate to $\operatorname{Re}A$. Then* (3.4) *holds; thus, condition* (2.2) *(i.e.,* (3.3)*) is necessary and sufficient for the weak well-posedness of the Cauchy problem for* (1) *on R_+ in $G = H \times H_{-1}$.*

Indeed, by virtue of Theorem 1.2: $D(\operatorname{Re} A) \subseteq D(\operatorname{Im} A) \Leftrightarrow \exists C < +\infty \forall \lambda \in \sigma(A) : |\operatorname{Im}\lambda| \leq C(1 + |\operatorname{Re}\lambda|)$ whence $\forall(\lambda, \mu) \in \sigma(A, B) \subseteq \sigma(A) \times \sigma(B) : |\operatorname{Im}\lambda| \leq C(1 + |\operatorname{Re}\lambda|)$.

Remark 1. Among particular cases of c.n.o. A and B where $D(\operatorname{Re} A) \subseteq D(\operatorname{Im} A)$ are the following:

1) c.n.o. A and B where A is self-adjoint;
2) c.n.o. A and B where A is bounded.

Corollary 3. *Let A and B be c.n.o. in H, and moreover, let $D(A) \subseteq D(B)$, i.e., B be subordinate to A. Then the conclusion of Corollary 2 is valid, and moreover, the necessary and sufficient condition for the weak well-posedness of the Cauchy problem for* (1) *on R_+ in $G = H \times H_{-1}$ can be rewritten in the following form:*

$$\inf \sigma(\operatorname{Re} A) > -\infty.$$

The proof is based on the following lemma.

Lemma 3.7. *Let A and B be c.n.o. in H. The following conditions are equivalent:*

1) $D(A) \subseteq D(B)$;
2) $\exists C < +\infty : \sigma(A, B) \subseteq \{(\lambda, \mu) |\, |\mu| \leq C^2\} \cup \{(\lambda, \mu) |\, |\mu| \leq C|\lambda|\}$;
3) $\exists C_1 < +\infty \forall(\lambda, \mu) \in \sigma(A, B) : \min(|\omega_1|, |\omega_2|) \leq C_1$.

Proof. The equivalence of conditions 1) and 2) follows from Theorem 1.2 for c.n.o. A and B. Furthermore, if 2) holds then, by virtue of Lemma 1.1, for each $(\lambda, \mu) \in \sigma(A, B)$:

$$\min(|\omega_1|, |\omega_2|) \leq 2 \min(\sqrt{|\mu|}, |\lambda|^{-1}|\mu|) \leq 2C,$$

i.e., 3) is valid. Conversely, if 3) holds then, by virtue of Lemma 1.1, for each $(\lambda, \mu) \in \sigma(A, B)$:

$$\min(\sqrt{|\mu|}, |\lambda|^{-1}|\mu|) \leq 2 \min(|\omega_1|, |\omega_2|) \leq 2C_1,$$

i.e., $\sqrt{|\mu|} \leq 2C_1$ or $|\mu| \leq 2C_1|\lambda|$; i.e., 2) is valid with $C = 2C_1$.

Proof of Corollary 3. We now complete the proof of Corollary 3. For definiteness, for each $(\lambda, \mu) \in \sigma(A, B)$ we denote by ω_2 a root of $\omega^2 + \lambda\omega + \mu$ such that $|\omega_2| = \min(|\omega_1|, |\omega_2|)$; this notation holds until the end of this proof. By Lemma 3.7, $\exists C < +\infty \forall(\lambda, \mu) \in \sigma(A, B) : |\omega_2| \leq C$. If $(\lambda, \mu) \in \sigma(A, B)$ is such that $|\omega_1| \leq 2C$, then $|\omega_i| \leq 2C$, $i = 1, 2$; otherwise

$$|\omega_1| > 2C \Rightarrow |\omega_1| > 2|\omega_2| \Rightarrow \left|\frac{\omega_1 - \omega_2}{\omega_1 + \omega_2}\right| \geq \frac{|\omega_1| - |\omega_2|}{|\omega_1| + |\omega_2|} \geq \frac{\frac{1}{2}|\omega_1|}{\frac{3}{2}|\omega_1|} = \frac{1}{3} > 0.$$

Summarizing, (3.2) is valid with $\varepsilon = \min(\frac{1}{2C}, \frac{1}{3})$. Moreover,

$$-C \leq \operatorname{Re}\omega_2(\lambda,\mu) \leq C \text{ for all } (\lambda,\mu) \in \sigma(A,B),$$

therefore condition (2.2) can be reformulated:

$$\sup_{(\lambda,\mu)\in\sigma(A,B),\, i=1,2} \operatorname{Re}\omega_i(\lambda,\mu) < +\infty \Leftrightarrow \sup_{(\lambda,\mu)\in\sigma(A,B)} \operatorname{Re}\omega_1(\lambda,\mu) < +\infty$$

$$\Leftrightarrow \sup_{(\lambda,\mu)\in\sigma(A,B)} (\operatorname{Re}\omega_1(\lambda,\mu) + \operatorname{Re}\omega_2(\lambda,\mu)) < +\infty$$

$$\Leftrightarrow \sup_{(\lambda,\mu)\in\sigma(A,B)} (-\lambda_1) < +\infty \Leftrightarrow \inf\sigma(\operatorname{Re}A) > -\infty.$$

Corollary 4. *Let A and B be c.n.o. in H, and moreover, let B be strongly subordinate to the operator $\frac{A^2}{4}$. Then the conclusion of Corollary 2 is valid.*

Proof. We conclude, from Theorem 1.2 for c.n.o. A and B and the remark after it, that B is strongly subordinate to $\frac{A^2}{4} \Leftrightarrow \exists 0 < R < +\infty, 0 < \varepsilon < 1 : \sigma(A,B) \subseteq \{(\lambda,\mu) |\, |\lambda|^2 + |\mu|^2 \leq R^2\} \cup \{(\lambda,\mu) |\, |\mu| \leq (1-\varepsilon)\frac{|\lambda|^2}{4}\}$, whence $\exists 0 < R < +\infty$, $0 < \varepsilon < 1 \forall (\lambda,\mu) \in \sigma(A,B) : |\lambda| \leq R$ or $|\mu - \frac{\lambda^2}{4}| \geq \varepsilon|\frac{\lambda^2}{4}| \Leftrightarrow |\frac{4\mu}{\lambda^2} - 1| \geq \varepsilon$; thus, (3.4) is valid, and with it (3.2) (it is equivalent to (3.4) by Lemma 3.3). The proof of Corollary 4 is over.

Thus, for «incomplete» equations of the form (1) with commuting normal A and B, i.e., if $A = 0$ (and even in the more general case of A bounded) or $B = 0$ (and even in the more general case of B subordinate to A), condition (3.2) holds automatically and, consequently, condition (2.2) is necessary and sufficient for the weak well-posedness of the Cauchy problem for (1) on R_+ in $G = H \times H_{-1}$. Notice that precisely these cases, where B is subordinate to A or A is bounded, are considered in [248] (without the additional assumption that A and B are commuting normal operators).

But in the general case of equation (1) with commuting normal A and B one cannot omit condition (3.2) in Theorem 3.1. Consider, for instance, A and B from Example 3.2. It is shown in Example 3.2 that condition (2.2) holds here. But for any $(\lambda,\mu) \in \sigma(A,B)$ such that $\lambda \neq 0$ we have: $\frac{|\lambda_2|}{|\lambda_1|} = +\infty$, $|\frac{4\mu}{\lambda^2} - 1| = 0$; taking into account that $\sup\limits_{(\lambda,\mu)\in\sigma(A,B)} |\lambda| = +\infty$, we obtain that (3.4) does not hold. Consequently, condition (3.2) is not satisfied.

It is shown in Example 3.2 that for these A and B not only the weak well-posedness in $H \times H_{-1}$ does not hold (which follows from Theorem 3.1), but even the weak well-posedness in $H \times H$ does not hold.

By the following example, consider how condition (2.2) is changed by the additional demand (3.2).

Example 3.3. Let A and B be c.n.o. in H, and moreover, let $\operatorname{Re} A = \operatorname{Im} B = 0$, i.e., $\forall(\lambda,\mu) \in \sigma(A,B) : \lambda_1 = \mu_2 = 0$. Then condition (3.3) may be written in the form:

$$\exists 0 < \gamma < +\infty \forall(\lambda,\mu) \in \sigma(A,B) : \mu_1 \geq -\frac{\lambda_2^2}{4} - \gamma^2,$$

whereas conditions (3.3) and (3.4) may be written in the form:

$$\exists 0 < \gamma,\ \varepsilon < +\infty \forall(\lambda,\mu) \in \sigma(A,B) : \mu_1 \geq (-1+\varepsilon)\frac{\lambda_2^2}{4} - \gamma^2.$$

6 Spectrum of the associated operator pencil

In order to study the Cauchy problem for equation (1), many authors considered the spectrum of the operator pencil $P(z) = z^2 + Az + B$, $z \in C^1$, associated with the equation (see, for instance, [332]). We denote this spectrum by $\sigma(P)$. By definition, $\sigma(P) \subseteq C^1$, and $C^1 \setminus \sigma(P)$ is the set of all $z \in C^1$ such that the linear operator $z^2 + Az + B$ in H with the domain $D(A) \cap D(B)$ has an inverse one in H which is densely defined in H and bounded.

For a *first order* equation $y'(t) + Ay(t) = 0$ where A is a normal operator, the associated operator pencil is of the form $P_1(z) = z + A$, $z \in C^1$. It is easy to see that $\sigma(P_1) = \{z \in C^1 | -z \in \sigma(A)\}$, and the following propositions are equivalent:

a) the Cauchy problem for the equation $y'(t) + Ay(t) = 0$ on R_+ is weakly well-posed in $G = H$ over $D(A^\infty)$;

b) $\inf_{\lambda \in \sigma(A)} \operatorname{Re} \lambda > -\infty$;

c) $\sup_{z \in \sigma(P_1)} \operatorname{Re} z < +\infty$.

For a *second order* equation $y''(t) + Ay'(t) + By(t) = 0$ with commuting normal operators A and B, Theorem 3.1 gives a necessary and sufficient condition for the weak well-posedness of the Cauchy problem for (1) on R_+ in $G = H \times H_{-1}$ in terms of the position of $\sigma(A,B)$ in C^2.

Statement 3.2. It is impossible to give a necessary and sufficient condition for the weak well-posedness of the Cauchy problem for equation (1) with commuting normal operators A and B on R_+ in $G = H \times H_{-1}$ in terms of the position in C^1 of the spectrum $\sigma(P)$ of the operator pencil $z^2 + Az + B$.

To prove this, we need the following lemma.

Lemma 3.8. *Let A and B be c.n.o. in H, and S stand for the set of all roots of polynomials $\omega^2 + \lambda\omega + \mu$ such that $(\lambda,\mu) \in \sigma(A,B)$. Then $S \subseteq \sigma(P) \subseteq \bar{S}$, where $\bar{S}$ denotes the closure of the set S in C^1.*

Lemma 3.8 follows immediately from the fact that in the case of c.n.o. A and B, $C^1 \setminus \sigma(P)$ is the set of all $z \in C^1$ such that $(z^2 + \lambda z + \mu)^{-1}(A, B)$ is a bounded operator on H, i.e.

$$\sup_{(\lambda,\mu)\in\sigma(A,B)} |(z^2 + \lambda z + \mu)^{-1}| < +\infty,$$

i.e.

$$\inf_{(\lambda,\mu)\in\sigma(A,B)} |(z - \omega_1(\lambda, \mu))\,(z - \omega_2(\lambda, \mu))| > 0.$$

Proof of Statement 3.2. To prove Statement 3.2, we give the following two examples. Spectrum of an operator pencil $z^2 + Az + B$ is one and the same in both cases, but the weak well-posedness of the Cauchy problem for (1) on R_+ in $G = H \times H_{-1}$ holds in the first case and does not hold in the second one.

Example 3.4. Let $H = l_2(C^1)$, $A = 0$, $B = (b_{kl})_{k,l=1}^{\infty}$ where $b_{kl} = 0$ for $k \neq l$, $b_{kl} = k^2$ for $k = l$.

Then $\sigma(A, B) = \{(\lambda_n, \mu_n) | n \in \mathbb{N},\ \lambda_n = 0,\ \mu_n = n^2\}$. For any $n \in \mathbb{N}$, we have for $(\lambda_n, \mu_n) \in \sigma(A, B) : \{\omega_1, \omega_2\} = \{in, -in\}$ (here i denotes the imaginary unit), thus $\operatorname{Re}\omega_1 = \operatorname{Re}\omega_2 = 0$, $|\frac{\omega_1 - \omega_2}{\omega_1 + \omega_2}| = +\infty$. Both conditions (2.2), (3.2) hold and so the weak well-posedness in $G = H \times H_{-1}$ holds by virtue of Theorem 3.1.

Here $S = \{in | n \in \mathbb{Z} \setminus \{0\}\}$; $\bar{S} = S$; so by Lemma 3.8: $\sigma(P) = \{in | n \in \mathbb{Z} \setminus \{0\}\}$.

Example 3.5. Let $H = l_2(C^1)$, and let $A = (a_{kl})_{k,l=1}^{\infty}$, $B = (b_{kl})_{k,l=1}^{\infty}$ be linear operators in H defined by diagonal matrices: $a_{kl} = b_{kl} = 0$ for $k \neq l$; $a_{kl} = -i(2k + 1)$, $b_{kl} = -k(k + 1)$ for $k = l$, if k is odd; $a_{kl} = i(2k - 1)$, $b_{kl} = -k(k - 1)$ for $k = l$, if k is even.

Then $\sigma(A, B) = \{(\lambda_n, \mu_n) | n \in \mathbb{N},\ \lambda_n = a_{nn},\ \mu_n = b_{nn}\}$. For an arbitrary odd $n \in \mathbb{N}$, we have for $(\lambda_n, \mu_n) \in \sigma(A, B) : \{\omega_1, \omega_2\} = \{in, i(n + 1)\}$, and so $\operatorname{Re}\omega_2 = \operatorname{Re}\omega_1 = 0$, $|\frac{\operatorname{Im}\omega_i}{\operatorname{Re}\omega_i}| = +\infty$ $(i = 1, 2)$, $\min\limits_{i=1,2} |\omega_i| = n$, $|\frac{\omega_1 - \omega_2}{\omega_1 + \omega_2}| = \frac{1}{2n+1}$. For an arbitrary even $n \in \mathbb{N}$, we have for $(\lambda_n, \mu_n) \in \sigma(A, B) : \{\omega_1, \omega_2\} = \{-i(n - 1), -in\}$, and so $\operatorname{Re}\omega_1 = \operatorname{Re}\omega_2 = 0$, $|\frac{\operatorname{Im}\omega_i}{\operatorname{Re}\omega_i}| = +\infty$ $(i = 1, 2)$, $\min\limits_{i=1,2} |\omega_i| = n - 1$, $|\frac{\omega_1 - \omega_2}{\omega_1 + \omega_2}| = \frac{1}{2n-1}$.

Summarizing, $\forall(\lambda_n, \mu_n) \in \sigma(A, B) : \operatorname{Re}\omega_1 = \operatorname{Re}\omega_2 = 0$; but $|\frac{\operatorname{Im}\omega_i}{\operatorname{Re}\omega_i}| = +\infty$ $(i = 1, 2)$, $\max\limits_{i=1,2} |\omega_i| \xrightarrow[n \to +\infty]{} +\infty$, $|\frac{\omega_1 - \omega_2}{\omega_1 + \omega_2}| \xrightarrow[n \to +\infty]{} 0$.

Thus, condition (2.2) holds, but condition (3.2) does not hold and so the weak well-posedness in $G = H \times H_{-1}$ does not hold, by virtue of Theorem 3.1.

Here $S = \{in | n \in \mathbb{Z} \setminus \{0\}\}$; $\bar{S} = S$; so by Lemma 3.8: $\sigma(P) = \{in | n \in \mathbb{Z} \setminus \{0\}\}$.

Statement 3.2 is proved.

Remark 1. Statement 3.2 remains valid if one replaces $G = H \times H_{-1}$ by an arbitrary G such that $H \times H \subseteq G \subseteq H \times H_{-1}$ (see Examples 3.4, 3.5).

Notice that condition (2.2) itself *can* be expressed in terms of the position of $\sigma(P)$ in C^1.

Lemma 3.9. *For c.n.o. A and B, condition (2.2) is equivalent to the following one:*

$$\sup_{z \in \sigma(P)} \operatorname{Re} z < +\infty. \tag{3.10}$$

Lemma 3.9 follows immediately from Lemma 3.8.

In particular, we deduce the following statement from Corollaries 2–4 of Theorem 3.1 and from Lemma 3.9.

Statement 3.3. Let A and B be c.n.o. in H, and moreover, let one of the following additional assumptions hold:

a) $D(\operatorname{Re} A) \subseteq D(\operatorname{Im} A)$;

b) $D(A) \subseteq D(B)$;

c) B is strongly subordinate to $\frac{A^2}{4}$.

The Cauchy problem for equation (1) on R_+ is weakly well-posed in $G = H \times H_{-1}$ if and only if (3.10) holds.

7 Fattorini's definitions of well-posedness of the Cauchy problem

In conclusion of Chapter 3, we discuss the relation between our definitions of well-posedness in $G \subseteq H \times H$ and weak well-posedness in $G \subseteq H \times \Phi'$ of the Cauchy problem for equation (1) on R_+, on one hand, — and the definitions of uniform well-posedness and strong well-posedness of the Cauchy problem for (1) on R_+ that are given by H.O. Fattorini in [93, Chapter VIII], on the other hand.

First of all, we restate those definitions.

The Cauchy problem for equation (1) on R_+ is called uniformly well-posedif the following two conditions are satisfied:

1) there exist linear subspaces D_0, D_1 of H which are dense in H and such that for any $(f_0, f_1) \in D_0 \times D_1$ there exists a solution $y(t)$ of the Cauchy problem for (1) on R_+ with i.c. $y(0) = f_0$, $y'(0) = f_1$;

2) there exists a nonnegative nondecreasing function $C(t)$ on R_+ such that for every solution $y(t)$ of (1) on R_+:

$$\| y(t) \| \leq C(t) (\| y(0) \| + \| y'(0) \|) \text{ for all } t \in R_+.$$

For each $f_1 \in D_1$ we set: $S(t)f_1 = y(t)$ for all $t \in R_+$, where $y(t)$ is the solution of (1) on R_+ having $y(0) = 0$, $y'(0) = f_1$. For each $t \geq 0$: $S(t)$ is a linear bounded operator in H, defined on the set D_1, which is dense in H. Thus, $S(t)$ can be extended up to a linear bounded operator on the whole H, which will be denoted by the same symbol.

But, as H.O. Fattorini showed in [84], [93, Chapter VIII], use of this definition of uniform well-posedness leads to paradoxical situations entailing loss of exponential growth of solutions and nonexistence of phase spaces; moreover, the Cauchy problem may be uniformly well-posed on a finite segment $[0, T]$ but not be uniformly well-posed on $[0, +\infty)$. This motivates the introduction of the additional Assumption 3.1.

Assumption 3.1 [93, Chapter VIII]:

1) For each $f \in H : S(t)f$ is continuously differentiable on R_+.
2) $S(t)H \subseteq D(A)$ for every $t \in R_+$, and for each $f \in H : AS(t)f$ is continuous on R_+.

The Cauchy problem for equation (1) on R_+ is said to be strongly well-posed if it is uniformly well-posed and Assumption 3.1 is satisfied.

Statement 3.4. Let A and B be commuting normal operators in H.

1. The following propositions are equivalent:

   a) the Cauchy problem for equation (1) on R_+ is weakly well-posed in $G = H \times H$;

   b) the Cauchy problem for equation (1) on R_+ is uniformly well-posed in the sense of H.O. Fattorini.

2. The following propositions are equivalent:

   a) the Cauchy problem for equation (1) on R_+ is well-posed in $G = D(B) \times (D(A) \cap D(|B|^{1/2}))$;

   b) the Cauchy problem for (1) on R_+ is weakly well-posed in $G = H \times H_-(|A| + 1)$;

   c) the Cauchy problem for (1) on R_+ is weakly well-posed in $G = H \times H_-(|A| + |B|^{1/2} + 1) = H \times H_{-1}$;

   d) the Cauchy problem for (1) on R_+ is strongly well-posed in the sense of H.O. Fattorini.

Proof. Taking into account Lemma 3.1, in order to prove point 1 it suffices to show that 1b) is valid if and only if $\forall 0 < T < +\infty \exists C_T < +\infty$ such that (3.1) holds.

Let $\forall 0 < T < +\infty \exists C_T < +\infty$ such that (3.1) holds.

Setting $D_0 = D_1 = \bigcup_{k=1}^{\infty} E_k H$, we see that for any $f_0 \in D_0$, $f_1 \in D_1$ there exists a solution $y(t) = \psi_0(A, B, t)f_0 + \psi_1(A, B, t)f_1$ of (1) on R_+ such that $y(0) = f_0$, $y'(0) = f_1$.

According to the proof of Theorem 3.2, if a solution $y(t)$ of (1) on R_+ such that $y(0) = f_0$, $y'(0) = f_1$ exists, then it can be written in the form: $y(t) = \psi_0(A,B,t)f_0 + \psi_1(A,B,t)f_1$ $(t \in R_+)$. Fix an arbitrary $0 < T < +\infty$. For any solution $y(t)$ of (1) on R_+ and for any $t \in [0,T]$ we have:

$$\begin{aligned}\| y(t) \| &= \| \psi_0(A,B,t)y(0) + \psi_1(A,B,t)y'(0) \| \\ &\leq \| \psi_0(A,B,t) \| \cdot \| y(0) \| + \| \psi_1(A,B,t) \| \cdot \| y'(0) \| \\ &= \sup_{(\lambda,\mu)\in\sigma(A,B)} |\psi_0(\lambda,\mu,t)| \cdot \| y(0) \| + \sup_{(\lambda,\mu)\in\sigma(A,B)} |\psi_1(\lambda,\mu,t)| \cdot \| y'(0) \| \\ &\leq C_T(\| y(0) \| + \| y'(0) \|).\end{aligned}$$

Thus, 1b) is valid.

Conversely, let 1b) be valid. Fix an arbitrary $0 < T < +\infty$. For any $f_0 \in D_0$, according to the proof of Theorem 3.2, a solution $y(t)$ of (1) on R_+ such that $y(0) = f_0$, $y'(0) = 0$ has the form: $y(t) = \psi_0(A,B,t)f_0$ $(t \in R_+)$. By virtue of condition 2) of the definition of the uniformly well-posed Cauchy problem for (1), for any $t \in [0,T]$: $\| y(t) \| = \| \psi_0(A,B,t)f_0 \| \leq C(T) \cdot \| f_0 \|$. Since $\| \psi_0(A,B,t)f_0 \| \leq C(T) \cdot \| f_0 \|$ on the set D_0 which is dense in H,

$$\| \psi_0(A,B,t) \| = \sup_{(\lambda,\mu)\in\sigma(A,B)} |\psi_0(\lambda,\mu,t)| \leq C(T),$$

for an arbitrary $t \in [0,T]$. Similarly,

$$\sup_{(\lambda,\mu)\in\sigma(A,B)} |\psi_1(\lambda,\mu,t)| \leq C(T), \quad \forall t \in [0,T].$$

Thus, $\forall 0 < T < +\infty \exists C_T = C(T) < +\infty$ such that (3.1) is valid.

We go on to prove point 2. By virtue of Theorems 3.1 and 3.2, conditions 2a), 2b), and 2c) are equivalent. Taking into account part IV of the proof of Theorem 3.1 and Lemma 3.6, we conclude that 2c) is equivalent to the following proposition: for every $0 < T < +\infty$ there exists $C'_T < +\infty$ such that (3.7) holds. To prove point 2, it now suffices to show that 2d) holds if and only if for every $0 < T < +\infty$ there exists $C'_T < +\infty$ such that (3.7) is valid. One can prove this similarly to point 1.

Chapter 4
Boundary-value problems on a finite segment

Let $-\infty < a < b < +\infty$. We now go on to study other classical boundary-value problems for equation (1) with c.n.o. A and B on the finite segment $[a, b]$.

1 The (weak) well-posedness of the Dirichlet problem

Theorem 4.1. *Let A and B be c.n.o. in H. The Dirichlet problem for equation* (1) *on $[a, b]$ is weakly well-posed in $G = H \times H$ if and only if one of the following equivalent conditions* 1), 2), 3) *holds:*

1) (2.8) *and* (2.9) *are valid;*
2) *a) there exists $0 < \gamma < +\infty$ such that $\sigma(A, B)$ is situated in C^2 outside the set*

$$\{(\lambda, \mu) \in C^2 | \lambda_1 > 2\gamma,$$
$$\mu_1 > \frac{1}{(\lambda_1 - 2\gamma)^2}(\mu_2^2 - \lambda_1\lambda_2 \cdot \mu_2 + \gamma(\lambda_1 - \gamma)(\lambda_2^2 + (\lambda_1 - 2\gamma)^2))\}$$
$$\cup \{(\lambda, \mu) \in C^2 | \lambda_1 < -2\gamma,$$
$$\mu_1 > \frac{1}{(\lambda_1 + 2\gamma)^2} \cdot (\mu_2^2 - \lambda_1\lambda_2 \cdot \mu_2 - \gamma(\lambda_1 + \gamma)\,(\lambda_2^2 + (\lambda_1 + 2\gamma)^2))\}; \tag{4.1}$$

b) there exists $0 < \varepsilon < +\infty$ such that $\sigma(A, B)$ is situated in C^2 outside the set

$$\bigcup_{l=1}^{\infty}\{(\lambda, \mu) \in C^2 | \left|\frac{\lambda^2}{4} - \mu + \frac{\pi^2 l^2}{(b-a)^2}\right| < l\varepsilon\}; \tag{4.2}$$

3) *there exist $0 < \gamma$, $\varepsilon < +\infty$ such that for each $(\lambda, \mu) \in \sigma(A, B)$:*
a) (μ_1, μ_2) lies in R^2 outside the union over all $l \in \mathbb{N}$ of the circles of radius εl centered at $(\frac{\lambda_1^2 - \lambda_2^2}{4} + \frac{\pi^2 l^2}{(b-a)^2}, \frac{\lambda_1\lambda_2}{2})$; (4.3)
b) if $|\lambda_1| > 2\gamma$ then (μ_1, μ_2) lies in R^2 outside the interior of the parabola $\mu_1 > \frac{1}{(|\lambda_1| - 2\gamma)^2} \cdot \mu_2^2$ transferred by $\frac{\lambda_1\lambda_2}{2}$ along the axis $0\mu_2$ and by $-\frac{\lambda_2^2}{4} + \gamma|\lambda_1| - \gamma^2$ along the axis $0\mu_1$. (4.4)

In this case, for every $(f_0, f_1) \in H \times H$ the weak solution of the Dirichlet problem for (1) *on $[a, b]$ with boundary conditions $y(a) = f_0$, $y(b) = f_1$ (or, what is the same, with b.d. (f_0, f_1)) has the form:*

$$y(t) = \sum_{i=0}^{1} \varphi_i(A, B, t) f_i \quad (t \in [a, b]).$$

Proof. The equivalence of conditions 1) and 2) follows from Lemma 3.2, similarly to the proof of Lemma 1.2. The equivalence of conditions 2) and 3) is obvious.

Lemma 4.1. *For the existence of a* $0 < C < +\infty$ *such that* $|\varphi_i(\lambda, \mu, t)| \leq C$ *for all* $(\lambda, \mu, t) \in \sigma(A, B) \times [a, b]$, $i = 0, 1$, *it is necessary and sufficient that conditions* (2.8) *and* (2.9) *hold.*

Proof. Necessity. Let $\exists 0 < C < +\infty \forall(\lambda, \mu, t) \in \sigma(A, B) \times [a, b], i = 0, 1 :$ $|\varphi_i(\lambda, \mu, t)| \leq C$. Since $|1 - e^{-(\omega_1-\omega_2)(b-a)}| \leq 2$ for all $(\lambda, \mu) \in C^2$, this implies that

$$|e^{\omega_2(t-a)}(1 - e^{-(\omega_1-\omega_2)(b-t)})| \leq 2C, \quad |e^{\omega_1(t-b)}(1 - e^{-(\omega_1-\omega_2)(t-a)})| \leq 2C,$$
$$\forall(\lambda, \mu) \in \sigma(A, B), t \in [a, b]. \tag{4.5}$$

Fix a $0 < \delta < \min(b - a, \frac{2\pi}{b-a})$ such that for all $z \in C^1$: if $|z| \leq 2\delta|b - a|$ then $\frac{1}{2} \leq |\frac{1-e^z}{z}| \leq \frac{3}{2}$. Further, fix a $0 < \delta_1 < \min(\frac{\pi}{2}, \delta(b - a - \delta))$. We set $-\pi \leq \arg z < \pi$, $\forall z \in C^1$.

For each $(\lambda, \mu) \in \sigma(A, B)$ one of the following cases holds.

1. $\operatorname{Re}\omega_1 - \operatorname{Re}\omega_2 > \delta$. Setting $t = \frac{a+b}{2}$, we have:

$$|1-e^{-(\omega_1-\omega_2)(b-t)}| = |1-e^{-(\omega_1-\omega_2)(t-a)}| = |1-e^{-(\omega_1-\omega_2)\frac{b-a}{2}}| \geq 1-e^{-\delta\cdot\frac{b-a}{2}} > 0;$$

now (4.5) implies that

$$\left|e^{\omega_2\cdot\frac{b-a}{2}}\right| \leq 2C\left(1 - e^{-\delta\cdot\frac{b-a}{2}}\right)^{-1}, \quad \left|e^{\omega_1\cdot\frac{a-b}{2}}\right| \leq 2C\left(1 - e^{-\delta\cdot\frac{b-a}{2}}\right)^{-1}.$$

Hence $\operatorname{Re}\omega_2 \leq \gamma_1$, $\operatorname{Re}\omega_1 \geq -\gamma_1$, where

$$\gamma_1 = \frac{2}{b-a}\ln\left(2C\left(1 - e^{-\delta\cdot\frac{b-a}{2}}\right)^{-1}\right).$$

Furthermore,

$$|(\omega_1 - \omega_2)(b - a) - 2\pi i l| \geq |\operatorname{Re}(\omega_1 - \omega_2)(b - a)| \geq \delta(b - a), \quad \forall l \in \mathbb{Z}.$$

2. $0 \leq \operatorname{Re}\omega_1 - \operatorname{Re}\omega_2 \leq \delta$, $|\operatorname{Im}\omega_1 - \operatorname{Im}\omega_2| > \delta$.

Set $\alpha = -\operatorname{Im}(\omega_1 - \omega_2)(b - a - \delta)$. Let, for definiteness, $\operatorname{Im}\omega_1 - \operatorname{Im}\omega_2 > \delta$. Then $\alpha < 0$. If $-\pi < \alpha < 0$ then values of $\arg(e^{-(\omega_1-\omega_2)(b-t)})$ for $t \in [a + \delta, b]$ cover, in particular, the whole segment $[\alpha, 0]$. If $\alpha \leq -\pi$ then these values cover, in particular, $(-\pi, 0]$. In any case, for a certain $t_1 \in [a+\delta, b]$: $\arg(e^{-(\omega_1-\omega_2)(b-t_1)}) = -\delta_1$. Therefore $|1 - e^{-(\omega_1-\omega_2)(b-t_1)}| \geq \sin\delta_1$ (see Fig. 4.1). In the same way, one can treat the case of $\operatorname{Im}\omega_1 - \operatorname{Im}\omega_2 < -\delta$.

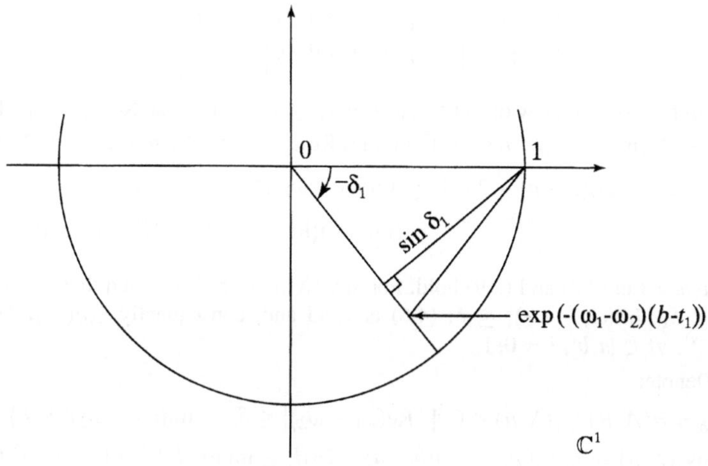

Fig. 4.1

Now (4.5) implies that $e^{\operatorname{Re}\omega_2\cdot\delta} \leq |e^{\omega_2(t_1-a)}| \leq \frac{2C}{\sin\delta_1}$ unless $\operatorname{Re}\omega_2 < 0$. Hence $\operatorname{Re}\omega_2 \leq \gamma_2$ where $\gamma_2 = |\delta^{-1}\ln\frac{2C}{\sin\delta_1}|$.

Similarly, one can prove that $\operatorname{Re}\omega_1 \geq -\gamma_2$.

Since $|\operatorname{Re}\omega_1 - \operatorname{Re}\omega_2| \leq \delta$ by the assumption, $\operatorname{Re}\omega_1 \geq -\gamma_2$, and $\operatorname{Re}\omega_2 \leq \gamma_2$, we conclude that $-\gamma_3 \leq \operatorname{Re}\omega_2 \leq \operatorname{Re}\omega_1 \leq \gamma_3$ where $\gamma_3 = \gamma_2 + \delta$.

So,

$$|e^{\omega_2(t_1-a)}| \geq e^{-\gamma_3(b-a)}, \quad |1 - e^{-(\omega_1-\omega_2)(b-t_1)}| \geq \sin\delta_1.$$

Now $|\varphi_0(\lambda, \mu, t_1)| \leq C$ implies that $|1 - e^{-(\omega_1-\omega_2)(b-a)}| \geq C^{-1}e^{-\gamma_3(b-a)}\sin\delta_1$.

Let $0 < \varepsilon_1 < 1$ be such that $\max\limits_{z\in C^1, |z|\leq\varepsilon_1} |1 - e^z| < C^{-1}e^{-\gamma_3(b-a)}\sin\delta_1$. We have proved that $|(\omega_1 - \omega_2)(b-a) - 2\pi i l| \geq \varepsilon_1$, $\forall l \in \mathbb{Z}$.

3. $0 \leq \operatorname{Re}\omega_1 - \operatorname{Re}\omega_2 \leq \delta$, $0 \leq |\operatorname{Im}\omega_1 - \operatorname{Im}\omega_2| \leq \delta$. Then $|\omega_1 - \omega_2| \leq 2\delta$. For any $t \in [a, b]$:

$$\frac{1}{2} \leq \left|\frac{1 - e^{-(\omega_1-\omega_2)(t-a)}}{(\omega_1 - \omega_2)(t-a)}\right| \leq \frac{3}{2}, \quad \frac{1}{2} \leq \left|\frac{1 - e^{-(\omega_1-\omega_2)(b-t)}}{(\omega_1 - \omega_2)(b-t)}\right| \leq \frac{3}{2}.$$

In particular, this is true for $t = a$ and $t = b$. Consequently, $\forall t \in [a,b]$:

$$\begin{aligned} \frac{1}{3}\left|\frac{t-a}{b-a}\right| &\le \left|\frac{1-e^{-(\omega_1-\omega_2)(t-a)}}{1-e^{-(\omega_1-\omega_2)(b-a)}}\right| \le 3\left|\frac{t-a}{b-a}\right|, \\ \frac{1}{3}\left|\frac{b-t}{b-a}\right| &\le \left|\frac{1-e^{-(\omega_1-\omega_2)(b-t)}}{1-e^{-(\omega_1-\omega_2)(b-a)}}\right| \le 3\left|\frac{b-t}{b-a}\right|. \end{aligned} \tag{4.6}$$

Set $t = \frac{a+b}{2}$. In view of (4.6), $|\varphi_0(\lambda,\mu,t)| \le C$ implies that $\operatorname{Re}\omega_2 \le \gamma_4$ where $\gamma_4 = \frac{2}{b-a}\ln(6C)$; $|\varphi_1(\lambda,\mu,t)| \le C$ implies $\operatorname{Re}\omega_1 \ge -\gamma_4$. Moreover, $\forall l \in \mathbb{Z}\setminus\{0\}$:

$$\begin{aligned} |(\omega_1-\omega_2)(b-a) - 2\pi i l| &\ge |\operatorname{Im}(\omega_1-\omega_2)(b-a) - 2\pi i l| \\ &\ge 2\pi|l| - \delta(b-a) \ge 2\pi - \delta(b-a) > 0. \end{aligned}$$

Sufficiency. Let (2.8) and (2.9) hold. For any $(\lambda,\mu) \in \sigma(A,B)$ such that $|\operatorname{Re}(\omega_1 - \omega_2)| \le \delta$, $|\operatorname{Im}(\omega_1-\omega_2)| \le \delta$: (4.6) is valid and, consequently, $|\varphi_i(\lambda,\mu,t)| \le 3e^{\gamma(b-a)}$, $\forall t \in [a,b]$, $i = 0,1$.

Denote:

$$\sigma_\delta = \sigma(A,B) \setminus \{(\lambda,\mu) \in C^2 \mid |\operatorname{Re}(\omega_1-\omega_2)| \le \delta, \quad |\operatorname{Im}(\omega_1-\omega_2)| \le \delta\}.$$

For any $(\lambda,\mu) \in \sigma_\delta : |(\omega_1-\omega_2)(b-a) - 2\pi i l| \ge \min(\varepsilon, \delta(b-a)) > 0$, $\forall l \in \mathbb{Z}$. Therefore $\exists \varepsilon_0 > 0 \forall (\lambda,\mu) \in \sigma_\delta : |1 - e^{-(\omega_1-\omega_2)(b-a)}| \ge \varepsilon_0$. Taking into account that $\forall (\lambda,\mu) \in \sigma_\delta$, $t \in [a,b] : |1 - e^{-(\omega_1-\omega_2)(t-a)}| \le 2$, $|1 - e^{-(\omega_1-\omega_2)(b-t)}| \le 2$, we obtain that $\forall (\lambda,\mu) \in \sigma_\delta : |\varphi_i(\lambda,\mu,t)| \le 2\varepsilon_0^{-1} e^{\gamma(b-a)}$, $\forall t \in [a,b]$, $i = 0,1$.

The lemma is proved.

Lemma 4.2. *Denote:*

$$\begin{aligned} N_D &= \bigcup_{l \in \mathbb{Z}\setminus\{0\}} \{(\lambda,\mu) \in C^2 | (\omega_1-\omega_2)(b-a) = 2\pi i l\} \\ &= \bigcup_{l \in \mathbb{N}} \left\{(\lambda,\mu) \in C^2 | 4\mu - \lambda^2 = \frac{4\pi^2 l^2}{(b-a)^2}\right\}. \end{aligned}$$

For each $k \in \mathbb{N}$, denote:

$$V_k = \square_k \cap \left\{(\lambda,\mu) \in C^2 | \inf_{l \in \mathbb{Z}\setminus\{0\}} |(\omega_1-\omega_2)(b-a) - 2\pi i l| \ge \frac{1}{k}\right\}.$$

1. *The Dirichlet problem for equation* (1) *on* $[a,b]$ *with boundary conditions* $y(a) = 0$, $y(b) = 0$ *has a unique weak solution if and only if* $E(N_D) = 0$.

2. *Let* $E(N_D) = 0$, $f_0 \in H$, $f_1 \in H$. *If the Dirichlet problem for* (1) *on* $[a,b]$ *with boundary conditions* $y(a) = f_0$, $y(b) = f_1$ *has a weak solution, then it is unique and has the form:*

$$y(t) = \lim_{k \to +\infty} \sum_{i=0}^{1} \varphi_i(A,B,t) E(V_k) f_i \quad (t \in [a,b]).$$

Proof. Let $E(N_D) \neq 0$. Then for a certain $l \in \mathbb{Z} \setminus \{0\} : E(\{(\lambda, \mu) \in C^2 | (\omega_1 - \omega_2)(b-a) = 2\pi i l\}) \neq 0$. This means that for a certain $k \in \mathbb{N} : E(\square_k \cap \{(\lambda, \mu) | (\omega_1 - \omega_2)(b-a) = 2\pi i l\}) \neq 0$. Take an $f \in E(\square_k \cap \{(\lambda, \mu) | (\omega_1 - \omega_2)(b-a) = 2\pi i l\})H$ such that $f \neq 0$. Then $u(t) = (e^{\omega_1(t-a)} - e^{\omega_2(t-a)})(A, B)f$ is a solution of equation (1) on $[a, b]$ such that $u(a) = u(b) = 0$, but $u(t_0) \neq 0$ for a certain $t_0 \in (a, b)$. Thus, a weak solution $y(t)$ of (1) on $[a, b]$ with $y(a) = y(b) = 0$ is not unique.

Conversely, let $E(N_D) = 0$. Obviously $(\bigcup_{k \in \mathbb{N}} V_k) \cup N_D = C^2$, and so $E(\bigcup_{k \in \mathbb{N}} V_k) = I$. Let $f_0 \in H$, $f_1 \in H$, and let $y(t)$ be a weak solution of (1) on $[a, b]$ such that $y(a) = f_0$, $y(b) = f_1$. Then $\forall t \in [a, b] : y(t) = \lim_{k \to \infty} E(V_k) y(t)$.

Fix an arbitrary $k \in \mathbb{N}$. $E(V_k)y(t)$ is a weak solution of (1) in $E(V_k)H$ on $[a, b]$ such that $(E(V_k)y)(a) = E(V_k)f_0$, $(E(V_k)y)(b) = E(V_k)f_1$. But equation (1) in $E(V_k)H$ has a usual (and, consequently, weak) solution on $[a, b]$ $u_k(t) = \sum_{i=0}^{1} \varphi_i(A, B, t) E(V_k) f_i$ such that $u_k(a) = E(V_k)f_0$, $u_k(b) = E(V_k)f_1$.

It remains to prove that this weak solution is unique. In fact, we have to prove that if $u(t)$ is a weak solution of equation (1) in $E(V_k)H$ on $[a, b]$ such that $u(a) = u(b) = 0$, then $u(t) = 0$ $(t \in [a, b])$.

Indeed, a weak solution $u(t)$ of equation (1) in $E_k H$ on $[a, b]$ is a usual solution and has the form:

$$u(t) = \psi_0(A, B, t)u(a) + \psi_1(A, B, t)u'(a) \quad (t \in [a, b]) \tag{4.7}$$

(see the proofs of Lemma 2.1 and Theorem 3.1).

Here $u(a) = 0$, and so

$$u(b) = \frac{e^{\omega_1(b-a)} - e^{\omega_2(b-a)}}{\omega_1 - \omega_2}(A, B)u'(a).$$

As follows from the definition of V_k, $\exists \varepsilon_k > 0 \forall (\lambda, \mu) \in V_k$:

$$\left| \frac{e^{\omega_1(b-a)}(1 - e^{-(\omega_1 - \omega_2)(b-a)})}{\omega_1 - \omega_2} \right| \geq \varepsilon_k.$$

Therefore, $u(b) = 0$ implies that $u'(a) = 0$. In view of (4.7), $u(t) = 0$ $(t \in [a, b])$. The lemma is proved.

Lemma 4.3. *Let* (2.8) *and* (2.9) *hold. Then*

$$\begin{aligned} &\exists 0 < C < +\infty : \\ &|\varphi_i^{(j)}(\lambda, \mu, t)| \ (j = 0, 1, 2),\ |\lambda \varphi_i'(\lambda, \mu, t)|, |\mu \varphi_i(\lambda, \mu, t)| \leq C(|\lambda|^2 + |\mu| + 1), \\ &\forall (\lambda, \mu, t) \in \sigma(A, B) \times [a, b], \quad i = 0, 1. \end{aligned} \tag{4.8}$$

Proof. Take δ and σ_δ from the proof of Lemma 4.1. It suffices to obtain the estimates (4.8) for $\sigma(A,B)$ replaced by σ_δ and for $\sigma(A,B)$ replaced by $\sigma(A,B)\setminus\sigma_\delta = \{(\lambda,\mu)\in\sigma(A,B) \mid |\operatorname{Re}(\omega_1-\omega_2)|\le\delta,\ |\operatorname{Im}(\omega_1-\omega_2)|\le\delta\}$. We prove the estimates for $\varphi_0(\lambda,\mu,t)$; one can prove them for $\varphi_1(\lambda,\mu,t)$ similarly.

For $(\lambda,\mu,t)\in\sigma_\delta\times[a,b]$, we may write:

$$\varphi_0(\lambda,\mu,t) = \left(1-e^{-(\omega_1-\omega_2)(b-a)}\right)^{-1} e^{\omega_2(t-a)} - \left(1-e^{-(\omega_1-\omega_2)(b-a)}\right)^{-1} e^{\omega_2(b-a)}e^{\omega_1(t-b)}.$$

Taking into account that $\exists 0<\varepsilon_0,\ \gamma<+\infty \forall(\lambda,\mu)\in\sigma_\delta$:

$$|1-e^{-(\omega_1-\omega_2)(b-a)}|\ge\varepsilon_0,\ \operatorname{Re}\omega_2\le\gamma,\ \text{and}\ \operatorname{Re}\omega_1\ge-\gamma,$$

Lemma 1.1 and the Vieta formulas $\lambda=-(\omega_1+\omega_2)$, $\mu=\omega_1\omega_2$ imply the estimates (4.8) with $\sigma(A,B)$ replaced by σ_δ.

For $(\lambda,\mu,t)\in(\sigma(A,B)\setminus\sigma_\delta)\times[a,b]$, we write:

$$\varphi_0(\lambda,\mu,t) = e^{\omega_2(t-a)}\frac{1-e^{-(\omega_1-\omega_2)(b-t)}}{1-e^{-(\omega_1-\omega_2)(b-a)}},$$

$$\begin{aligned}\varphi_0'(\lambda,\mu,t) &= \omega_2 e^{\omega_2(t-a)}\frac{1-e^{-(\omega_1-\omega_2)(b-t)}}{1-e^{-(\omega_1-\omega_2)(b-a)}}\\ &\quad-\frac{1}{b-a}e^{\omega_2(t-a)}e^{(\omega_1-\omega_2)(t-b)}\cdot\frac{(\omega_1-\omega_2)(b-a)}{1-e^{-(\omega_1-\omega_2)(b-a)}},\end{aligned}$$

$$\begin{aligned}\varphi_0''(\lambda,\mu,t) &= \omega_2^2 e^{\omega_2(t-a)}\frac{1-e^{-(\omega_1-\omega_2)(b-t)}}{1-e^{-(\omega_1-\omega_2)(b-a)}}\\ &\quad-\frac{2}{b-a}\omega_2 e^{\omega_2(t-a)}e^{(\omega_1-\omega_2)(t-b)}\cdot\frac{(\omega_1-\omega_2)(b-a)}{1-e^{-(\omega_1-\omega_2)(b-a)}}\\ &\quad-(\omega_1-\omega_2)\cdot\frac{1}{b-a}e^{\omega_2(t-a)}e^{(\omega_1-\omega_2)(t-b)}\cdot\frac{(\omega_1-\omega_2)(b-a)}{1-e^{-(\omega_1-\omega_2)(b-a)}}.\end{aligned}$$

Taking into account (2.8) and the fact that

$$\forall(\lambda,\mu,t)\in(\sigma(A,B)\setminus\sigma_\delta)\times[a,b]:\ \frac{1}{2}\le\left|\frac{1-e^{-(\omega_1-\omega_2)(b-a)}}{(\omega_1-\omega_2)(b-a)}\right|\le\frac{3}{2}$$

and (4.6) holds, Lemma 1.1 and the Vieta formulas imply the estimates (4.8) with $\sigma(A,B)$ replaced by $\sigma(A,B)\setminus\sigma_\delta$.

The lemma is proved.

Lemma 4.4. *Let* (2.8) *and* (2.9) *hold. Then for any* $g \in D(A^2) \cap D(B)$*:*

1) $y(t) = \varphi_0(A,B,t)g$ $(t \in [a,b])$ *is a solution of* (1) *on* $[a,b]$ *with* $y^{(j)}(t) = \varphi_0^{(j)}(A,B,t)g$ $(t \in [a,b])$, $j = 0,1,2$, *such that* $y(a) = g$, $y(b) = 0$;
2) $y(t) = \varphi_1(A,B,t)g$ $(t \in [a,b])$ *is a solution of* (1) *on* $[a,b]$ *with* $y^{(j)}(t) = \varphi_1^{(j)}(A,B,t)g$ $(t \in [a,b])$, $j = 0,1,2$, *such that* $y(a) = 0$, $y(b) = g$;
3) *for* $i = 0,1 : (\varphi_i(A,B,t)Ag)' = A(\varphi_i(A,B,t)g)' \in C([a,b],H)$.

The *proof* is based on Lemma 4.3 and is similar to the proof of Lemma 2.4.

The further proof of Theorem 4.1 is quite analogous to the proof of Theorem 3.1 and will be omitted here.

Theorem 4.2. *Let A and B be c.n.o. in H. The Dirichlet problem for equation* (1) *on* $[a,b]$ *is well-posed in* $G = (D(B) \cap D(A^2)) \times (D(B) \cap D(A^2))$ *if and only if one of the equivalent conditions* 1), 2), 3) *of Theorem* 4.1 *holds.*

Proof. The sufficiency follows from Lemmas 4.2 and 4.4. Let's prove the necessity. Let the Dirichlet problem for equation (1) on $[a,b]$ be well-posed in $(D(B) \cap D(A^2)) \times (D(B) \cap D(A^2))$. Using Lemma 4.2, we obtain similarly to the proof of Theorem 3.2 that

$$\begin{gathered}\exists 0 < C < +\infty : \frac{|\mu|+1}{|\lambda|^2+|\mu|+1}|\varphi_i(\lambda,\mu,t)| \le C,\\ \frac{|\lambda|+1}{|\lambda|^2+|\mu|+1}|\varphi_i'(\lambda,\mu,t)| \le C,\\ \forall(\lambda,\mu,t) \in \sigma(A,B) \times [a,b], \quad i = 0,1.\end{gathered} \tag{4.9}$$

It remains to show that (4.9) implies (2.8) and (2.9).

Set $P_1 = \{(\lambda,\mu) \in \mathbb{C}^2 | \frac{1}{2}|\omega_2| < |\omega_1| < 2|\omega_2|\} \cup \{(\lambda,\mu) \in \mathbb{C}^2 | |\omega_1| < 2, |\omega_2| < 2\}$, $P_2 = \{(\lambda,\mu) \in \mathbb{C}^2 | |\omega_1| \ge 2|\omega_2|, |\omega_1| \ge 2\}$, $P_3 = \{(\lambda,\mu) \in \mathbb{C}^2 | |\omega_2| \ge 2|\omega_1|, |\omega_2| \ge 2\}$. Obviously $P_1 \cup P_2 \cup P_3 = \mathbb{C}^2$.

It is easy to see from the Vieta formulas and the proof of Lemma 1.1 that $\forall(\lambda,\mu) \in P_1 : |\lambda|^2 \le 16(|\mu|+1)$. Now (4.9) implies that $|\varphi_i(\lambda,\mu,t)| \le 17C$, $\forall(\lambda,\mu,t) \in (\sigma(A,B) \cap P_1) \times [a,b]$, $i = 0,1$. Lemma 4.1 remains valid if $\sigma(A,B)$ is replaced by an arbitrary $P \subseteq \mathbb{C}^2$. Making use of this fact for $P = \sigma(A,B) \cap P_1$, we obtain: $\exists 0 < \gamma_1, \varepsilon_1 < +\infty \forall(\lambda,\mu) \in \sigma(A,B) \cap P_1 : \operatorname{Re}\omega_1 \ge -\gamma_1, \operatorname{Re}\omega_2 \le \gamma_1$, $|(\omega_1-\omega_2)(b-a) - 2\pi i l| \ge \varepsilon_1$, $\forall l \in \mathbb{Z} \setminus \{0\}$.

For any $(\lambda,\mu) \in P_2 : |\mu| \le \frac{1}{2}|\omega_1|^2 \le 2|\lambda|^2$, $|\lambda| \ge \frac{1}{2}|\omega_1| \ge 1$, and hence $|\omega_1 - \omega_2| \ge \frac{|\omega_1|}{2} \ge \frac{|\omega_1+\omega_2|}{3} = \frac{|\lambda|}{3} \ge \frac{1}{12} \cdot \frac{|\lambda|^2+|\mu|+1}{|\lambda|}$. Now (4.9) implies that

$$\left|\frac{\varphi_i'(\lambda,\mu,t)}{\omega_1-\omega_2}\right| \le 12C, \quad \forall(\lambda,\mu,t) \in (\sigma(A,B) \cap P_2) \times [a,b], \quad i = 0,1.$$

But $\forall(\lambda,\mu)\in\sigma(A,B)\cap P_2$:

$$\left|\frac{\varphi_0'(\lambda,\mu,b)}{\omega_1-\omega_2}\right|=\frac{e^{\operatorname{Re}\omega_2(b-a)}}{1-e^{-(\omega_1-\omega_2)(b-a)}},\qquad \left|\frac{\varphi_1'(\lambda,\mu,a)}{\omega_1-\omega_2}\right|=\frac{e^{\operatorname{Re}\omega_1(a-b)}}{1-e^{-(\omega_1-\omega_2)(b-a)}}.$$

Thus,

$$\forall(\lambda,\mu)\in\sigma(A,B)\cap P_2:\quad \frac{e^{\operatorname{Re}\omega_2(b-a)}}{1-e^{-(\omega_1-\omega_2)(b-a)}}\le 12C,\quad \frac{e^{\operatorname{Re}\omega_1(a-b)}}{1-e^{-(\omega_1-\omega_2)(b-a)}}\le 12C. \tag{4.10}$$

Since $|1-e^{-(\omega_1-\omega_2)(b-a)}|\le 2$, it follows from (4.10) that

$$\forall(\lambda,\mu)\in\sigma(A,B)\cap P_2:\ \operatorname{Re}\omega_2\le\gamma_2,\ \operatorname{Re}\omega_1\ge-\gamma_2\ \text{where}\ \gamma_2=\frac{\ln(24C)}{b-a}. \tag{4.11}$$

Taking into account (4.10) and (4.11), we obtain similarly to the proof of Lemma 4.1 that $\exists\varepsilon_2>0\,\forall(\lambda,\mu)\in\sigma(A,B)\cap P_2:|(\omega_1-\omega_2)(b-a)-2\pi il|\ge\varepsilon_2$, $\forall l\in\mathbb{Z}\setminus\{0\}$.

One can consider $\sigma(A,B)\cap P_3$ quite similarly.

The theorem is proved.

2 The (weak) well-posedness of the Dirichlet problem in particular cases

Let us now analyze condition 1) of Theorem 4.1, which is necessary and sufficient for the weak well-posedness of the Dirichlet problem for (1) on $[a,b]$ in $G=H\times H$. In addition to a traditional condition, — now this is the second condition of 1), i.e. (2.9), and the equivalent one (4.2), — there arise condition (2.8) and the equivalent one (4.1). In a series of particular cases, this condition (2.8) holds automatically, and so (2.9) (i.e. (4.2)) becomes a necessary and sufficient condition for the weak well-posedness of the Dirichlet problem for (1) on $[a,b]$ in $G=H\times H$.

Corollary 1. *Let A and B be c.n.o. in H, and moreover, let* $\operatorname{Re}A$ *be bounded. Then* (2.8) *holds. The Dirichlet problem for* (1) *on* $[a,b]$ *is weakly well-posed in* $G=H\times H$ *if and only if* (2.9) *(or, what is equivalent,* (4.2)*) holds.*

Indeed, in this case there exists $0<\gamma<+\infty$ such that for any $(\lambda,\mu)\in\sigma(A,B):\operatorname{Re}\omega_1(\lambda,\mu)+\operatorname{Re}\omega_2(\lambda,\mu)=-\operatorname{Re}\lambda\in[-\gamma,\gamma]$; this would be impossible if $\max_{i=1,2}\operatorname{Re}\omega_i(\lambda,\mu)<-\gamma$ or $\min_{i=1,2}\operatorname{Re}\omega_i(\lambda,\mu)>\gamma$.

Corollary 2. *Let A and B be c.n.o. in H, and moreover, let $D(A)\subseteq D(B)$. Then* (2.8) *holds. The Dirichlet problem for* (1) *on* $[a,b]$ *is weakly well-posed in* $G=H\times H$ *if and only if* (2.9) *(or, what is equivalent,* (4.2)*) holds.*

Indeed, let for each $(\lambda,\mu)\in\sigma(A,B)$ $\omega(\lambda,\mu)$ be such a root of $\omega^2+\lambda\omega+\mu$ that $|\omega(\lambda,\mu)| = \min_{i=1,2}|\omega_i(\lambda,\mu)|$. By virtue of Lemma 3.7, $\exists 0<\gamma<+\infty \forall(\lambda,\mu)\in\sigma(A,B) : |\omega(\lambda,\mu)|\le\gamma$, and so $-\gamma\le \operatorname{Re}\omega(\lambda,\mu)\le\gamma$ and consequently

$$\max_{i=1,2}\operatorname{Re}\omega_i(\lambda,\mu)\ge\operatorname{Re}\omega(\lambda,\mu)\ge-\gamma,\quad \min_{i=1,2}\operatorname{Re}\omega_i(\lambda,\mu)\le\operatorname{Re}\omega(\lambda,\mu)\le\gamma.$$

Thus, for «incomplete» equations of the form (1) with commuting normal A and B, i.e., if $A=0$ (and even in the more general case of A bounded) or if $B=0$ (and even in the more general case of B subordinate to A), condition (4.1) holds automatically and, consequently, condition (4.2) is necessary and sufficient for the weak well-posedness of the Dirichlet problem for (1) on $[a,b]$ in $G=H\times H$.

In the general case of equation (1) with c.n.o. A and B, even in the case of self-adjoint A and B one cannot omit condition (4.1) in Theorem 4.1. Indeed, let A be self-adjoint, $\sigma(A)=R^1$, $B=\frac{A^2}{4}$ (for instance: $A=-2i\frac{d}{dx}$, $B=-\frac{d^2}{dx^2}$ in $H=L_2(R^1)$). Then $\sigma(A,B)=\{(\lambda,\mu)\in R^2|\mu=\frac{\lambda^2}{4}\}$. Condition (4.2) is satisfied. But condition (4.1) which may be written here as the following one:

$$\exists 0<\gamma<+\infty : \sigma(A,B)\subseteq\{(\lambda,\mu)\in R^2|\,|\lambda|\le 2\gamma\}$$
$$\cup\{(\lambda,\mu)\in R^2|\,|\lambda|>2\gamma,\mu\le\gamma|\lambda|-\gamma^2\}$$

does not hold, and so condition 2) of Theorem 4.1 does not hold.

3 Boundary conditions for the Dirichlet problem

In this section, we assume that A and B are c.n.o. in H and condition 1) of Theorem 4.1 is satisfied. Theorem 4.1 implies that in this case for every $(f_0,f_1)\in H\times H$ there exists a unique weak solution of the Dirichlet problem for (1) on $[a,b]$ with boundary conditions (b.c.) $y(a)=f_0$, $y(b)=f_1$. On the other hand, if for a certain (f_0,f_1) there exists a weak solution $y(t)$ of the Dirichlet problem for (1) on $[a,b]$ with b.c. $y(a)=f_0$, $y(b)=f_1$, then by virtue of Definition 1.4: $f_0=y(a)\in H$, $f_1=y(b)\in H$. We have proved the following statements.

Corollary 3. *Let A and B be c.n.o. in H, and let condition* 1) *of Theorem* 4.1 *be satisfied. Then $H\times H$ is the set of all (f_0,f_1) such that the Dirichlet problem for* (1) *on $[a,b]$ with b.c. $y(a)=f_0$, $y(b)=f_1$ has a unique weak solution.*

Corollary 4. *Let A and B be c.n.o. in H such that condition* 1) *of Theorem* 4.1 *is valid. Denote by Y the set of all weak solutions of equation* (1) *on $[a,b]$. There exists a one-to-one mapping of $H\times H$ onto Y given by the following formula:*

$$H\times H\ni(f_0,f_1)\leftrightarrow y(t)=\varphi_0(A,B,t)f_0+\varphi_1(A,B,t)f_1\in Y\subseteq C([a,b],H).$$

Further, Theorem 4.2 implies that in this case for every $(f_0, f_1) \in (D(B) \cap D(A^2)) \times (D(B) \cap D(A^2))$ there exists a unique solution of the Dirichlet problem for (1) on $[a, b]$ with b.c. $y(a) = f_0$, $y(b) = f_1$. Now a natural question arises: what is the set of *all* (f_0, f_1) such that the Dirichlet problem for (1) on $[a, b]$ with b.c. $y(a) = f_0$, $y(b) = f_1$ is uniquely solvable?

One may also approach this question in another way. Corollary 3 asserts that in this case $H \times H$ is the set of all (f_0, f_1) such that the Dirichlet problem for (1) on $[a, b]$ with b.c. $y(a) = f_0$, $y(b) = f_1$ has a unique weak solution. Now the question is: what is the set of all $(f_0, f_1) \in H \times H$ such that this weak solution is *usual*?

Denote this set by F_D. Just as previously, Theorem 4.2 and Definition 1.1 imply that

$$(D(B) \cap D(A^2)) \times (D(B) \cap D(A^2)) \subseteq F_D \subseteq D(B) \times D(B).$$

If $D(B) \subseteq D(A^2)$ then $F_D = D(B) \times D(B) = (D(B) \cap D(A^2)) \times (D(B) \cap D(A^2))$. Otherwise $D(B) \cap D(A^2)$ does not coincide with $D(B)$ and the question is not trivial. It is answered by the following theorem.

We index $\omega_i(\lambda, \mu)$, $i = 1, 2$ so that for every $(\lambda, \mu) \in \mathbb{C}^2 : \operatorname{Re} \omega_1(\lambda, \mu) \geq \operatorname{Re} \omega_2(\lambda, \mu)$ and, for definiteness, if $\operatorname{Re} \omega_1(\lambda, \mu) = \operatorname{Re} \omega_2(\lambda, \mu)$ then $\operatorname{Im} \omega_1(\lambda, \mu) \geq \operatorname{Im} \omega_2(\lambda, \mu)$. Denote: $\Delta_1 = \{(\lambda, \mu) \in \mathbb{C}^2 |\ |\omega_1| \geq |\omega_2|\}$; $\Delta_2 = \{(\lambda, \mu) \in \mathbb{C}^2 |\ |\omega_1| < |\omega_2|\}$.

Theorem 4.3. *Let A and B be c.n.o. in H, and let condition* 1) *of Theorem* 4.1 *hold, i.e., the Dirichlet problem for* (1) *on* $[a, b]$ *be weakly well-posed in* $G = H \times H$. *The Dirichlet problem for* (1) *on* $[a, b]$ *with b.c.* $y(a) = f_0$, $y(b) = f_1$ *has a unique solution if and only if* $f_0 \in D(B)$, $f_1 \in D(B)$, *and moreover,*

$$E(\Delta_1)f_1 - e^{\omega_2(b-a)}(A, B)E(\Delta_1)f_0 \in D(A^2),$$
$$E(\Delta_2)f_0 - e^{\omega_1(a-b)}(A, B)E(\Delta_2)f_1 \in D(A^2).$$

Remark. In view of (2.8), $e^{\omega_2(b-a)}(A, B)$ and $e^{\omega_1(a-b)}(A, B)$ are bounded operators on H.

Proof. Let P_1, P_2, P_3 be from the proof of Theorem 4.2. For each of $i = 1, 2, 3$, denote: $H_i = E(P_i)H$; let A_i and B_i be the restrictions of A and B to H_i. Then A_i and B_i are c.n.o. in H_i; $\sigma(A_i, B_i) \subseteq \overline{P_i}$ (here $\overline{}$ denotes the closure of a set).

Let $f_0 \in H$, $f_1 \in H$. Obviously, $y(t)$ is a solution of equation (1) in H on $[a, b]$ such that $y(a) = f_0$, $y(b) = f_1$ if and only if for each of $i = 1, 2, 3 : E(P_i)y(t)$ is a solution of the equation

$$y_i''(t) + A_i y_i'(t) + B_i y_i(t) = 0$$

in H_i on $[a, b]$ such that $(E(P_i)y)(a) = E(P_i)f_0$, $(E(P_i)y)(b) = E(P_i)f_1$.

So, it suffices to prove Theorem 4.3 for each of the following three cases:

1) $H = E(P_1)H$;
2) $H = E(P_2)H$;
3) $H = E(P_3)H$.

1. Let 1) hold. Then $\forall(\lambda,\mu) \in \sigma(A,B) \subseteq \overline{P_1} : |\lambda|^2 \leq 16(|\mu|+1)$. By virtue of Theorem 1.2 for c.n.o. A and B and the remark to it, $D(B) \subseteq D(A^2)$. Thus, $F_D = D(B)\times D(B)$. Further, $\forall(f_0,f_1) \in D(B)\times D(B) : f_0 \in D(A^2), f_1 \in D(A^2)$, and consequently,

$$E(\Delta_1)f_1 - e^{\omega_2(b-a)}(A,B)E(\Delta_1)f_0 \in D(A^2),$$
$$E(\Delta_2)f_0 - e^{\omega_1(a-b)}(A,B)E(\Delta_2)f_1 \in D(A^2).$$

2. Let 2) hold. For any $(\lambda,\mu) \in P_2 : |\omega_1 - \omega_2| \geq \frac{1}{2}|\omega_1| \geq 1$. By virtue of this fact, the fact that $\sigma(A,B) \subseteq \overline{P_2}$, and (2.9),

$$\exists 0 < \varepsilon_0 < +\infty \forall(\lambda,\mu) \in \sigma(A,B) : \ \varepsilon_0 \leq |1 - e^{-(\omega_1-\omega_2)(b-a)}| \leq 2. \qquad (4.12)$$

Denote: $d(\lambda,\mu) = (1 - e^{-(\omega_1-\omega_2)(b-a)})^{-1}, \forall(\lambda,\mu) \in \sigma(A,B)$.

Since $P_2 \subseteq \Delta_1$, we may reformulate the statement of Theorem 4.3 in the following form:

There exists a solution $y(t)$ of equation (1) on $[a,b]$ such that $y(a) = f_0$, $y(b) = f_1$ if and only if $f_0 \in D(B), f_1 \in D(B)$, and $f_1 - e^{\omega_2(b-a)}(A,B)f_0 \in D(A^2)$.

We now prove this statement. Let $y(t)$ be a solution of equation (1) on $[a,b]$, $y(a) = f_0$, $y(b) = f_1$. By Definition 1.1, $f_0 = y(a) \in D(B)$, $f_1 = y(b) \in D(B)$.

In particular, $y(t)$ is a weak solution of (1) on $[a,b]$. By Theorem 4.1: $y(t) = \varphi_0(A,B,t)f_0 + \varphi_1(A,B,t)f_1$ $(t \in [a,b])$. Let

$$h_1 = d(A,B)(f_1 - e^{\omega_2(b-a)}(A,B)f_0), \quad h_2 = d(A,B)(f_0 - e^{\omega_1(a-b)}(A,B)f_1).$$

In view of (4.12), $f_0 \in D(B)$ and $f_1 \in D(B)$ imply that $h_1 \in D(B)$ and $h_2 \in D(B)$, and we may rewrite:

$$\begin{aligned} y(t) &= \varphi_0(A,B,t)f_0 + \varphi_1(A,B,t)f_1 \\ &= e^{\omega_1(t-b)}(A,B)h_1 + e^{\omega_2(t-a)}(A,B)h_2 \quad (t \in [a,b]). \end{aligned}$$

Here $e^{\omega_2(t-a)}(A,B)h_2$ is a solution of (1) on $[a,b]$. Indeed, $\forall(\lambda,\mu,t) \in P_2 \times [a,b] : |e^{\omega_2(t-a)}| \leq e^{\gamma(b-a)}$; further, $|\omega_2|^2 \leq |\omega_2|\,|\omega_1| = |\mu|$, hence $|\omega_2| \leq \frac{1}{2}(1+|\mu|)$, $|\lambda\omega_2| = |(\omega_1+\omega_2)\omega_2| \leq 2|\mu|$; summarizing, $|(e^{\omega_2(t-a)})^{(j)}|$ $(j = 0,1,2)$, $|\lambda(e^{\omega_2(t-a)})'|$, $|\mu e^{\omega_2(t-a)}| \leq 2e^{\gamma(b-a)}(|\mu|+1)$. Since $h_2 \in D(B)$, $e^{\omega_2(t-a)}(A,B)h_2$ is a solution of (1) on $[a,b]$.

Thus, $e^{\omega_1(t-b)}(A,B)h_1 = y(t) - e^{\omega_2(t-a)}(A,B)h_2$ is a solution of (1) on $[a,b]$ as well. In particular, at every $t \in [a,b]$ there exists $(e^{\omega_1(t-b)}(A,B)h_1)' \in D(A)$; analogously to the proof of Theorem 3.2,

$$(e^{\omega_1(t-b)}(A,B)h_1)' = \omega_1 e^{\omega_1(t-b)}(A,B)h_1.$$

Setting here $t = b$, we have: $\omega_1(A,B)h_1 \in D(A)$, whence $\int_{P_2} |\lambda\omega_1|^2 d(Eh_1,h_1) < +\infty$. But $\forall(\lambda,\mu) \in P_2 : |\omega_1| \geq \frac{2}{3}|\lambda|$; so, $\int_{P_2} |\lambda|^4 d(Eh_1,h_1) < +\infty$. We conclude that $h_1 \in D(A^2)$.

Conversely, if $h_1 \in D(A^2)$, then $e^{\omega_1(t-b)}(A,B)h_1$ is a solution of (1) on $[a,b]$. Indeed, $\forall(\lambda,\mu,t) \in P_2 \times [a,b] : |e^{\omega_1(t-b)}| \leq e^{\gamma(b-a)}$; further, $|\omega_1|^2 \leq 4|\lambda|^2$, hence $|\mu| \leq \frac{1}{2}|\omega_1|^2 \leq 2|\lambda|^2$, $|\omega_1| \leq \frac{1}{2}(1+4|\lambda|^2)$, $|\lambda\omega_1| = |(\omega_1+\omega_2)\omega_1| \leq 2|\omega_1|^2 \leq 8|\lambda|^2$. The further proof is analogous to the previous proof for $e^{\omega_2(t-a)}(A,B)h_2$.

We have proved that if $y(t)$ is a solution of (1) on $[a,b]$, $y(a) = f_0$, and $y(b) = f_1$, then $f_0 \in D(B)$, $f_1 \in D(B)$, and $h_1 \in D(A^2)$. Moreover, we have proved that if $f_0 \in D(B)$, $f_1 \in D(B)$, and $h_1 \in D(A^2)$, then

$$y(t) = \varphi_0(A,B,t)f_0 + \varphi_1(A,B,t)f_1 = e^{\omega_1(t-b)}(A,B)h_1 + e^{\omega_2(t-a)}(A,B)h_2$$

is a solution of (1) on $[a,b]$ with $y(a) = f_0$, $y(b) = f_1$.

By virtue of (4.12): $h_1 \in D(A^2)$ if and only if $f_1 - e^{\omega_2(b-a)}(A,B)f_0 \in D(A^2)$.

3. The proof for the case 3) is quite analogous.

The theorem is proved.

In fact, in the case under study we have found the set of all usual solutions of equation (1) on $[a,b]$. Namely, there exists a one-to-one mapping of this set onto the set of all (f_0,f_1) satisfying conditions of Theorem 4.3, given by the formula:

$$(f_0,f_1) \leftrightarrow y(t) = \varphi_0(A,B,t)f_0 + \varphi_1(A,B,t)f_1 \quad (t \in [a,b]).$$

For each $(\lambda,\mu) \in R^2$ set: $\sqrt{\frac{\lambda^2}{4} - \mu} \geq 0$ if $\frac{\lambda^2}{4} \geq \mu$; $\operatorname{Im}\sqrt{\frac{\lambda^2}{4} - \mu} \geq 0$ if $\frac{\lambda^2}{4} \leq \mu$.

Corollary 1. *Let A and B be commuting self-adjoint operators in H, and let the Dirichlet problem for* (1) *on* $[a,b]$ *be weakly well-posed in* $G = H \times H$. *The Dirichlet problem for* (1) *on* $[a,b]$ *with b.c.* $y(a) = f_0$, $y(b) = f_1$ *has a unique solution if and only if* $f_0 \in D(B)$, $f_1 \in D(B)$, *and moreover,*

$$E_A([0,+\infty))f_0 - e^{(-\frac{\lambda}{2}+\sqrt{\frac{\lambda^2}{4}-\mu})(a-b)}(A,B)E_A([0,+\infty))f_1 \in D(A^2),$$

$$E_A((-\infty,0))f_1 - e^{(-\frac{\lambda}{2}-\sqrt{\frac{\lambda^2}{4}-\mu})(b-a)}(A,B)E_A((-\infty,0))f_0 \in D(A^2).$$

Remark 1. In Corollary 1, one may replace f_i in the last two lines by $E(\{(\lambda,\mu)|\mu \leq \frac{\lambda^2}{4}\})f_i$, since $f_i \in D(B) \Rightarrow E(\{(\lambda,\mu)|\mu > \frac{\lambda^2}{4}\})f_i \in D(B) \Rightarrow E(\{(\lambda,\mu)|\mu > \frac{\lambda^2}{4}\})f_i \in D(A^2)$.

Remark 2. The following example shows that in the general case F_D may be strictly wider than $(D(B)\cap D(A^2)) \times (D(B)\cap D(A^2))$ but strictly narrower than $D(B) \times D(B)$.

Example 4.1. Let A be an unbounded self-adjoint operator in H, $B = 0$. Then the Dirichlet problem for (1) on $[a,b]$ is weakly well-posed in $G = H \times H$. Furthermore, $(D(B)\cap D(A^2)) \times (D(B)\cap D(A^2)) = D(A^2) \times D(A^2)$; $D(B) \times D(B) = H \times H$. The Dirichlet problem for (1) on $[a,b]$ with b.c. $y(a) = f_0$, $y(b) = f_1$ has a unique solution if and only if $f_0 \in H$, $f_1 \in H$, and $(f_0 - f_1) \in D(A^2)$.

Let $f_1 \in H$, $f_1 \notin D(A^2)$. If $f_0 = 0$ then $(f_0, f_1) \in D(B)\times D(B)$ but $(f_0, f_1) \notin F_D$. If $f_0 = f_1$ then $(f_0, f_1) \in F_D$ but $(f_0, f_1) \notin (D(B)\cap D(A^2))\times(D(B)\cap D(A^2))$.

One can prove the following corollaries of Theorem 4.3 quite similarly to Example 4.1.

Corollary 2. *Let A and B be c.n.o. in H, and let condition* 1) *of Theorem* 4.1 *hold.* $F_D = D(B) \times D(B)$ *if and only if* $D(B) \subseteq D(A^2)$ *(i.e.,* $\exists 0 < c < +\infty : \sigma(A,B) \subseteq \{(\lambda,\mu) \in C^2 |\, |\mu| + 1 \geq c|\lambda|^2\}$*).*

Corollary 3. *Let A and B be c.n.o. in H, and let condition* 1) *of Theorem* 4.1 *hold.* $F_D = (D(B)\cap D(A^2)) \times (D(B)\cap D(A^2))$ *if and only if* $D(B) \subseteq D(A^2)$.

4 The weak well-posedness and the well-posedness of the Neumann problem

Theorem 4.4. *Let A and B be c.n.o. in H. The Neumann problem for equation* (1) *on* $[a,b]$ *is weakly well-posed in* $G = H_{-2} \times H_{-2}$ *if and only if one of the following equivalent conditions* 1), 2), 3), 4) *holds:*

1) (2.8) *and* (2.9) *are valid, and moreover,*

$$\exists 0 < \varepsilon_1 < +\infty \forall (\lambda,\mu) \in \sigma(A,B) : \min_{i=1,2} |\omega_i| \geq \varepsilon_1; \tag{4.13}$$

2) (4.1) *and* (4.2) *are valid, and moreover, there exists* $0 < \varepsilon_1 < +\infty$ *such that* $\sigma(A,B)$ *is situated in* C^2 *outside the set*

$$\{(\lambda,\mu) \in C^2 |\, |\mu| \leq \varepsilon_1\} \cup \{(\lambda,\mu) \in C^2 |\, |\mu| \leq \varepsilon_1|\lambda|\}; \tag{4.14}$$

3) *there exist* $0 < \gamma, \varepsilon < +\infty$ *such that for each* $(\lambda,\mu) \in \sigma(A,B)$: (4.3) *and* (4.4) *are valid, and moreover,* (μ_1,μ_2) *lies in* R^2 *outside the circle of radius* $\max(\varepsilon, \varepsilon|\lambda|)$ *centered at* $(0,0)$*;*

4) *the Dirichlet problem for* (1) *on* $[a,b]$ *is weakly well-posed in* $G = H \times H$, *and moreover,* B^{-1} *is bounded and* $D(B) \subseteq D(A)$.

If 1) *holds then the Neumann problem for* (1) *on* $[a,b]$ *is weakly well-posed in* $G = H_{-2} \times H_{-2}$ *not only over* $D(A^2) \cap D(B)$, *but over the whole* $D(A) \cap D(B)$ *as well. For every* $(f_0, f_1) \in H_{-2} \times H_{-2}$ *the weak solution of the Neumann problem for* (1) *on* $[a,b]$ *with boundary data* (f_0, f_1) *has the form:* $y(t) = \chi_0(A,B,t)f_0 + \chi_1(A,B,t)f_1$ $(t \in [a,b])$.

Proof. Conditions (4.13) and (4.14) are equivalent by virtue of Lemma 1.1.

Lemma 4.5. *For the existence of a* $0 < C < +\infty$ *such that* $(\min_{j=1,2} |\omega_j| + 1) \cdot |\chi_i(\lambda,\mu,t)| \le C$, $\forall (\lambda,\mu,t) \in \sigma(A,B) \times [a,b]$, $i = 0,1$, *it is necessary and sufficient that* (2.8), (2.9), *and* (4.13) *hold.*

The *proof* is similar to the proof of Lemma 4.1. In the proof of the necessity, we first show that $(\min_{j=1,2} |\omega_j| + 1) \cdot |\chi_i(\lambda,\mu,t)| \le C$, $\forall (\lambda,\mu,t) \in \sigma(A,B) \times [a,b]$, $i = 0,1$, implies (2.8) and (2.9). Then we show that (2.8), (2.9), and the fact that $|\chi_i(\lambda,\mu,t)| \le C$, $\forall (\lambda,\mu,t) \in \sigma(A,B) \times [a,b]$, $i = 0,1$, imply (4.13).

The further proof of Theorem 4.4 is similar to the proof of Theorem 4.1.

Notice that if (2.8), (2.9), and (4.13) hold, then for an arbitrary $f \in H_{-2}$, $i = 0,1$: $\chi_i(A,B,t)f$ is a weak solution of (1) on $[a,b]$ not only over $D(A^2) \cap D(B)$, but over the whole $D(A) \cap D(B)$ as well. (By Definition 1.4, $D(A) \cap D(B)$ is the maximal possible Φ.)

Indeed, in view of (4.13): $f' = (\min_{j=1,2} |\omega_j|)^{-1}(A,B)f \in H$. For any $g \in D(A) \cap D(B)$, $t \in [a,b]$: $(\chi_i(A,B,t)f, g) =$

$$(\chi_i(A,B,t)f', (\min_{j=1,2} |\omega_j|)(A,B)g) = (f', \chi_i(A^*,B^*,t)(\min_{j=1,2} |\omega_j|)(A,B)g).$$

Denote: $g_1 = (\min_{j=1,2} |\omega_j|)\,(A,B)g$. For $\chi_i(A^*,B^*,t)g_1 \in C^2([a,b],H)$ (and, in general, for $\chi_i(A^*,B^*,t)g_1$ to be a solution of the equation $y''(t) + A^*y'(t) + B^*y(t) = 0$ on $[a,b]$), it suffices that $g_1 \in D(A) \cap D(|B|^{1/2})$. But this means that

$$(\min_{j=1,2} |\omega_j|)(A,B)g \in D((\max_{j=1,2} |\omega_j|)(A,B)),$$

i.e.,

$$g \in D((\max_{j=1,2} |\omega_j| \cdot \min_{j=1,2} |\omega_j|)(A,B)) = D(|\mu|(A,B)) = D(B).$$

This is the case, since $g \in D(A) \cap D(B)$.

This and similar arguments show that $\chi_i(A,B,t)f$ is a weak solution of (1) on $[a,b]$ over $D(A) \cap D(B)$.

Theorem 4.5. *Let A and B be c.n.o. in H. The Neumann problem for* (1) *on* $[a,b]$ *is well-posed in* $G = (D(A) \cap D(|B|^{1/2})) \times (D(A) \cap D(|B|^{1/2})) = H_1 \times H_1$ *if and only if one of the following equivalent conditions* 1), 2), 3), 4) *holds:*

1) (2.8) *and* (2.9) *are valid, and moreover,*

$$\exists 0 < \varepsilon_1 < +\infty \forall(\lambda,\mu) \in \sigma(A,B): \ |\mu| = |\omega_1| \cdot |\omega_2| \geq \varepsilon_1; \tag{4.15}$$

2) (4.1) *and* (4.2) *are valid, and moreover, there exists* $0 < \varepsilon_1 < +\infty$ *such that* $\sigma(A,B)$ *is situated in* C^2 *outside the set* $\{(\lambda,\mu) \in C^2 | \, |\mu| \leq \varepsilon_1\}$;
3) *there exist* $0 < \gamma,\ \varepsilon < +\infty$ *such that for each* $(\lambda,\mu) \in \sigma(A,B)$: (4.3) *and* (4.4) *are valid, and moreover,* (μ_1,μ_2) *lies in* R^2 *outside the circle of radius* ε *centered at* $(0,0)$;
4) *the Dirichlet problem for* (1) *on* $[a,b]$ *is well-posed in* $G = (D(A^2) \cap D(B)) \times (D(A^2) \cap D(B))$, *and moreover,* B^{-1} *is bounded.*

If 1) *holds then for every* $(f_0, f_1) \in H_1 \times H_1$ *the solution of the Neumann problem for* (1) *on* $[a,b]$ *with boundary conditions* $y'(a) = f_0$, $y'(b) = f_1$ *has the form:* $y(t) = \chi_0(A,B,t)f_0 + \chi_1(A,B,t)f_1$ $(t \in [a,b])$.

Proof. The proof of Theorem 4.5 is similar to the proof of Theorem 4.2.

The analogue of Lemma 4.3 is the following assertion:

Let (2.8), (2.9), and (4.15) hold. Then

$$\begin{aligned}&\exists 0 < C < +\infty: \ |\chi_i^{(j)}(\lambda,\mu,t)| \quad (j = 0,1,2), \ |\lambda\chi_i'(\lambda,\mu,t)|,\\ &|\mu\chi_i(\lambda,\mu,t)| \leq C(|\lambda| + \sqrt{|\mu|} + 1),\\ &\qquad \forall(\lambda,\mu,t) \in \sigma(A,B) \times [a,b], \quad i = 0,1.\end{aligned} \tag{4.16}$$

To prove this, we first show that if (2.8) and (2.9) hold then $\exists 0 < C < +\infty \forall(\lambda,\mu,t) \in \sigma(A,B) \times [a,b], i = 0,1: |\chi_i^{(j)}(\lambda,\mu,t)|$ $(j = 1,2)$, $|\lambda\chi_i'(\lambda,\mu,t)|$, $|\mu\chi_i(\lambda,\mu,t)| \leq C(|\lambda| + \sqrt{|\mu|} + 1)$. Then (4.15) (which is a weaker condition than (4.13)) implies: $|\chi_i(\lambda,\mu,t)| \leq \frac{|\mu|}{\varepsilon_1}|\chi_i(\lambda,\mu,t)| \leq \varepsilon_1^{-1}C(|\lambda| + \sqrt{|\mu|} + 1)$, $\forall(\lambda,\mu,t) \in \sigma(A,B) \times [a,b], i = 0,1$.

To prove the necessity in Theorem 4.5, we have to prove that (4.16) implies (2.8), (2.9), and (4.15).

Let (4.16) hold. Denote:

$$P_1 = \{(\lambda,\mu) \in \sigma(A,B) | \max(|\lambda|, \sqrt{|\mu|}, 1) = |\lambda|\};$$

$$P_2 = \{(\lambda,\mu) \in \sigma(A,B) | \max(|\lambda|, \sqrt{|\mu|}, 1) = \sqrt{|\mu|}\};$$

$$P_3 = \{(\lambda,\mu) \in \sigma(A,B) | \max(|\lambda|, \sqrt{|\mu|}, 1) = 1\}.$$

Consider separately each of the sets P_1, P_2, P_3. When considering a certain P_i, by conditions (2.8), (2.9), (4.13), and (4.15) we mean the same conditions with $\sigma(A,B)$ replaced by P_i.

For any $(\lambda,\mu,t) \in P_1 \times [a,b]$, $i = 0,1$: $|\lambda\chi_i'(\lambda,\mu,t)| \le 3C|\lambda|$, i.e., $|\chi_i'| = |\varphi_i| \le 3C$. According to Lemma 4.1, with $\sigma(A,B)$ replaced by P_1, (2.8) and (2.9) hold. Further, $|\chi_i(\lambda,\mu,t)| \le 3C|\lambda| \Rightarrow \frac{1}{|\lambda|}|\chi_i| \le 3C$, $\forall(\lambda,\mu,t) \in P_1 \times [a,b]$, $i = 0,1$; hence, similarly to the proof of Lemma 4.5, we find in view of (2.8) and (2.9) that $\exists 0 < \varepsilon_1 < +\infty \forall(\lambda,\mu) \in P_1 : \min_{j=1,2} |\omega_j| \cdot |\lambda| \ge \varepsilon_1$. By Lemma 1.1, $\forall(\lambda,\mu) \in P_1 : \frac{|\mu|}{2|\lambda|} \le \min_{j=1,2} |\omega_j| \le \frac{2|\mu|}{|\lambda|}$. Thus, $|\mu| \ge \frac{1}{2}\varepsilon_1$, $\forall(\lambda,\mu) \in P_1$.

For any $(\lambda,\mu,t) \in P_2 \times [a,b]$, $i = 0,1$: $|\mu\chi_i(\lambda,\mu,t)| \le 3C\sqrt{|\mu|}$, hence $\min_{j=1,2} |\omega_j| \cdot |\chi_i| \le \sqrt{|\omega_1| \cdot |\omega_2|} \cdot |\chi_i| \le 3C$. By Lemma 4.5, with $\sigma(A,B)$ replaced by P_2, (2.8) and (2.9) hold; (4.15) follows from the definition of P_2.

For any $(\lambda,\mu,t) \in P_3 \times [a,b]$, $i = 0,1$: $\max_{j=1,2} |\omega_j| \le 2$, $|\chi_i(\lambda,\mu,t)| \le 3C$, hence $\min_{j=1,2} |\omega_j| \cdot |\chi_i| \le 6C$. By Lemma 4.5, with $\sigma(A,B)$ replaced by P_3, conditions (2.8), (2.9), and (4.13) (which implies (4.15)) hold.

5 Boundary conditions for the Neumann problem

Let A and B be c.n.o. in H, and let condition 1) of Theorem 4.4 be satisfied. Theorem 4.4 asserts that in this case for any $(f_0, f_1) \in H_{-2} \times H_{-2}$ there exists a unique weak solution of the Neumann problem for (1) on $[a,b]$ with boundary data (f_0, f_1). The question now arises: what is the set of *all* (f_0, f_1) such that there exists a unique weak solution of the Neumann problem for (1) on $[a,b]$ with b.d. (f_0, f_1)? It is answered by the following theorem. (We make use of the notations introduced before Theorem 4.3.)

Theorem 4.6. *Let A and B be c.n.o. in H, and let condition* 1) *of Theorem* 4.4 *be valid, i.e., the Neumann problem for* (1) *on* $[a,b]$ *be weakly well-posed in* $G = H_{-2} \times H_{-2}$. *The Neumann problem for* (1) *on* $[a,b]$ *with b.d.* (f_0, f_1) *has a unique weak solution if and only if* $f_0 \in H_{-1}$, $f_1 \in H_{-1}$, *and moreover,*

$$E(\Delta_1)f_0 - e^{\omega_1(a-b)}(A,B)E(\Delta_1)f_1 \in H_{-2},$$

$$E(\Delta_2)f_1 - e^{\omega_2(b-a)}(A,B)E(\Delta_2)f_0 \in H_{-2}.$$

For every such an (f_0, f_1), *the weak solution of the Neumann problem for* (1) *on* $[a,b]$ *with b.d.* (f_0, f_1) *has the form:*

$$y(t) = \lim_{k\to\infty} \sum_{i=0}^{1} \chi_i(A,B,t)E_k f_i \quad (t \in [a,b]).$$

In the *proof* of the necessity, we use identification of f_0, f_1 with $\mathbb{E}$-sequences $\{f_0^{(k)}\}_{k=1}^{\infty}, \{f_1^{(k)}\}_{k=1}^{\infty}$ (see Section 4 of Chapter 6) and the fact that

$$y(a) = \lim_{k\to\infty} \sum_{i=0}^{1} \chi_i(A,B,a) f_i^{(k)} \in H,$$

$$y(b) = \lim_{k\to\infty} \sum_{i=0}^{1} \chi_i(A,B,b) f_i^{(k)} \in H.$$

The further proof is analogous to the proofs of Theorems 4.3 and 4.7 and will be omitted here.

Denote the set of all such (f_0, f_1) by F'_N. Theorem 4.6 implies that if A and B are c.n.o. in H and condition 1) of Theorem 4.4 is valid, then

$$H_{-2} \times H_{-2} \subseteq F'_N \subseteq H_{-1} \times H_{-1}.$$

It follows from Lemma 1.1 that for c.n.o. A and B in H, $H_{-1} = H_{-2}$ if and only if $D(B) \subseteq D(A^2)$. We have obtained the following corollary.

Corollary 1. *Let A and B be c.n.o. in H, let condition* 1) *of Theorem* 4.4 *be valid, and let, moreover, $D(B) \subseteq D(A^2)$. Then $F'_N = H_{-1} \times H_{-1} = H_{-2} \times H_{-2} = H_-(|B|^{1/2}+1) \times H_-(|B|^{1/2}+1)$.*

Similarly to Corollaries 2 and 3 of Theorem 4.3, one can prove the following corollaries.

Corollary 2. *Let A and B be c.n.o. in H, and let condition* 1) *of Theorem* 4.4 *hold. $F'_N = H_{-2} \times H_{-2}$ if and only if $D(B) \subseteq D(A^2)$ (i.e., $\exists 0 < c < +\infty$: $\sigma(A,B) \subseteq \{(\lambda,\mu) \in C^2 \mid |\mu| + 1 \geq c|\lambda|^2\}$).*

Corollary 3. *Let A and B be c.n.o. in H, and let condition* 1) *of Theorem* 4.4 *hold. $F'_N = H_{-1} \times H_{-1}$ if and only if $D(B) \subseteq D(A^2)$.*

Set for each $(\lambda,\mu) \in R^2$: $\sqrt{\frac{\lambda^2}{4} - \mu} \geq 0$ if $\frac{\lambda^2}{4} \geq \mu$, $\operatorname{Im}\sqrt{\frac{\lambda^2}{4} - \mu} \geq 0$ if $\frac{\lambda^2}{4} \leq \mu$.

Corollary 4. *Let A and B be commuting self-adjoint operators in H, and let the Neumann problem for* (1) *on $[a,b]$ be weakly well-posed in $G = H_{-2} \times H_{-2}$. The Neumann problem for* (1) *on $[a,b]$ with b.d. (f_0, f_1) has a unique weak solution if and only if $f_0 \in H_{-1}$, $f_1 \in H_{-1}$, and moreover,*

$$E_A([0,+\infty))f_1 - e^{(-\frac{\lambda}{2} - \sqrt{\frac{\lambda^2}{4} - \mu})(b-a)}(A,B)E_A([0,+\infty))f_0 \in H_{-2},$$

$$E_A((-\infty,0))f_0 - e^{(-\frac{\lambda}{2} + \sqrt{\frac{\lambda^2}{4} - \mu})(a-b)}(A,B)E_A((-\infty,0))f_1 \in H_{-2}.$$

Remark 1. In Corollary 4, one may replace f_i in the last two lines by $E(\{(\lambda,\mu)|\mu \leq \frac{\lambda^2}{4}\})f_i$, since $f_i \in H_{-1} \Rightarrow E(\{(\lambda,\mu)|\mu > \frac{\lambda^2}{4}\})f_i \in H_{-1} \Rightarrow E(\{(\lambda,\mu)|\mu > \frac{\lambda^2}{4}\})f_i \in H_{-2}$.

We go on to study the usual solvability of the Neumann problem.

Let A and B be c.n.o. in H, and let condition 1) of Theorem 4.5 be satisfied. Theorem 4.5 shows that in this case for any $(f_0, f_1) \in (D(A) \cap D(|B|^{1/2})) \times (D(A) \cap D(|B|^{1/2}))$ there exists a unique solution of the Neumann problem for (1) on $[a,b]$ with b.c. $y'(a) = f_0$, $y'(b) = f_1$. This brings up the question: what is the set of *all* (f_0, f_1) such that there exists a unique solution of the Neumann problem for (1) on $[a,b]$ with b.c. $y'(a) = f_0$, $y'(b) = f_1$?

Denote this set by F_N. Just as for the Dirichlet problem, Theorem 4.5 and Definition 1.1 now imply that

$$(D(A) \cap D(|B|^{1/2})) \times (D(A) \cap D(|B|^{1/2})) \subseteq F_N \subseteq D(A) \times D(A).$$

We see that if A and B are c.n.o. in H and condition 1) of Theorem 4.5 is satisfied and, moreover, $D(A^2) \subseteq D(B)$, then $F_N = D(A) \times D(A) = (D(A) \cap D(|B|^{1/2})) \times (D(A) \cap D(|B|^{1/2}))$.

Otherwise $D(A) \cap D(|B|^{1/2})$ does not coincide with $D(A)$ and the question is not trivial. It is answered by the following theorem.

Theorem 4.7. *Let A and B be c.n.o. in H, and let condition* 1) *of Theorem* 4.5 *be valid, i.e., the Neumann problem for* (1) *on* $[a,b]$ *be well-posed in* $G = (D(A)\cap D(|B|^{1/2})) \times (D(A)\cap D(|B|^{1/2})) = H_1 \times H_1$. *Then* $(D(A)\cap D(|B|^{1/2})) \times (D(A) \cap D(|B|^{1/2}))$ *is the set of all* (f_0, f_1) *such that the Neumann problem for* (1) *on* $[a,b]$ *with b.c.* $y'(a) = f_0$, $y'(b) = f_1$ *has a unique solution.*

Proof. In view of Theorem 4.5, it remains to prove that if the Neumann problem for (1) on $[a,b]$ with b.d. (f_0, f_1) has a unique solution, then $(f_0, f_1) \in H_1 \times H_1$.

Let $y(t)$ be a solution of equation (1) on $[a,b]$ such that $y'(a) = f_0$, $y'(b) = f_1$. By Definition 1.1, $f_i \in D(A)$, $i = 0, 1$. We have to prove that, what is more, $f_i \in D(A) \cap D(|B|^{1/2})$, $i = 0, 1$.

If $\sigma(A,B) \subseteq \{(\lambda,\mu) \in \mathbb{C}^2 | \, |\mu| \leq |\lambda|^2 + 1\}$, then by virtue of Theorem 1.2 for c.n.o A and B and the remark to it: $D(A) \subseteq D(|B|^{1/2})$ and, consequently, $D(A) \cap D(|B|^{1/2}) = D(A)$.

Similarly to the proof of Theorem 4.3, it now remains to consider the case of $\sigma(A,B) \subseteq \{(\lambda,\mu) \in \mathbb{C}^2 | \, |\mu| \geq |\lambda|^2 + 1\}$. We have to prove that if (2.8), (2.9), and (4.15) hold and $y(t)$ is a solution of equation (1) on $[a,b]$ with $y'(a) = f_0$, $y'(b) = f_1$, then $f_0 \in D(|B|^{1/2})$, $f_1 \in D(|B|^{1/2})$.

For any $(\lambda,\mu) \in \sigma(A,B) : |\omega_1 - \omega_2|^2 = |4\mu - \lambda^2| \geq 4|\mu| - |\lambda|^2 \geq 4$. This and (2.9) imply (4.12).

Denote: $d(\lambda,\mu) = (1 - e^{-(\omega_1 - \omega_2)(b-a)})^{-1}$, $\forall(\lambda,\mu) \in \sigma(A,B)$.

By Lemma 1.1, $\forall(\lambda,\mu)\in\sigma(A,B):\frac{1}{2}\sqrt{|\mu|}\le\min_{j=1,2}|\omega_j|\le 2\sqrt{|\mu|}$. Therefore, (4.15) implies (4.13).

Since (2.8), (2.9), and (4.13) hold, we have by virtue of Theorem 4.4: $y(t)=\sum_{i=0}^{1}\chi_i(A,B,t)f_i$ $(t\in[a,b])$. In view of (2.8), (4.12), and (4.13): $f_i'=d(A,B)f_i\in H$, $i=0,1$; $h_1'=f_1'-e^{\omega_2(b-a)}(A,B)f_0'\in H$, $h_2'=f_0'-e^{\omega_1(a-b)}(A,B)f_1'\in H$; $h_1=\omega_1^{-1}(A,B)h_1'\in H$, $h_2=\omega_2^{-1}(A,B)h_2'\in H$, and we may rewrite:

$$y(t)=\sum_{i=0}^{1}\chi_i(A,B,t)f_i=e^{\omega_2(t-a)}(A,B)h_2+e^{\omega_1(t-b)}(A,B)h_1\quad(t\in[a,b]).$$

Now $y(a)=h_2+e^{\omega_1(a-b)}(A,B)h_1$, $y(b)=e^{\omega_2(b-a)}(A,B)h_2+h_1$, whence

$$\begin{aligned}h_2&=d(A,B)(y(a)-e^{\omega_1(a-b)}(A,B)y(b)),\\ h_1&=d(A,B)(y(b)-e^{\omega_2(b-a)}(A,B)y(a)).\end{aligned}\tag{4.17}$$

By Definition 1.1: $y(a)\in D(B)$, $y(b)\in D(B)$. By virtue of (4.17): $h_2\in D(B)$, $h_1\in D(B)$.

By Lemma 1.1, $\forall(\lambda,\mu)\in\sigma(A,B):\frac{1}{2}\le\frac{1}{2}\sqrt{|\mu|}\le\min_{j=1,2}|\omega_j|\le\max_{j=1,2}|\omega_j|\le 2\sqrt{|\mu|}$. Thus, $h_2=\omega_2^{-1}(A,B)h_2'\in D(B)$, $h_1=\omega_1^{-1}(A,B)h_1'\in D(B)\Leftrightarrow h_2'\in D(|B|^{1/2})$, $h_1'\in D(|B|^{1/2})$.

Here

$$\begin{aligned}h_2'&=f_0'-e^{\omega_1(a-b)}(A,B)f_1',\\ -h_1'&=e^{\omega_2(b-a)}(A,B)f_0'-f_1'.\end{aligned}$$

Just as previously (with $y(a)$, $y(b)$, h_2, h_1 replaced by h_2', $-h_1'$, f_0', $-f_1'$ respectively), $h_2'\in D(|B|^{1/2})$ and $-h_1'\in D(|B|^{1/2})$ imply that $f_0'\in D(|B|^{1/2})$, $-f_1'\in D(|B|^{1/2})$.

In view of (4.12), this means that $f_0\in D(|B|^{1/2})$, $f_1\in D(|B|^{1/2})$.

The proof is over.

Corollary 1. *Let A and B be c.n.o. in H, and let condition* 1) *of Theorem* 4.5 *be valid.* $F_N=D(|B|^{1/2})\times D(|B|^{1/2})$ *if and only if* $D(B)\subseteq D(A^2)$.

6 The inverse Cauchy problem

Let now $-\infty\le a<b<+\infty$.

For the inverse Cauchy problem on $[a,b]$ for equation (1) with c.n.o. A and B, all the results obtained in Chapter 3 for the Cauchy problem on R_+ remain valid, with the corresponding replacements in the necessary and sufficient conditions. Namely, λ, λ_1, λ_2, ω_1, ω_2 are replaced by $(-\lambda)$, $(-\lambda_1)$, $(-\lambda_2)$, $(-\omega_1)$, $(-\omega_2)$ respectively; $\max_{i=1,2} \operatorname{Re}\omega_i(\lambda,\mu)$ is replaced by $(-\min_{i=1,2} \operatorname{Re}\omega_i(\lambda,\mu))$; $\inf\sigma(\operatorname{Re} A)$ is replaced by $(-\sup\sigma(\operatorname{Re} A))$; ψ_0, ψ_1 are replaced by η_0, η_1 respectively.

As an example, we give some statements.

Theorem 4.8. *Let A and B be c.n.o. in H, and let $H \times H_-(|A|+1) \subseteq G \subseteq H \times H_{-1}$. The inverse Cauchy problem for equation* (1) *on* $[a,b]$ *is weakly well-posed in G if and only if one of the following equivalent conditions* 1), 2), 3) *holds:*

1) *a)* (2.13) *is valid;*
   b) (3.2) *is valid;*
2) *a)* $\exists 0<\gamma<+\infty : \sigma(A,B) \subseteq \{(\lambda,\mu)\in \mathbb{C}^2 | \lambda_1 < 2\gamma,$

$$\mu_1 > \frac{1}{(\lambda_1-2\gamma)^2}(\mu_2^2 - \lambda_1\lambda_2\cdot\mu_2 + \gamma(\lambda_1-\gamma)(\lambda_2^2+(\lambda_1-2\gamma)^2))\}; \quad (4.18)$$

   b) (3.4) *is valid;*
3) *there exist* $0<\gamma,\ \varepsilon<+\infty$ *such that for each* $(\lambda,\mu)\in\sigma(A,B)$ *one of the following conditions is satisfied:*
   a) $\lambda_1 < 2\gamma$, $|\lambda_2| \le \frac{1}{\varepsilon}(1+|\lambda_1|)$, (μ_1,μ_2) *lies in* $\mathbb{R}^2$ *to the right of the parabola*

$$\mu_1 - \left(-\frac{\lambda_2^2}{4}+\lambda_1\gamma-\gamma^2\right) = \frac{1}{(\lambda_1-2\gamma)^2}\left(\mu_2 - \frac{\lambda_1\lambda_2}{2}\right)^2; \quad (4.19)$$

   b) $\lambda_1 < 2\gamma$, $|\lambda_2| > \frac{1}{\varepsilon}(1+|\lambda_1|)$, (μ_1,μ_2) *lies in* $\mathbb{R}^2$ *to the right of the parabola* (4.19) *and outside the circle of radius* $\varepsilon|\lambda|^2$ *the center of which is situated on the axis of the parabola* (4.19) *to the right of its vertex by* $(\frac{\lambda_1}{2}-\gamma)^2$.

In this case, for every $(f_0,f_1)\in H\times H_{-1}$ the weak solution of the inverse Cauchy problem for (1) *on* $[a,b]$ *with b.d.* (f_0,f_1) *can be written in the following form:* $y(t) = \eta_0(A,B,t)f_0 + \eta_1(A,B,t)f_1 (t\in[a,b])$.

Theorem 4.9. *Let A and B be c.n.o. in H. The inverse Cauchy problem for equation* (1) *on* $[a,b]$ *is well-posed in* $G = D(B)\times H_1$ *if and only if one of the equivalent conditions* 1), 2), 3) *of Theorem* 4.8 *holds.*

Corollary 3. *Let A and B be c.n.o. in H, and moreover, let* $D(A)\subseteq D(B)$, *i.e., B be subordinate to A. Then* (3.2) *is valid; thus, condition* (2.13) *(i.e.,* (4.18)*) is necessary and sufficient for the weak well-posedness of the inverse Cauchy problem for* (1) *on* $[a,b]$ *in* $G = H\times H_{-1}$. *Moreover, this condition can be rewritten in the following form:*

$$\sup\sigma(\operatorname{Re} A) < +\infty.$$

Part II
Initial Data of Solutions

Introduction to Part II

In Sections 3, 5 of Chapter 4, we consider the following question. Let A and B be c.n.o. in H. We fix a certain boundary-value problem for equation (1) on $[a, b]$ and a certain space of boundary data $G \subseteq H \times H (G \subseteq \Phi' \times \Phi')$, which may depend on A and B. Further, we assume that this boundary-value problem is well-posed (weakly well-posed) in G, i.e., for any $(f_0, f_1) \in G$ there exists a unique solution (weak solution) of this boundary-value problem for equation (1) on $[a, b]$ with b.d. (f_0, f_1). The question is: what is the set of *all* (f_0, f_1) such that this boundary-value problem for equation (1) on $[a, b]$ with b.d. (f_0, f_1) has a unique solution (weak solution)? (It is obvious that this set contains G.)

We have studied this question for the Dirichlet and Neumann problems. In the second part of the work, we go on to study this question for the Cauchy problem (and, consequently, for the inverse Cauchy problem, since for that problem all the results are preserved with the replacements indicated in Section 6 of Chapter 4).

We treat all the questions for $R_+ = [0, +\infty)$. For any other $[a, b]$ such that $-\infty < a < b \leq +\infty$, all the statements remain valid.

Denote for each $(\lambda, \mu) \in C^2 : a = -\frac{1}{2} \operatorname{Re}(\omega_1 + \omega_2) = \frac{1}{2} \operatorname{Re} \lambda$, $b_1 = \frac{1}{2} |\operatorname{Re}(\omega_1 - \omega_2)| = |\operatorname{Re} \sqrt{\frac{\lambda^2}{4} - \mu}|$, $b_2 = \frac{1}{2} |\operatorname{Im}(\omega_1 - \omega_2)| = |\operatorname{Im} \sqrt{\frac{\lambda^2}{4} - \mu}|$, $d = -\max_{i=1,2} \operatorname{Re} \omega_i = a - b_1$. Taking into account Theorem 3.1, Lemmas 3.2 and 3.3, Theorem 3.2, and Statement 3.4, we obtain the following statement.

Statement 5.1. Let A and B be c.n.o. in H, and let $P = \sigma(A, B)$. The following conditions are equivalent:

1) the Cauchy problem for equation (1) on R_+ is strongly well-posed in the sense of H.O. Fattorini's definition (Section 7 of Chapter 3; [93, Chapter VIII]);

2) the Cauchy problem for (1) on R_+ is well-posed in $G = D(B) \times H_1 = D(B) \times (D(A) \cap D(|B|^{1/2}))$;

3) the Cauchy problem for (1) on R_+ is weakly well-posed in $G = H \times H_{-1} = H \times H_-(|A| + |B|^{1/2} + 1)$;

4) $\exists 0 \leq \gamma, C < +\infty : P \subseteq \{(\lambda, \mu) \in C^2 | d \geq -\gamma, |\operatorname{Im} \lambda|^2 \leq C(1 + a^2 + b_1^2 + b_2^2)\}$. (2)

Here (2) for $P = \sigma(A, B)$ means that both conditions (2.2) (i.e., (3.3)) and (3.4) hold.

Let A and B be c.n.o. in H such that $P = \sigma(A, B)$ satisfies (2).

The Cauchy problem for equation (1) on R_+ is well-posed in $G = D(B) \times H_1$, i.e., for any $(f_0, f_1) \in D(B) \times (D(A) \cap D(|B|^{1/2}))$: the Cauchy problem for (1) on R_+ with initial conditions (i.c.) $y(0) = f_0$, $y'(0) = f_1$ has a unique solution. Now a natural question arises:

1. What is the set of *all* (f_0, f_1) such that the Cauchy problem for (1) on R_+ with i.c. $y(0) = f_0$, $y'(0) = f_1$ has a unique solution?

We denote this set by F_C. By virtue of Definition 1.1, for every solution $y(t)$ of (1) on $R_+ : y(0) \in D(B)$, $y'(0) \in D(A)$. Taking into account Statement 5.1, we have:

$$D(B) \times (D(A) \cap D(|B|^{1/2})) \subseteq F_C \subseteq D(B) \times D(A).$$

Thus, $F_C \subseteq D(B) \times D(A) \subseteq H \times H_{-1}$. But Statement 5.1 implies that for every $(f_0, f_1) \in H \times H_{-1}$ there exists a unique weak solution of the Cauchy problem for (1) on R_+ with b.d. (f_0, f_1). So the question 1 is:

– for what $(f_0, f_1) \in H \times H_{-1}$ this weak solution is *usual*?

Since $F_C \subseteq H \times H_{-1}$, we conclude from Statement 5.1 that there exists a one-to-one mapping of F_C onto the set Y of all solutions of (1) on R_+; this mapping is given by the formula: $Y \ni y(t) \leftrightarrow (y(0), y'(0)) \in F_C$. Therefore, the question 1 may also be considered as the following one:

– to indicate *all* solutions of equation (1) on R_+.

In order to answer this question, we need to introduce a new integral transform $R(t)$ $(t \geq 0)$ for σ-finite measures and to study its boundary behaviour as $t \to 0$ depending on the sub-integral measure. Let us explain this in more detail.

At first, let us say some words on definitions. Generally speaking, a measure ρ defined on Borel subsets of a closed $P \subseteq C^2$ is said to be σ-finite if there exists a sequence $\{P_k\}_{k=1}^{\infty}$ such that $P = \bigcup\limits_{k=1}^{\infty} P_k$ and $\rho(P_k) < +\infty$ for all $k \geq 1$. In this monograph, we use the term «σ-finite» in a more restricted sense. Namely, we demand that as this sequence one can take $P_k = P \cap \Box_k = \{(\lambda, \mu) \in P \| \lambda| \leq k,\ |\mu| \leq k\}$, $k \geq 1$. In other words, we call a measure ρ σ-finite if $\rho(\Delta) < +\infty$ for all bounded Borel sets $\Delta \subseteq P$.

Let's attempt to answer the question 1. By Theorem 3.1, for any $(f_0, f_1) \in H \times H_{-1}$: if there exists a solution of (1) on R_+ such that $y(0) = f_0$, $y'(0) = f_1$, then it is unique and has the form: $y(t) = \psi_0(A, B, t) f_0 + \psi_1(A, B, t) f_1$ $(t \in R_+)$. Thus, we have to indicate all $(f_0, f_1) \in D(B) \times D(A)$ such that $y(t) = \psi_0(A, B, t) f_0 + \psi_1(A, B, t) f_1$ $(t \in R_+)$ is a solution of the Cauchy problem for (1) on R_+ with i.d. (f_0, f_1).

Let us assume that (f_0, f_1) is such a vector.

Here $f_0 \in D(B)$. On the other hand, by virtue of Theorem 3.2, for $f_0 \in D(B)$: $y_0(t) = \psi_0(A, B, t) f_0$ $(t \in R_+)$ is a solution of (1) on R_+ such that $y_0(0) = f_0$,

$y_0'(0) = 0$. So, $y_1(t) = y(t) - y_0(t) = \psi_1(A, B, t)f_1$ $(t \in R_+)$ is a solution of (1) on R_+ such that $y_1(0) = 0$, $y_1'(0) = f_1$. Moreover, $f_1 \in D(A)$.

Conversely, if $f_0 \in D(B)$ and $f_1 \in D(A)$ is such that $y_1(t) = \psi_1(A, B, t)f_1$ $(t \in R_+)$ is a solution of (1) on R_+ with $y_1(0) = 0$, $y_1'(0) = f_1$, then $y(t) = \psi_0(A, B, t)f_0 + \psi_1(A, B, t)f_1$ $(t \in R_+)$ is a solution of the Cauchy problem for (1) on R_+ with i.d. (f_0, f_1).

Summarizing, F_C is the set of all $(f_0, f_1) \in D(B) \times D(A)$ such that $y_1(t) = \psi_1(A, B, t)f_1$ $(t \in R_+)$ is a solution of (1) on R_+ with $y_1(0) = 0$ and $y_1'(0) = f_1$.

In view of Definition 1.1, the last condition on $f_1 \in D(A)$ contains the following one: for all $t \in R_+$, $y_1(t) = \psi_1(A, B, t)f_1 \in D(B)$, and $By_1(t) = B\psi_1(A, B, t)f_1$ is continuous in H on R_+. In particular,

$$\| By_1(t) \|^2 = \| B\psi_1(A, B, t)f_1 \|^2$$

is continuous on R_+. In other words,

$$R(t) = \int_{\sigma(A,B)} |\psi_1(\lambda, \mu, t)|^2 |\mu|^2 d(Ef_1, f_1)$$

is finite for all $t \in R_+$ and continuous on R_+.

We arrive at the following problem.

One has a set $P = \sigma(A, B)$ which satisfies (2). Furthermore, one has a σ-finite measure ρ on $P = \sigma(A, B)$ (here $d\rho = |\mu|^2 d(Ef_1, f_1)$). The question is: what conditions must this σ-finite measure ρ on P satisfy for its integral transform $R(t) = \int_P |\psi_1(\lambda, \mu, t)|^2 d\rho_{\lambda,\mu}$ to be finite for all $t \in R_+$ and continuous on R_+?

In Chapter 5, we answer this question in the general case where P is a closed subset of C^2 which satisfies (2) and ρ is a σ-finite measure on P. As well, we answer some other questions related to this one.

In Chapter 6, based on these results, we answer the question 1 on the set of all initial data (f_0, f_1) that ensure the unique solvability of the Cauchy problem for (1) on R_+. Moreover, we answer some other interesting questions related to the question 1.

Chapter 5
Boundary behaviour of an integral transform $R(t)$ as $t \to 0$ depending on the sub-integral measure

In what follows, we understand «a measure» as «a nonnegative countably additive measure defined on Borel subsets».

Let P be a closed subset of C^2 such that (2) holds, ρ be a σ-finite measure on P, and $0 < T \leq +\infty$. In this chapter, we consider a new integral transform

$$R(t) = \int_P |\psi_1(\lambda, \mu, t)|^2 d\rho_{\lambda,\mu} \quad (t \in [0, T])$$

and study the connection between the boundary behaviour of $R(t)$ as $t \to 0$, on one hand, and the behaviour of the sub-integral measure ρ on P, on the other hand.

Note that for all $(\lambda, \mu) \in P$, $t \in R_+$:

$$|\psi_1(\lambda, \mu, t)|^2 = e^{-2at} \frac{\cosh^2 b_1 t \cdot \sin^2 b_2 t + \sinh^2 b_1 t \cdot \cos^2 b_2 t}{b_1^2 + b_2^2}.$$

1 Analogy with Tauberian theorems

There exists a certain analogy between these considerations and Tauberian theorems. We now explain this analogy.

Let $P^0 = R_+ = [0, +\infty)$, ρ^0 be a σ-finite measure on P^0, and for all $\lambda \in P^0$: $\psi^0(\lambda, t)$ $(t \in R_+)$ stand for the fundamental solution of the first order ordinary differential equation $\psi'(t) + \lambda\psi(t) = 0$, i.e., for a solution of this equation such that $\psi(0) = 1$.

Consider the integral transform

$$L(t) = \int_{P^0} |\psi^0(\lambda, t)|^2 d\rho^0 \quad (t \in R_+).$$

Here $\psi^0(\lambda, t) = e^{-\lambda t}$ for $\lambda \in P^0$, $t \in R_+$. Thus,

$$L(t) = \int_{R_+} e^{-2t\lambda} d\rho^0 \quad (t \in R_+).$$

We see that $L(t)$ is the Laplace transform of the σ-finite measure ρ^0 on R_+, up to the replacement of t by $2t$. The connection between the boundary behaviour of $L(t)$ (or, what is the same, of the Laplace transform) as $t \to 0$, on one hand, and the behaviour of the σ-finite measure ρ^0 on $P^0 = R_+$, on the other hand, has been studied extensively. Theorems on this connection are known as Tauberian theorems [106 (Section 5 of Chapter XIII), 278].

In Chapter 5, we meet with a similar situation. Now P is a closed subset of C^2 which satisfies (2), ρ is a σ-finite measure on P, and for all $(\lambda, \mu) \in P$: $\psi_1(\lambda, \mu, t)$ $(t \in R_+)$ stands for the fundamental solution of the second order ordinary differential equation

$$\psi''(t) + \lambda\psi'(t) + \mu\psi(t) = 0,$$

i.e., for a solution of this equation such that $\psi(0) = 0$, $\psi'(0) = 1$.

We consider the integral transform

$$R(t) = \int_P |\psi_1(\lambda, \mu, t)|^2 d\rho \quad (t \in R_+)$$

and study the connection between the boundary behaviour of $R(t)$ as $t \to 0$, on one hand, and the behaviour of the σ-finite measure ρ on P, on the other hand.

In this sense, the integral transform $R(t)$ and, accordingly, our results in Chapter 5 correspond to second order equation $\psi''(t)+\lambda\psi'(t)+\mu\psi(t) = 0$ in the same manner as the Laplace transform $L(t)$ and Tauberian theorems correspond to first order equation $\psi'(t)+\lambda\psi(t) = 0$. From this point of view, our results may be considered as analogues of simplest Tauberian theorems.

This point is the end of the analogy. The boundary behaviour of $R(t)$ as $t \to 0$ differs from the behaviour of $L(t)$. Indeed, $R(0) = 0$ in the general case, whereas $L(0) = \rho^0(P^0) = \rho^0(R_+)$. Therefore, the conditions arising in our theorems differ essentially from those contained in Tauberian theorems.

2 A model example

First of all, we explain the main ideas of this chapter considering a model example.

Let $P \subseteq \{(\lambda, \mu) \in R^2 | \lambda \geq 2,\ \mu = \frac{\lambda^2}{2}\}$, P be a closed set. Here for all $(\lambda, \mu) \in P : d = \frac{\lambda}{2} \geq 1$, $\operatorname{Im} \lambda = 0$. Thus, P satisfies (2) with $C = \gamma = 0$.

The question is: what must be a σ-finite measure ρ on P for the integral transform $R(t) = \int_P |\psi_1(\lambda, \mu, t)|^2 d\rho$ to be continuous on R_+? (Note that $R(t) \in C(R_+)$ implies, in particular, $R(t) < +\infty$ for all $t \in R_+$.)

Now we rewrite this question. For any $(\lambda, \mu) \in P : a = d = b_2 = \frac{\lambda}{2}$, $b_1 = 0$. Therefore, for all $(\lambda, \mu) \in P$, $t \in R_+ : |\psi_1(\lambda, \mu, t)|^2 = e^{-2dt}\frac{\sin^2 dt}{d^2}$. Furthermore, there exists a one-to-one mapping of P onto a closed set $\Omega \subseteq [1, +\infty)$ which is given by the formula: $P \ni (\lambda, \mu) \leftrightarrow d = \frac{\lambda}{2} \in \Omega$. Denote this mapping of P onto Ω by κ.

By a σ-finite measure ρ on P, we introduce a σ-finite measure ν on Ω as the following: for each Borel set $\Delta \subseteq \Omega$,

$$\nu(\Delta) \stackrel{\text{def}}{=} \rho(\kappa^{-1}(\Delta)) = \rho(\{(\lambda, \mu) \in P | d \in \Delta\}) = \rho\left(\left\{(\lambda, \mu) \in P | \frac{\lambda}{2} \in \Delta\right\}\right)$$

Now one has for all $t \in R_+ : R(t) = \int_\Omega e^{-2dt} \frac{\sin^2 dt}{d^2} d\nu$, and the question can be rewritten so: what must be a σ-finite measure ν on Ω for $R(t) = \int_\Omega e^{-2dt} \frac{\sin^2 dt}{d^2} d\nu$ to be continuous on R_+?

Set for all Borel sets $\Delta \subseteq \Omega : \nu_1(\Delta) = \int_\Delta \frac{1}{d^2} d\nu$. Now ν_1 is a σ-finite measure on Ω, and for all $t \in R_+ : R(t) = \int_\Omega e^{-2dt}(\sin^2 dt) d\nu_1$.

It is obvious that

$$\nu_1(\Omega) < +\infty \tag{5.1}$$

is a sufficient condition for $R(t) \in C(R_+)$. This follows immediately from the Lebesgue theorem on majorized convergence.

If one omits the multiplier $\sin^2 dt$ in the sub-integral function and considers the integral transform $L_1(t) = \int_\Omega e^{-2dt} d\nu_1$, which is, in fact, the Laplace transform of ν_1 for $2t$, then, what is more, (5.1) is a necessary and sufficient condition for $L_1(t) \in C(R_+)$.

In the present situation, where the sub-integral function of $R(t)$ contains the multiplier $\sin^2 dt$, (5.1) is sufficient but not necessary for $R(t) \in C(R_+)$.

A necessary and sufficient condition for $R(t) \in C(R_+)$ is weaker, and it is given by the following lemma.

Lemma 5.1. *Let ν_1 be a σ-finite measure on a closed set $\Omega \subseteq [1, +\infty)$, and let for all $t \in R_+ : R(t) = \int_\Omega e^{-2dt}(\sin^2 dt) d\nu_1$. $R(t) \in C(R_+)$ if and only if*

$$\lim_{x \to +\infty} \nu_1(\Omega \cap [x, 2x]) = 0. \tag{5.2}$$

Note that for any $x > 0 : \Omega \cap [x, 2x] = (\Omega \cap \{d | d \leq 2x\}) \setminus \{d | d < x\}$.

Remark 1. Fix an arbitrary $c > 1$. Lemma 5.1 remains valid if one replaces 2 by c in (5.2).

For definiteness, we understand ν_1 as a σ-finite measure on R_+ such that $\nu_1(R_+ \setminus \Omega) = 0$. One may express (5.1) in the form:

$$\nu_1(R_+) < +\infty, \tag{5.1}$$

whereas (5.2) may be expressed in the form:

$$\lim_{x\to+\infty} \nu_1([x,2x]) = 0. \tag{5.2}$$

Let us compare these two conditions. By the Cauchy criterion for convergence of an integral, one may rewrite (5.1) as the following condition:

$$\forall\varepsilon > 0 \exists X_\varepsilon > 0 \forall x \geq X_\varepsilon, \quad \forall y \geq x : \nu_1([x,y]) < \varepsilon, \tag{5.1}$$

whereas (5.2) may be rewritten as:

$$\forall\varepsilon > 0 \exists X_\varepsilon > 0 \forall x \geq X_\varepsilon, \quad y = 2x : \nu_1([x,y]) < \varepsilon. \tag{5.2}$$

Denote for each $n \geq 1 : a_n = \nu_1([2^{n-1}, 2^n))$. Then (5.1) may be written as:

$$\sum_{n=1}^{\infty} a_n < +\infty, \tag{5.1}$$

whereas condition (5.2) is equivalent to:

$$\lim_{n\to+\infty} a_n = 0. \tag{5.2}$$

The last expressions for (5.1) and (5.2) show that in the general case (5.2) is a strictly weaker condition than (5.1).

3 Proof of Lemma 5.1

Necessity. The proof of the necessity is quite simple. Indeed, let $R(t) \in C(R_+)$. Taking into account that $R(0) = 0$ by the definition, one has: $\frac{1}{t}\int_t^{2t} R(\tau)d\tau \to 0$ as $t \to 0$.

By the Foubini theorem, for any $t > 0$:

$$\frac{1}{t}\int_t^{2t} R(\tau)d\tau = \frac{1}{t}\int_t^{2t}\left(\int_{R_+} e^{-2d\tau}\sin^2(d\tau)d\nu_1\right)d\tau$$

$$= \int_{R_+}\left(\frac{1}{t}\int_t^{2t} e^{-2d\tau}\sin^2(d\tau)d\tau\right)d\nu_1 \geq \int_{\frac{1}{t}\leq d\leq\frac{2}{t}} e^{-4dt}\left(\frac{1}{t}\int_t^{2t}\sin^2(d\tau)d\tau\right)d\nu_1$$

$$= \int_{\frac{1}{t}\leq d\leq\frac{2}{t}} e^{-4dt}\frac{1}{2}\left(1 - \frac{\sin(dt)\cos(3dt)}{dt}\right)d\nu_1 \geq K_1\cdot\nu_1\left(\left[\frac{1}{t},\frac{2}{t}\right]\right),$$

where

$$K_1 = \frac{1}{2}e^{-8}\left(1 - \max_{1\le z\le 2}\left|\frac{\sin z}{z}\right|\right) > 0.$$

Thus, $\frac{1}{t}\int_t^{2t} R(\tau)d\tau \underset{t\to 0}{\longrightarrow} 0$ implies $\nu_1([\frac{1}{t},\frac{2}{t}]) \underset{t\to 0}{\longrightarrow} 0$, i.e., $\nu_1([x,2x]) \to 0$ as $x \to +\infty$.

Sufficiency. The proof of the sufficiency is more complicated. Let (5.2) hold. We now show that $R(t) < +\infty$ for all $t > 0$, and moreover, $R(t) \to R(0) = 0$ as $t \to 0$.

The main idea of the proof is the following. For each $t > 0$, one divides the integral $R(t)$ into three parts: $R(t) = R_1(t) + R_2(t) + R_3(t)$, where

$$R_1(t) = \int\limits_{1\le d<\frac{1}{t}} e^{-2dt}\sin^2(dt)d\nu_1,$$

$$R_2(t) = \int\limits_{\frac{1}{t}\le d<\frac{2}{t}} e^{-2dt}\sin^2(dt)d\nu_1,$$

$$R_3(t) = \int\limits_{d\ge\frac{2}{t}} e^{-2dt}\sin^2(dt)d\nu_1.$$

Then one proves separately for $i = 1$, $i = 2$, and $i = 3$ that $R_i(t) < +\infty$ for all $t > 0$ and $R_i(t) \to 0$ as $t \to 0$.

For $i = 2$, this follows immediately from (5.2), since $R_2(t) \le \nu_1([\frac{1}{t},\frac{2}{t}])$ for any $t > 0$.

For any $t > 0$, $R_1(t) < +\infty$ since ν_1 is a σ-finite measure on R_+.

We now prove that $R_1(t) \to 0$ as $t \to 0$. Set for each $0 < t \le 1 : x = \frac{1}{t}$; $n \in \mathbb{N} \cup \{0\}$ be such a number that $x \in [2^n, 2^{n+1})$. Taking into account that $1 \le d \le \frac{1}{t} \Rightarrow \sin^2(dt) \le (dt)^2 \le dt \le 1$, we have:

$$R_1(t) \le \int\limits_{1\le d<\frac{1}{t}} \sin^2(dt)d\nu_1 \le \int\limits_{1\le d<\frac{1}{t}} (dt)d\nu_1 = \int\limits_{1\le d<2} \frac{d}{x}d\nu_1$$

$$+ \int\limits_{2\le d<4} \frac{d}{x}d\nu_1 + \cdots + \int\limits_{2^{n-1}\le d<2^n} \frac{d}{x}d\nu_1 + \int\limits_{2^n\le d<x} d\nu_1$$

$$\le \frac{1}{2^{n-1}}\nu_1([1,2]) + \frac{1}{2^{n-2}}\nu_1([2,4]) + \cdots + \nu_1([2^{n-1},2^n]) + \nu_1([2^n,2^{n+1}]).$$

Denote the last sum by S_n. Here $S_n = \sum\limits_{k=1}^{n} C_{nk}a_k + a_{n+1}$, where $a_k = \nu_1([2^{k-1},2^k])$ for any $k \ge 1$, $C_{nk} = \frac{1}{2^{n-k}}$ for any $n \ge 1$, $1 \le k \le n$.

Condition (5.2) implies that $a_k = \nu_1([2^{k-1}, 2^k]) \to 0$ as $k \to +\infty$. By the Toeplitz theorem on a regular transformation of a sequence [344, 145, 51], $S_n \to 0$ as $n \to +\infty$ as well. Since for any $0 < t \leq 1 : 0 \leq R_1(t) \leq S_n$, and $n \to +\infty$ as $t \to 0$, we have: $R_1(t) \to 0$ as $t \to 0$.

Finally, we consider $R_3(t)$. Fix an arbitrary $t > 0$. We have:

$$\begin{aligned}
R_3(t) &\leq \int\limits_{d \geq \frac{2}{t}} e^{-2dt} d\nu_1 = \int\limits_{\frac{2}{t} \leq d < \frac{4}{t}} e^{-2dt} d\nu_1 + \int\limits_{\frac{4}{t} \leq d < \frac{8}{t}} e^{-2dt} d\nu_1 \\
&+ \cdots + \int\limits_{\frac{2^n}{t} \leq d < \frac{2^{n+1}}{t}} e^{-2dt} d\nu_1 + \cdots \leq e^{-4} \nu_1 \left(\left[\frac{2}{t}, \frac{4}{t} \right] \right) \\
&+ e^{-8} \nu_1 \left(\left[\frac{4}{t}, \frac{8}{t} \right] \right) + \cdots + e^{-2^{n+1}} \nu_1 \left(\left[\frac{2^n}{t}, \frac{2^{n+1}}{t} \right] \right) + \cdots \\
&= \sum_{n=1}^{\infty} e^{-2^{n+1}} \nu_1 \left(\left[\frac{2^n}{t}, \frac{2^{n+1}}{t} \right] \right) \leq \sup_{n \geq 1} \nu_1 \left(\left[\frac{2^n}{t}, \frac{2^{n+1}}{t} \right] \right) \cdot \sum_{n=1}^{\infty} e^{-2^{n+1}} \\
&\leq \sup_{n \geq 1} \nu_1 \left(\left[\frac{2^n}{t}, \frac{2^{n+1}}{t} \right] \right) \text{ since } \sum_{n=1}^{\infty} e^{-2^{n+1}} \leq \sum_{n=1}^{\infty} e^{-2n} = \frac{e^{-2}}{1 - e^{-2}} < 1.
\end{aligned}$$

Here $\nu_1([\frac{2^n}{t}, \frac{2^{n+1}}{t}]) \to 0$ as $n \to +\infty$ by virtue of (5.2). In particular, this sequence is bounded, and, consequently,

$$R_3(t) \leq \int\limits_{d \geq \frac{2}{t}} e^{-2dt} d\nu_1 \leq \sup_{n \geq 1} \nu_1 \left(\left[\frac{2^n}{t}, \frac{2^{n+1}}{t} \right] \right) < +\infty.$$

(Notice that $R_1(t) < +\infty$ and $R_2(t) < +\infty$ hold automatically because ν_1 is a σ-finite measure on R_+, i.e., for any bounded Borel set $\Delta \subset R_+ : \nu_1(\Delta) < +\infty$.)

We have proved in passing that for an arbitrary $t > 0$:

$$\int\limits_{d \geq \frac{2}{t}} e^{-2dt} d\nu_1 < +\infty. \tag{5.3}$$

Now we show that $R_3(t) \to 0$ as $t \to 0$. Take an arbitrary $\varepsilon > 0$. By virtue of (5.2), $\exists X_\varepsilon \geq 1 \forall x > X_\varepsilon : \nu_1([x, 2x]) < \varepsilon$. Let $0 < t < \frac{2}{X_\varepsilon}$. Then for any $n \geq 1$: $x = \frac{2^n}{t} > X_\varepsilon$, and therefore $\nu_1([\frac{2^n}{t}, \frac{2^{n+1}}{t}]) < \varepsilon$. Thus, $R_3(t) \leq \sup_{n \geq 1} \nu_1([\frac{2^n}{t}, \frac{2^{n+1}}{t}]) \leq \varepsilon$ for any $0 < t < \frac{2}{X_\varepsilon}$. We have proved that $R_3(t) \to 0$ as $t \to 0$.

We have proved that for all $t \in R_+ : R(t) < +\infty$, and $R(t) \to R(0) = 0$ as $t \to 0$. It remains to prove that for any $t_0 > 0 : R(t)$ is continuous at t_0.

Fix an arbitrary $t_0 > 0$ and a $0 < \delta < t_0$. For all $t \in (t_0 - \delta, t_0 + \delta)$, the functions $e^{-2td} \sin^2 td$ $(d \in R_+)$ are majorized by the function $e^{-2(t_0-\delta)d}$. Setting in (5.3): $t = t_0 - \delta$, and taking into account that ν_1 is a σ-finite measure on R_+, one has: $\int_{R_+} e^{-2(t_0-\delta)d} d\nu_1 < +\infty$. Now the fact that $R(t) = \int_{R_+} e^{-2td} \sin^2(td) d\nu_1 \underset{t \to t_0}{\longrightarrow} R(t_0)$ follows immediately from the Lebesgue theorem on majorized convergence. The proof of Lemma 5.1 is now completed.

Remark 1. Fix an arbitrarily small $0 < \delta < +\infty$. From the proof of Lemma 5.1 we observe that (5.2) can be deduced from the fact that $R(t)$ is measurable on $[0, \delta]$ and continuous at the point $t = 0$, whereas (5.2) implies that $R(t)$ is continuous on the whole R_+.

4 Further results

We continue our treatment with the notations introduced in Section 2.

Fix an arbitrary $0 < T < +\infty$.

Corollary 1. *$R(t)$ is continuous on $[0, T]$ if and only if (5.2) holds.*

This follows from Remark 1 in the previous section.

Lemma 5.2. *$R(t)$ is continuous and bounded on $(0, T]$ if and only if*

$$\exists K < +\infty \forall x > 0 : \nu_1([x, 2x]) \leq K. \tag{5.4}$$

One can prove Lemma 5.2 just as Lemma 5.1.

Lemma 5.3. *$R(t)$ is integrable on $(0, T]$ if and only if*

$$\int_{R_+} \frac{d\nu_1}{d+1} < +\infty. \tag{5.5}$$

Moreover, if (5.5) holds then $R(t)$ is continuous on $(0, T]$.

The *proof* of Lemma 5.3 is based on the Foubini theorem. The last statement of Lemma 5.3 follows from the fact that if (5.5) holds then for any $t > 0$: (5.3) is valid.

We have proved the following statement.

Statement 5.2. Denote for each $n \geq 1 : a_n = \nu_1([2^{n-1}, 2^n))$; here $a_n \geq 0$ for all $n \geq 1$.

1) $\nu_1(R_+) < +\infty$ if and only if $\sum_{n=1}^{\infty} a_n < +\infty$;
2) $R(t) = \int_{R_+} e^{-2dt} \sin^2(dt) d\nu_1$ is continuous on $[0, T]$ if and only if $a_n \to 0$ as $n \to +\infty$;
3) $R(t)$ is continuous and bounded on $(0, T]$ if and only if $\{a_n\}_{n=1}^{\infty}$ is a bounded sequence;
4) $R(t)$ is continuous and integrable on $(0, T]$ if and only if $\sum_{n=1}^{\infty} \frac{a_n}{2^n} < +\infty$.

Fix an arbitrary closed set $\Omega \subseteq [1, +\infty)$. Consider four propositions about a σ-finite measure ν_1 on Ω:

$$1) \ \nu_1(\Omega) < +\infty; \tag{5.6}$$

$$2) \ R(t) = \int_\Omega e^{-2dt} \sin^2(dt) d\nu_1 \text{ is continuous on } [0, T]; \tag{5.7}$$

$$3) \ R(t) \text{ is continuous and bounded on } (0, T]; \tag{5.8}$$

$$4) \ R(t) \text{ is continuous and integrable on } (0, T]. \tag{5.9}$$

It is obvious that everyone of these conditions implies the following one. Statement 5.2 shows that in the general case the converse implications may not be valid. For instance, let $\Omega = [1, +\infty)$. If ν_1 is such that for every $n \geq 1 : a_n = \frac{1}{n}$, then (5.7) holds but (5.6) does not. If ν_1 is such that for every $n \geq 1 : a_n = 1$, then (5.8) holds but (5.7) does not. If ν_1 is such that for every $n \geq 1 : a_n = n$, then (5.9) holds but (5.8) does not.

This situation differs from the situation for the Laplace transform $L_1(t) = \int_\Omega e^{-2dt} d\nu_1$. In that case, for any closed $\Omega \subseteq [1, +\infty)$ the following conditions on a σ-finite measure ν_1 on Ω are equivalent:

1) $\nu_1(\Omega) < +\infty$;
2) $L_1(t) = \int_\Omega e^{-2dt} d\nu_1$ is continuous on $[0, T]$;
3) $L_1(t)$ is continuous and bounded on $(0, T]$.

Now natural questions arise:

1) to indicate *all* closed $\Omega \subseteq [1, +\infty)$ such that condition (5.7) on a σ-finite measure ν_1 on Ω is equivalent to (5.6);

2) to indicate *all* closed $\Omega \subseteq [1, +\infty)$ such that condition (5.8) on a σ-finite measure ν_1 on Ω is equivalent to (5.7);

3) to indicate *all* closed $\Omega \subseteq [1, +\infty)$ such that condition (5.9) on a σ-finite measure ν_1 on Ω is equivalent to (5.8).

These questions are answered by the following lemma.

Lemma 5.4. *Let Ω be a closed subset of $[1,+\infty)$.*

1) *Conditions (5.6) and (5.7) on a σ-finite measure ν_1 on Ω are equivalent if and only if Ω is bounded.*
2) *Conditions (5.7) and (5.8) on a σ-finite measure ν_1 on Ω are equivalent if and only if Ω is bounded.*
3) *Conditions (5.8) and (5.9) on a σ-finite measure ν_1 on Ω are equivalent if and only if Ω is bounded.*

Proof. Fix an arbitrary closed set $\Omega \subseteq [1,+\infty)$. Then condition (5.6) on a σ-finite measure ν_1 on Ω implies (5.7). Further, (5.7) implies (5.8). Finally, (5.8) implies (5.9). In view of Lemma 5.3, (5.9) is equivalent to (5.5).

These implications hold for any closed $\Omega \subseteq [1,+\infty)$. But if, moreover, Ω is bounded, i.e., d is bounded on Ω, then (5.5) may be written as: $\nu_1(\Omega) < +\infty$. In other words, (5.5) implies (5.6), and so conditions (5.5), (5.6), (5.7), (5.8), and (5.9) are equivalent.

The sufficiency is proved. Let us prove the necessity.

Let Ω be an unbounded closed subset of $[1,+\infty)$. Take the sequence of intervals $[2^{n-1},2^n)$, $n \geq 1$. In view of the unboundedness of Ω, one can choose such an infinite subsequence $[2^{n_k-1},2^{n_k})$, $k \geq 1$, of this sequence that for every $k \geq 1 : [2^{n_k-1},2^{n_k}) \cap \Omega \neq \emptyset$ (if this were not the case, then $\exists N < +\infty \forall n \geq N :$ $\Omega \cap [2^{n-1},2^n) = \emptyset$, i.e., $\Omega \subseteq [1,2^{N-1})$). Take $\{d_k\}_{k=1}^{\infty}$ such that $\forall k \geq 1 :$ $d_k \in [2^{n_k-1},2^{n_k}) \cap \Omega$.

Set for each $k \geq 1 : \nu_1(\{d_k\}) = q_k > 0$, and set $\nu_1([1,+\infty) \setminus \bigcup\limits_{k=1}^{\infty} \{d_k\}) = 0$. Then $\operatorname{supp} \nu_1 \subseteq \Omega$. For each $n \geq 1$: if $n = n_k$ for a certain $k \geq 1$ then $a_n = q_k$; otherwise $a_n = 0$.

In order to prove the necessity in 1), we set for every $k \geq 1 : q_k = \frac{1}{k}$. Then $\sum\limits_{n=1}^{\infty} a_n = \sum\limits_{k=1}^{\infty} \frac{1}{k} = +\infty$ while $a_n \to 0$ as $n \to +\infty$. Statement 5.2 shows that for this ν_1: (5.7) holds but (5.6) does not.

In order to prove the necessity in 2), we set for every $k \geq 1 : q_k = 1$. Then $\{a_n\}_{n=1}^{\infty}$ does not converge to 0 as $n \to +\infty$ while $\forall n \geq 1 : a_n \leq 1$. Statement 5.2 shows that for this ν_1: (5.8) holds but (5.7) does not.

In order to prove the necessity in 3), we set for every $k \geq 1 : q_k = k$. Then $\{a_n\}_{n=1}^{\infty}$ is unbounded while $\sum\limits_{n=1}^{\infty} \frac{a_n}{2^n} = \sum\limits_{k=1}^{\infty} \frac{k}{2^{n_k}} \leq \sum\limits_{k=1}^{\infty} \frac{k}{2^k} < +\infty$. Statement 5.2 shows that for this ν_1: (5.9) holds but (5.8) does not.

The proof of Lemma 5.4 is over.

Remark 1. We have shown that for any unbounded closed $\Omega \subseteq [1,+\infty)$ there exists such a σ-finite measure ν_1 on Ω that (5.7) holds but (5.6) does not. Similarly, one can prove the following statement, which is stronger:

Let Ω be an unbounded closed subset of $[1,+\infty)$, ν_0 be an arbitrary σ-finite measure on Ω such that $\operatorname{supp}\nu_0 = \Omega$. For this measure, there exists such a continuous function $\varphi \geq 0$ on Ω that for a σ-finite measure ν_1 on Ω defined as: $\nu_1(\Delta) = \int_\Delta \varphi(d)d\nu_0$, $\forall\Delta \in \mathcal{B}(\Omega)$, — (5.7) holds but (5.6) does not.

The same statements are valid: for conditions (5.8) and (5.7) respectively; for conditions (5.9) and (5.8) respectively.

Remark 2. Let us return to a closed $P \subseteq \{(\lambda,\mu) \in R^2 | \lambda \geq 2, \mu = \frac{\lambda^2}{2}\}$ and a σ-finite measure ρ on P. It is obvious that all the statements of Sections 2, 3, 4 may be reformulated in terms of P, $(\lambda,\mu) \in P$, a σ-finite measure ρ on P, $R(t) = \int_P |\psi_1(\lambda,\mu,t)|^2 d\rho$ on R_+.

For instance, $R(t) \in C(R_+)$ if and only if

$$\lim_{x\to+\infty} \int\limits_{\{(\lambda,\mu)\in P | x\leq\lambda\leq 2x\}} \frac{1}{\lambda^2} d\rho = 0.$$

Let $0 < T < +\infty$, and P be a closed subset of $\{(\lambda,\mu) \in R^2 | \lambda \geq 2, \mu = \frac{\lambda^2}{2}\}$ such that for a σ-finite measure ρ on $P : R(t) \in C([0,T])$ implies $\int_P \frac{d\rho}{\lambda^2} < +\infty$. This is the case if and only if $\exists 0 < \gamma_1 < +\infty \forall(\lambda,\mu) \in P : d = -\max\limits_{i=1,2} \operatorname{Re}\omega_i \leq \gamma_1$. In the present situation, this is the case if and only if P is bounded.

5 Continuity of $R(t)$ on R_+ in extreme cases

1. Let $0 \leq \gamma < +\infty$, $P = \{(\lambda,\mu) \in R^2 | \lambda \geq -2\gamma, \mu \geq \frac{\lambda^2}{4}\}$. Here $\forall(\lambda,\mu) \in P$: $\operatorname{Im}\lambda = 0$; $d = a = \frac{\lambda}{2} \geq -\gamma$; $b_1 = 0$, $b_2 = \sqrt{\mu - \frac{\lambda^2}{4}} \geq 0$. In particular, P satisfies (2) with $C = 0$. Note that $\forall(\lambda,\mu) \in P : d^2 + b_2^2 + 1 = \mu + 1$.

The question is: what must be a σ-finite measure ρ on P for

$$R(t) \stackrel{\text{def}}{=} \int\limits_P |\psi_1(\lambda,\mu,t)|^2 d\rho$$

to be continuous on R_+?

There exists a one-to-one mapping of P onto $\Omega = \{(d,b_2) | d \geq -\gamma, b_2 \geq 0\} \subset R^2$ which is given by the formula: $P \ni (\lambda,\mu) \leftrightarrow (d,b_2) = (\frac{\lambda}{2}, \sqrt{\mu - \frac{\lambda^2}{4}}) \in \Omega$. Denote this mapping of P onto Ω by κ. By a σ-finite measure ρ on P, we introduce a measure ν on Ω as the following: for each Borel set $\Delta \subseteq \Omega$, $\nu(\Delta) \stackrel{\text{def}}{=} \rho(\kappa^{-1}(\Delta)) = \rho(\{(\lambda,\mu) \in P | (d,b_2) \in \Delta\})$. The one-to-one mapping κ^{-1} maps any bounded subset of Ω onto a bounded subset of P, and therefore the condition that ρ is a σ-finite measure on P implies that ν is a σ-finite measure on Ω.

For any $(\lambda,\mu)\in P,\ t\in R_+ : |\psi_1(\lambda,\mu,t)|^2 = e^{-2dt\frac{\sin^2 b_2 t}{b_2^2}}$. Now one can write for all $t\in R_+$:

$$R(t)\overset{\text{def}}{=}\int\limits_P |\psi_1(\lambda,\mu,t)|^2 d\rho_{\lambda,\mu} = \int\limits_\Omega e^{-2dt\frac{\sin^2 b_2 t}{b_2^2}} d\nu_{d,b_2}.$$

Lemma 5.5. *Let* $0\le\gamma<+\infty$, $\Omega=\{(d,b_2)|d\ge-\gamma,\ b_2\ge 0\}$, *and* ν *be a* σ-*finite measure on* Ω. *Denote for each Borel set* $\Delta\subseteq\Omega : \nu_1(\Delta)=\int_\Delta \frac{1}{d^2+b_2^2+1}d\nu$. *Denote for each* $x>0 : \Omega_x\overset{\text{def}}{=}\Omega\cap(\{(d,b_2)|d\le 2x\}\setminus\{(d,b_2)|d<x,\ b_2<x\})$ (see Fig. 5.1). *The integral transform* $R(t)=\int_\Omega e^{-2dt\frac{\sin^2 b_2 t}{b_2^2}}d\nu$ *is continuous on* $R_+=[0,+\infty)$ *if and only if*

$$\nu_1(\Omega_x)<+\infty \text{ for all } x>0, \text{ and } \lim_{x\to+\infty}\nu_1(\Omega_x)=0. \qquad (5.10)$$

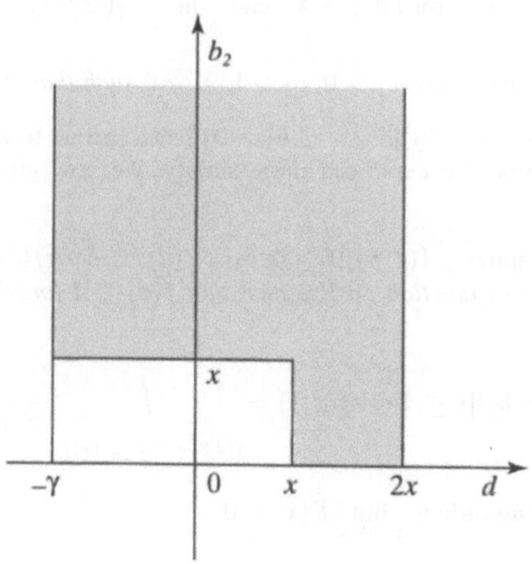

Fig. 5.1

The *proof of Lemma* 5.5 is similar to the proof of Lemma 5.1 (see Sections 2, 3 of this chapter) but substantially more complicated.

Note that we consider separately the integrals over $\{(d,b_2)|-\gamma\le d\le 1,\ 0\le b_2<1\}$, over $\{(d,b_2)|b_2\ge 1,\ -\gamma\le d\le b_2\}$, and over $\{(d,b_2)|d>1,$

$0 \le b_2 < d\}$. In the integrals over $\{(d,b_2)|d > 1, 0 \le b_2 < d\}$, we rewrite the expression for the sub-integral function in the following form:

$$e^{-2dt}\frac{\sin^2 b_2 t}{b_2^2} = (dt)^2 e^{-2dt}\left(\frac{\sin b_2 t}{b_2 t}\right)^2 \cdot \frac{1}{d^2},$$

for all $t > 0$.

Corollary 1. *Fix an arbitrarily small* $0 < \delta < +\infty$. *Condition* (5.10) *is necessary for* $R(t)$ *to be a measurable function on* $[0,\delta]$ *such that* $R(t) \to R(0) = 0$ *as* $t \to 0$. *On the other hand,* (5.10) *is sufficient for* $R(t)$ *to be continuous on the whole* R_+.

Corollary 2. *For arbitrary* $c_1 > 0$, $c_2 > 1$, $x > 0$, *we denote:*

$$\Omega_x^{c_1,c_2} \overset{\text{def}}{=} \Omega \cap (\{(d,b_2)|d \le c_2 x\} \setminus \{(d,b_2)|d < x,\, b_2 < c_1 x\})$$

(in particular, $\Omega_x^{1,2} = \Omega_x$*). For* $R(t)$ *to be continuous on* R_+, *it is necessary that for all* $c_1 > 0$, $c_2 > 1$, $X \ge 0$*:*

$$\nu_1(\Omega_x^{c_1,c_2}) < +\infty \text{ for all } x > X, \text{ and } \lim_{x\to+\infty} \nu_1(\Omega_x^{c_1,c_2}) = 0, \tag{5.11}$$

and sufficient that there exist $c_1 > 0$, $c_2 > 1$, $X \ge 0$ *such that* (5.11) *holds.*

If $\operatorname{supp}\nu \subseteq \Omega = \{(d,b_2)|d \ge -\gamma, b_2 \ge 0\}$ has a special form, then condition (5.10) may sometimes be expressed more simply. We now give an example of this.

Corollary 3. *Let* $\operatorname{supp}\nu \subseteq \{(d,b_2)|d \ge 0, b_2 = f(d)\}$, *where* $f(d)$ *is a real-valued increasing continuous function on* R_+ *such that* $f(d) \ge d$ *for all* $d \ge 0$. *Denote for each* $x > 0$*:*

$$F(x) \overset{\text{def}}{=} \nu_1(\{(d,b_2)|x \le d \le 2f(x)\}) = \int\limits_{\{(d,b_2)|x\le d\le 2f(x)\}} \frac{d\nu}{d^2 + f^2(d) + 1}.$$

$R(t) \in C(R_+)$ *if and only if* $\lim_{x\to+\infty} F(x) = 0$.

The corollaries that are similar to Corollaries 1–3 of Lemma 5.5 hold for Lemma 5.6 and Theorems 5.1–5.3 below as well.

2. Let $0 \le \gamma < +\infty$, $P = \{(\lambda,\mu) \in R^2|\lambda \ge -2\gamma, -\gamma\lambda - \gamma^2 \le \mu \le \frac{\lambda^2}{4}\}$. In view of Lemma 1.2, $\forall(\lambda,\mu) \in R^2 : d \ge -\gamma \Leftrightarrow -\max_{i=1,2} \operatorname{Re}\omega_i(\lambda,\mu) \ge -\gamma \Leftrightarrow \lambda \ge -2\gamma, \mu \ge -\gamma\lambda - \gamma^2$. Taking into account this inference, for any $(\lambda,\mu) \in P$: $\operatorname{Im}\lambda = 0$; $d = \frac{\lambda}{2} - \sqrt{\frac{\lambda^2}{4} - \mu} \ge -\gamma$, $a = \frac{\lambda}{2} \ge -\gamma$, $b_1 = a - d = \sqrt{\frac{\lambda^2}{4} - \mu} \ge 0$,

$b_2 = 0$. In particular, P satisfies (2) with $C = 0$. Note that $\forall(\lambda, \mu) \in P$: $a^2 + 1 = \frac{\lambda^2}{4} + 1$.

There exists a one-to-one mapping of P onto $\Omega = \{(d, a) | d \geq -\gamma,\ a \geq d\} \subset R^2$ which is given by the formula:

$$P \ni (\lambda, \mu) \leftrightarrow (d, a) = \left(\frac{\lambda}{2} - \sqrt{\frac{\lambda^2}{4} - \mu}, \frac{\lambda}{2}\right) \in \Omega.$$

Note that $\forall(d, a) \in \Omega : b_1 = a - d \geq 0$. Just as previously, we introduce a measure ν on Ω by a σ-finite measure ρ on P, and show that ν is a σ-finite measure on Ω since ρ is σ-finite.

For any $(\lambda, \mu) \in P$, $t \in R_+ : |\psi_1(\lambda, \mu, t)|^2 = e^{-2at} \frac{\sinh^2 b_1 t}{b_1^2}$. Now one can write for all $t \in R_+$:

$$R(t) = \int\limits_P |\psi_1(\lambda, \mu, t)|^2 d\rho_{\lambda,\mu} = \int\limits_\Omega e^{-2at} \frac{\sinh^2 b_1 t}{b_1^2} d\nu_{d,a}.$$

Lemma 5.6. *Let* $0 \leq \gamma < +\infty$, $\Omega = \{(d, a) | d \geq -\gamma,\ a \geq d\} \subset R^2$, *for all* $(d, a) \in \Omega : b_1 = a - d$, *and* ν *be a* σ*-finite measure on* Ω. *Denote for each Borel set* $\Delta \subseteq \Omega : \nu_1(\Delta) = \int_\Delta \frac{d\nu}{a^2+1}$. *Denote for each* $x > 0$:

$$\Omega_x \overset{\text{def}}{=} \Omega \cap (\{(d, a) | d \leq 2x\} \setminus \{(d, a) | d < x,\ a < x\})$$

(see Fig. 5.2). *The integral transform* $R(t) = \int_\Omega e^{-2at} \frac{\sinh^2 b_1 t}{b_1^2} d\nu$ *is continuous on* $R_+ = [0, +\infty)$ *if and only if*

$$\nu_1(\Omega_x) < +\infty \text{ for all } x > 0, \text{ and } \lim_{x \to +\infty} \nu_1(\Omega_x) = 0.$$

The *proof* is similar to the proof of Lemma 5.5. Note that for all $a \geq 1$, $-\gamma \leq d \leq a$, $t > 0$ we rewrite the expression for the sub-integral function in the following form:

$$e^{-2at} \frac{\sinh^2 b_1 t}{b_1^2} = (at)^2 e^{-2at} \left(\frac{\sinh b_1 t}{b_1 t}\right)^2 \cdot \frac{1}{a^2}.$$

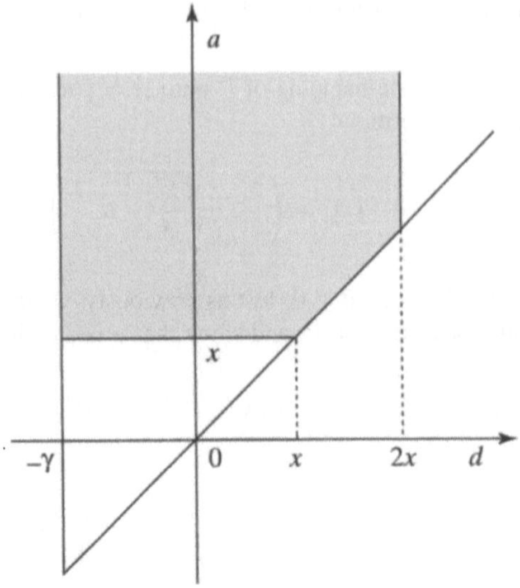

Fig. 5.2

6 Continuity of $R(t)$ on R_+

Let now P be an arbitrary closed subset of C^2 such that (2) holds. In particular,

$$\exists 0 < C < +\infty \forall (\lambda,\mu) \in P : \ |\operatorname{Im}\lambda|^2 \le C(a^2 + d^2 + b_2^2 + 1). \tag{5.12}$$

We study the same question: what must be a σ-finite measure ρ on P for $R(t) = \int_P |\psi_1(\lambda,\mu,t)|^2 d\rho \in C(R_+)$?

1. Recall that $\forall(\lambda,\mu) \in P$:

$$a = \frac{1}{2}\operatorname{Re}\lambda \ge -\gamma, \quad d = \frac{1}{2}\operatorname{Re}\lambda - \left|\operatorname{Re}\sqrt{\frac{\lambda^2}{4} - \mu}\right| = -\max_{i=1,2}\operatorname{Re}\omega_i \ge -\gamma,$$

$$b_1 = a - d = \left|\operatorname{Re}\sqrt{\frac{\lambda^2}{4} - \mu}\right| \ge 0, \quad b_2 = \left|\operatorname{Im}\sqrt{\frac{\lambda^2}{4} - \mu}\right| \ge 0.$$

There exists a continuous mapping κ of P into $\Omega = \{(d,a,b_2) \in R^3 | d \ge -\gamma, a \ge d, b_2 \ge 0\}$ which is given by the formula: $P \ni (\lambda,\mu) \mapsto (d,a,b_2) \in \Omega$. Note that $\forall (d,a,b_2) \in \Omega : b_1 = a - d \ge 0$. Just as previously, to any σ-finite measure ρ on P there corresponds a measure ν on Ω which is defined by the following formula: for each Borel set $\Delta \subseteq \Omega$, $\nu(\Delta) \overset{\text{def}}{=} \rho(\kappa^{-1}(\Delta)) = \rho(\{(\lambda,\mu) \in P | (d,a,b_2) \in \Delta\})$.

But now the situation differs from the previous section. Let us explain this. First, κ may not be a one-to-one mapping.

Further, for an arbitrary closed $P \subseteq C^2$ and a measure ρ on P, define a mapping κ of P into $R^3 = \{(d, a, b_2)\}$ and a measure ν on R^3 just as previously. If P is an arbitrary closed subset of C^2, then κ may map an unbounded Borel subset of P onto a bounded subset of R^3. For instance, if $P = C^2$ then κ maps $\{(\lambda, \mu) | \operatorname{Re} \lambda = 0, \operatorname{Im} \lambda \in R^1, \mu = \frac{\lambda^2}{4}\} \subset C^2$ onto $\{(0,0,0)\} \subset R^3$.

If $P \subseteq C^2$ is such that a certain unbounded Borel set $\Delta \subseteq P$ is mapped by κ into a bounded subset of R^3, then a measure ν on R^3 may not be σ-finite while ρ is a σ-finite measure on P. Indeed, let a Borel set $\Delta \subseteq R^3$ be bounded but $\kappa^{-1}(\Delta) = \{(\lambda, \mu) \in P | (d, a, b_2) \in \Delta\}$ be unbounded. Then $\nu(\Delta) = \rho(\kappa^{-1}(\Delta))$ may be equal to $+\infty$, since $\kappa^{-1}(\Delta) \subseteq C^2$ is unbounded and the fact that ρ is σ-finite does not control $\rho(\kappa^{-1}(\Delta))$.

Nevertheless, we now establish that in the case which is of concern in this section, a measure ν on Ω is σ-finite if the original measure ρ on P is σ-finite. To do this, we need the following lemma.

Lemma 5.7. *Let $P \subseteq C^2$ satisfy* (5.12). *Then*

$$\begin{aligned}&\exists 0 < c_1 < c_2 < +\infty \forall (\lambda, \mu) \in P :\\ &c_1(a^2 + d^2 + b_2^2 + 1) \leq |\lambda|^2 + |\mu| + 1 \leq c_2(a^2 + d^2 + b_2^2 + 1).\end{aligned}$$

Corollary 1. *Let P be such a closed subset of C^2 that* (5.12) *holds. There exist $0 < c_1 < c_2 < +\infty$ such that for any σ-finite measure ρ on P and Borel set $\Delta \subseteq R^3$:*

$$\frac{1}{c_2} \int_{\Delta} \frac{d\nu}{a^2 + d^2 + b_2^2 + 1} \leq \int_{\kappa^{-1}(\Delta)} \frac{d\rho}{|\lambda|^2 + |\mu| + 1} \leq \frac{1}{c_1} \int_{\Delta} \frac{d\nu}{a^2 + d^2 + b_2^2 + 1}$$

or all of these integrals are equal to $+\infty$.

Corollary 2. *Let P be such a closed subset of C^2 that* (5.12) *holds. Then a set $\Delta \subseteq \kappa(P)$ is bounded in R^3 if and only if $\kappa^{-1}(\Delta) \subseteq P$ is bounded in C^2. Accordingly, a measure ν on R^3 is σ-finite if and only if the original measure ρ on P is σ-finite.*

Remark 1. Notice that if one replaces P which satisfies (5.12) by C^2, then Lemma 5.7 and Corollaries 1, 2 fail to be true. To show this, it suffices to consider $\Delta = \{(0,0,0)\}$; then $\kappa^{-1}(\Delta) = \{(\lambda, \mu) \in C^2 | \operatorname{Re} \lambda = 0, \operatorname{Im} \lambda \in R^1, \mu = \frac{\lambda^2}{4}\}$.

We have seen that in the present situation, where P satisfies (2), to a σ-finite measure ρ on P there corresponds a measure ν on Ω which is σ-finite as well.

2. For any $t \in R_+$, $(\lambda, \mu) \in P$:

$$|\psi_1(\lambda, \mu, t)|^2 = e^{-2at} \frac{\sin^2 b_2 t \cosh^2 b_1 t + \cos^2 b_2 t \sinh^2 b_1 t}{b_1^2 + b_2^2}.$$

Now one can write for all $t \in R_+$:

$$R(t) \stackrel{\text{def}}{=} \int_P |\psi_1(\lambda,\mu,t)|^2 d\rho_{\lambda,\mu} = \int_\Omega e^{-2at} \frac{\sin^2 b_2 t \cosh^2 b_1 t + \cos^2 b_2 t \sinh^2 b_1 t}{b_1^2 + b_2^2} d\nu_{d,a,b_2}.$$

Lemma 5.8. *Let* $0 \le \gamma < +\infty$, $\Omega = \{(d,a,b_2) | d \ge -\gamma,\ a \ge d,\ b_2 \ge 0\} \subset R^3$, *for all* $(d,a,b_2) \in \Omega : b_1 = a - d$, *and* ν *be a* σ*-finite measure on* Ω. *For each Borel set* $\Delta \subseteq \Omega$, *denote:* $\nu_1(\Delta) = \int_\Delta \frac{1}{d^2+a^2+b_2^2+1} d\nu$. *For each* $x > 0$, *denote:* $\Omega_x = \Omega \cap (\{(d,a,b_2) | d \le 2x\} \setminus \{(d,a,b_2) | d < x,\ a < x,\ b_2 < x\})$ (see Fig. 5.3). *The integral transform*

$$R(t) = \int_\Omega e^{-2at} \frac{\sin^2 b_2 t \cosh^2 b_1 t + \cos^2 b_2 t \sinh^2 b_1 t}{b_1^2 + b_2^2} d\nu$$

is continuous on R_+ *if and only if*

$$\nu_1(\Omega_x) < +\infty \text{ for all } x > 0, \text{ and } \lim_{x \to +\infty} \nu_1(\Omega_x) = 0.$$

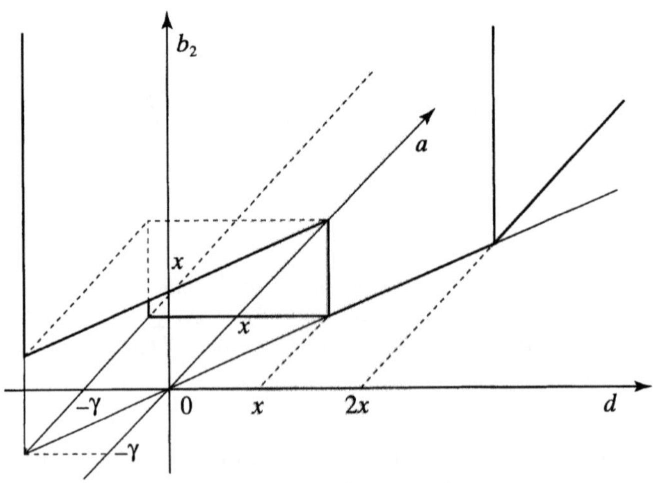

Fig. 5.3

In order to *prove* Lemma 5.8, we write for each $t > 0 : R(t) = R_1(t) + R_2(t)$, where

$$R_1(t) = \int\limits_{\Omega} e^{-2dt} \frac{\sin^2 b_2 t}{b_2^2} \left(\frac{1+e^{-2b_1 t}}{2}\right)^2 \frac{b_2^2}{b_1^2 + b_2^2} d\nu,$$

$$R_2(t) = \int\limits_{\Omega} e^{-2at} \frac{\sinh^2 b_1 t}{b_1^2} \cos^2 b_2 t \frac{b_1^2}{b_1^2 + b_2^2} d\nu.$$

Then we make use of Lemmas 5.5 and 5.6 and treatments which are similar to the proofs of these lemmas.

3. We now return to $P \subseteq C^2$. Taking into account Lemma 5.7 and its corollaries, Lemma 5.8 implies the following theorem.

Theorem 5.1. *Let $P \subseteq C^2$ be a closed set such that* (2) *holds, ρ be a σ-finite measure on P. Denote for each Borel set $\Delta \subseteq P$:*

$$\rho_1(\Delta) = \int\limits_{\Delta} \frac{1}{|\lambda|^2 + |\mu| + 1} d\rho.$$

Set for each $x > 0$: $P_x = P \cap (\{(\lambda, \mu) | d \leq 2x\} \setminus \{(\lambda, \mu) | d < x,\ a < x,\ b_2 < x\})$. The integral transform $R(t) = \int_P |\psi_1(\lambda, \mu, t)|^2 d\rho$ is continuous on R_+ if and only if

$$\rho_1(P_x) < +\infty \text{ for all } x > 0, \text{ and } \lim_{x \to +\infty} \rho_1(P_x) = 0. \tag{5.13}$$

Remark 1. Note that $\forall x > 0 : P_x = P \cap (C^2 \setminus \{(\lambda, \mu) | d > 2x\}) \cap (C^2 \setminus \{(\lambda, \mu) | a < x,\ b_2 < x\}) = P \cap (C^2 \setminus \{(\lambda, \mu) | \lambda_1 > 4x,\ \mu_1 > \frac{1}{(\lambda_1 - 4x)^2}(\mu_2^2 - \lambda_1\lambda_2 \cdot \mu_2 + 2x(\lambda_1 - 2x)(\lambda_2^2 + (\lambda_1 - 4x)^2))\}) \cap (C^2 \setminus \{(\lambda, \mu) | \lambda_1 < 2x,\ |\operatorname{Im}\sqrt{\lambda^2 - 4\mu}| < 2x\})$.

Corollary 1. *If $\rho_1(P) < +\infty$ then $R(t) \in C(R_+)$.*

Proof. Let $\rho_1(P) < +\infty$. Fix an arbitrary $x > 0$. $P_x \subseteq P$; hence $\rho_1(P_x) \leq \rho_1(P) < +\infty$. Moreover, $P_x \subseteq P \setminus \{(\lambda, \mu) | d < x,\ a < x,\ b_2 < x\}$, and hence $\rho_1(P_x) \leq \rho_1(P \setminus \{(\lambda, \mu) | d < x,\ a < x,\ b_2 < x\})$. $\rho_1(P) < +\infty$ implies that $\rho_1(P \setminus \{(\lambda, \mu) | d < x,\ a < x,\ b_2 < x\}) \to 0$ as $x \to +\infty$. This completes the proof.

7 Continuity, boundedness, and integrability of $R(t)$ on a finite segment $[0, T]$

Fix an arbitrary $0 < T < +\infty$. Assume that P is a closed subset of C^2 such that (2) holds, ρ is a σ-finite measure on P. We make use of the notations which were introduced in Theorem 5.1. Notice that ρ_1 is a σ-finite measure on P as is ρ.

Corollary 2. *The integral transform $R(t)$ is continuous on $[0, T]$ if and only if* (5.13) *is valid.*

The following two theorems can be proved analogously.

Theorem 5.2. *The integral transform $R(t)$ is continuous and bounded on $(0, T]$ if and only if*

$$\exists K < +\infty \forall x > 0 : \rho_1(P_x) \le K. \tag{5.14}$$

Theorem 5.3. *The integral transform $R(t)$ is integrable on $(0, T]$ if and only if*

$$\int_P \frac{1}{|d|+1} d\rho_1 < +\infty. \tag{5.15}$$

Moreover, if (5.15) *holds then $R(t)$ is continuous on $(0, T]$.*

The proof is based on the Foubini theorem.

Note that here

$$\int_P \frac{1}{|d|+1} d\rho_1 = \int_P \frac{1}{(|\lambda|^2 + |\mu| + 1)(|d| + 1)} d\rho.$$

Remark 1. Fix an arbitrary closed $P \subseteq C^2$ such that (2) holds. By definition, for a σ-finite measure ρ on $P : R(t) = \int_P |\psi_1(\lambda, \mu, t)|^2 d\rho$ for all $t \in R_+$. Now let us write four conditions on a σ-finite measure ρ on P:

$$1)\ \rho_1(P) < +\infty; \tag{5.16}$$

$$2)\ R(t) = \int_P |\psi_1(\lambda, \mu, t)|^2 d\rho \text{ is continuous on } [0, T]; \tag{5.17}$$

$$3)\ R(t) \text{ is continuous and bounded on } (0, T]; \tag{5.18}$$

$$4)\ R(t) \text{ is continuous and integrable on } (0, T]. \tag{5.19}$$

It is obvious that everyone of these conditions implies the following one. In the general case the converse implications may not be valid. For instance, we have shown in Section 4 that for $P = \{(\lambda, \mu) \in R^2 | \lambda \ge 2,\ \mu = \frac{\lambda^2}{2}\}$, everyone of these conditions is strictly weaker than the previous one (i.e., follows from the previous one but is not equivalent to it).

8 Equivalence of conditions on a sub-integral measure

Let $0 < T < +\infty$.

1. What must be a closed $P \subseteq C^2$ satisfying (2) for conditions (5.16) and (5.17) on a σ-finite measure ρ on P to be equivalent?

One may pose the same question for conditions (5.17) and (5.18) and for conditions (5.18) and (5.19) as well.

Theorem 5.4. *Let P be a closed subset of C^2 such that* (2) *holds.*

1) *Conditions* (5.16) *and* (5.17) *on a σ-finite measure ρ on P are equivalent if and only if*

$$\exists 0 < \gamma_1 < +\infty : \ P \subseteq \{(\lambda, \mu) \in C^2 | d \leq \gamma_1\}, \tag{5.20}$$

or, what is the same,

$$\exists 0 < \gamma_1 < +\infty \forall (\lambda, \mu) \in P : \ \max_{i=1,2} \operatorname{Re} \omega_i \geq -\gamma_1.$$

2) *The statement* 1), *with* (5.16) *and* (5.17) *replaced by* (5.17) *and* (5.18) *respectively.*

3) *The statement* 1), *with* (5.16) *and* (5.17) *replaced by* (5.18) *and* (5.19) *respectively.*

The proof is similar to the proof of Lemma 5.4.

Moreover, one can prove the following statement.

Statement 5.3. Let P be a closed subset of C^2 such that (2) holds but (5.20) does not; ρ_0 be an arbitrary σ-finite measure on P such that $\operatorname{supp} \rho_0 = P$.

1) For this measure, there exists such a nonnegative continuous function φ on P that for a σ-finite measure ρ on P such that $\rho_1(\Delta) = \int_\Delta \varphi(\lambda, \mu) d\rho_0$, $\forall \Delta \in \mathcal{B}(P)$, — (5.17) holds but (5.16) does not.

2) The statement 1), with (5.16) and (5.17) replaced by (5.17) and (5.18) respectively.

3) The statement 1), with (5.16) and (5.17) replaced by (5.18) and (5.19) respectively.

Remark 1. In Theorem 5.4 and Statement 5.3, one can replace conditions (5.16), (5.17), (5.18), and (5.19) by conditions (5.16), (5.13), (5.14), and (5.15) respectively. This follows from Corollary 2 of Theorem 5.1 and Theorems 5.2, 5.3.

Remark 2. Let $0 < \gamma_1 < +\infty$. Then:

1) $\{(\lambda, \mu) \in R^2 | d \leq \gamma_1\} = R^2 \setminus \{(\lambda, \mu) | \lambda > 2\gamma_1, \ \mu > \gamma_1 \lambda - \gamma_1^2\}$;

2) $$\{(\lambda, \mu) \in C^2 | d \leq \gamma_1\} = C^2 \setminus \{(\lambda, \mu) | \lambda_1 > 2\gamma_1, \ \mu_1 > \frac{1}{(\lambda_1 - 2\gamma_1)^2} \cdot (\mu_2^2 - \lambda_1 \lambda_2 \cdot \mu_2 + \gamma_1(\lambda_1 - \gamma_1)(\lambda_2^2 + (\lambda_1 - 2\gamma_1)^2))\}.$$

This follows from Lemma 1.2 and the proof of Lemma 3.2.

2. Let P be a closed subset of C^2 such that (2) holds, f be a nonnegative continuous function on P. For each σ-finite measure ρ on P we denote: $R_f(t) = \int_P |\psi_1(\lambda, \mu, t)|^2 f(\lambda, \mu) d\rho$, for all $t \in R_+$; ρ_f is a σ-finite measure on P such that for each Borel set $\Delta \subseteq P$:

$$\rho_f(\Delta) = \int\limits_{\Delta} f(\lambda, \mu) d\rho_1 = \int\limits_{\Delta} \frac{f(\lambda, \mu)}{|\lambda|^2 + |\mu| + 1} d\rho.$$

If $f(\lambda, \mu) \equiv 1$ on P, then for any $t \in R_+ : R_f(t) = R(t)$; for any Borel set $\Delta \subseteq P : \rho_f(\Delta) = \rho_1(\Delta)$. We have proved (see Theorem 5.1 and Corollary 1 of it) that in this case condition

$$\rho_1(P) < +\infty \tag{5.16}$$

on a σ-finite measure ρ on P implies that

$$\rho_f(P_x) < +\infty \text{ for all } x > 0, \text{ and } \rho_f(P_x) \to 0 \text{ as } x \to +\infty, \tag{5.21}$$

or, what is the same,

$$R_f(t) \in C([0, T]). \tag{5.22}$$

Theorem 5.5. *Let P be a closed subset of C^2 such that* (2) *holds, let $f \in C(P)$, and let $f(\lambda, \mu) \geq 0$ for all $(\lambda, \mu) \in P$. Condition* (5.16) *on a σ-finite measure ρ on P implies* (5.21) *(or, what is the same,* (5.22)*) if and only if*

$$\exists K < +\infty \forall (\lambda, \mu) \in P : f(\lambda, \mu) \leq K \tag{5.23}$$

(i.e., f is a bounded function on P).

Moreover, one can prove the following statement.

Statement 5.4. Let P be a closed subset of C^2 such that (2) holds; f be a nonnegative continuous function on P such that (5.23) does not hold (i.e., f is an unbounded function on P); and ρ_0 be an arbitrary σ-finite measure on P such that $\operatorname{supp} \rho_0 = P$. For this measure, there exists such a nonnegative continuous function φ on P that for a σ-finite measure ρ on P such that $\rho_1(\Delta) = \int_\Delta \varphi(\lambda, \mu) d\rho_0$, $\forall \Delta \in \mathcal{B}(P)$, — (5.16) holds but (5.21) (or, what is the same, (5.22)) does not.

3. If P is a closed subset of C^2 such that (2) holds, and (5.20) is valid, then for a σ-finite measure ρ on P and a continuous nonnegative function f on P: (5.21) (or, what is the same, (5.22)) implies that

$$\rho_f(P) < +\infty. \tag{5.24}$$

This follows from Theorem 5.4 where ρ_1 is replaced by ρ_f and $R(t)$ by $R_f(t)$ respectively.

Further, if P is a closed subset of C^2 such that (2) holds, f is a nonnegative continuous function on P, and (5.23) holds, then for a σ-finite measure ρ on P: (5.16) implies (5.24).

This follows immediately from the definition of ρ_f.

From these two statements, we deduce the following one:

Let P be a closed subset of C^2 such that (2) holds, $f \in C(P)$, $f(\lambda, \mu) \geq 0$ for all $(\lambda, \mu) \in P$, and

$$\exists 0 < \gamma_1,\ K < +\infty : P \subseteq \{(\lambda, \mu) \in C^2 | d \leq \gamma_1\} \cup \{(\lambda, \mu) \in C^2 | f(\lambda, \mu) \leq K\}. \tag{5.25}$$

Then for a σ-finite measure ρ on P the simultaneous fulfilment of conditions (5.16) and (5.21) implies (5.24).

Theorem 5.6. *Let P be a closed subset of C^2 such that* (2) *holds, f be a nonnegative continuous function on P. The simultaneous fulfilment of conditions* (5.16) *and* (5.21) *for a σ-finite measure ρ on P implies* (5.24) *if and only if* (5.25) *holds.*

Moreover, one can prove the following statement.

Statement 5.5. Let P be a closed subset of C^2 such that (2) holds, f be a nonnegative continuous function on P such that condition (5.25) does not hold. Further, let ρ_0 be an arbitrary σ-finite measure on P such that $\operatorname{supp} \rho_0 = P$. For this measure, there exists such a nonnegative continuous function φ on P that for a σ-finite measure ρ on P such that $\rho_1(\Delta) = \int_\Delta \varphi(\lambda, \mu) d\rho_0$, $\forall \Delta \in \mathfrak{B}(P)$, — (5.16) and (5.21) hold but (5.24) does not.

Chapter 6
Initial data of solutions

1 The set of initial data of solutions

Let A and B be c.n.o. in H such that $P = \sigma(A,B)$ satisfies (2). By virtue of Statement 5.1, this means that the Cauchy problem for equation (1) on R_+ is well-posed in $G = D(B) \times (D(A) \cap D(|B|^{1/2}))$. In other words, for any $(f_0, f_1) \in D(B) \times (D(A) \cap D(|B|^{1/2}))$ there exists a unique solution of equation (1) on R_+ such that $y(0) = f_0$, $y'(0) = f_1$.

The question is: to indicate the set of *all* (f_0, f_1) such that the Cauchy problem for equation (1) on R_+ with initial conditions $y(0) = f_0$, $y'(0) = f_1$ has a unique solution.

For equivalent formulations of this question see Introduction to Part II.

1. Denote the set of such (f_0, f_1) by F_C. Obviously F_C depends on A and B. We have shown in Introduction to Part II that

$$D(B) \times (D(A) \cap D(|B|^{1/2})) \subseteq F_C \subseteq D(B) \times D(A).$$

Theorem 6.1. *Let A and B be c.n.o. in H such that both conditions (3.3), (3.4) hold, i.e., all of the equivalent conditions 1), 2), 3), 4) of Statement 5.1 hold. Denote: $P = \sigma(A,B)$. The Cauchy problem for equation (1) on R_+ with initial conditions $y(0) = f_0$, $y'(0) = f_1$ admits a unique solution if and only if $f_0 \in D(B)$, $f_1 \in D(A)$, and*

$$\forall x > 0 : \; E(P_x)f_1 \in D(|B|^{1/2}), \text{ and } \lim_{x\to+\infty} |B|^{1/2}E(P_x)f_1 = 0. \tag{6.1}$$

If this is the case then a solution $y(t)$ of (1) on R_+ with $y(0) = f_0$, $y'(0) = f_1$ has the form: $y(t) = \sum_{i=0}^{1} \psi_i(A,B,t)f_i$ $(t \in R_+)$.

Remark 1. By the definitions, (6.1) may be rewritten in the following form:

$$\forall x > 0 : \int_{P_x} |\mu| d(Ef_1, f_1) < +\infty, \text{ and } \lim_{x\to+\infty} \int_{P_x} |\mu| d(Ef_1, f_1) = 0. \tag{6.2}$$

Remark 2. Obviously (6.1) is a weaker condition than $f_1 \in D(|B|^{1/2})$. Indeed, let $f_1 \in D(|B|^{1/2})$, i.e., $\int_P |\mu| d(Ef_1, f_1) < +\infty$. Then for any $x > 0$: $\int_{P_x} |\mu| d(Ef_1, f_1) \leq \int_P |\mu| d(Ef_1, f_1) < +\infty$. Furthermore, $P_x \subseteq P \setminus \{(\lambda, \mu) \in C^2 | d < x,\ a < x,\ b_2 < x\}$; so,

$$\int_{P_x} |\mu| d(Ef_1, f_1) \leq$$

$$\int_P |\mu| d(Ef_1, f_1) - \int_{P \cap \{(\lambda,\mu) | d<x, a<x, b_2<x\}} |\mu| d(Ef_1, f_1) \to 0 \text{ as } x \to +\infty.$$

Proof of Theorem 6.1. In fact, we have started to prove Theorem 6.1 in Introduction to Part II. We have shown that if a solution $y(t)$ of (1) on R_+ with $y(0) = f_0$, $y'(0) = f_1$ exists, then it has the form:

$$y(t) = \sum_{i=0}^{1} \psi_i(A, B, t) f_i \quad (t \in R_+).$$

We have proved that F_C consists of all (f_0, f_1) such that $f_0 \in D(B)$, $f_1 \in D(A)$, and $y_1(t) = \psi_1(A, B, t) f_1$ is a solution of equation (1) on R_+ with $y_1(0) = 0$, $y_1'(0) = f_1$.

Now we can complete this proof. To this end, we introduce a σ-finite measure ρ on P such that $d\rho = |\mu|^2 d(Ef_1, f_1)$. Theorem 5.1 implies that $\| By_1(t) \|^2$ is continuous on R_+, i.e.,

$$\| B\psi_1(A, B, t) f_1 \|^2 = \int_P |\psi_1(\lambda, \mu, t)|^2 |\mu|^2 d(Ef_1, f_1) \in C(R_+)$$

if and only if

$$\begin{aligned} &\forall x > 0 : \int_{P_x} \frac{|\mu|^2}{|\lambda|^2 + |\mu| + 1} d(Ef_1, f_1) < +\infty, \\ &\text{and } \lim_{x \to +\infty} \int_{P_x} \frac{|\mu|^2}{|\lambda|^2 + |\mu| + 1} d(Ef_1, f_1) = 0. \end{aligned} \tag{6.3}$$

Conversely, let $f_1 \in D(A)$ and (6.3) be valid. Since both conditions (2.2) and (3.2) hold, $f_1 \in D(A)$ implies that $y_1(t) = \psi_1(A, B, t) f_1 \in C^1(R_+, H)$ with $y_1'(t) = \psi_1'(A, B, t) f_1$ $(t \in R_+)$; $y_1(0) = 0$, $y_1'(0) = f_1$, $Ay_1'(t) = A\psi_1'(A, B, t) f_1 \in C(R_+, H)$. Furthermore, by virtue of Theorem 5.1, (6.3) implies that

$$\| B\psi_1(A, B, t) f_1 \|^2 = \int_P |\psi_1(\lambda, \mu, t)|^2 |\mu|^2 d(Ef_1, f_1) \to 0 \text{ as } t \to 0,$$

i.e., $By_1(t) = B\psi_1(A,B,t)f_1$ is continuous at $t = 0$. Moreover, by virtue of Theorems 5.1 and 5.3, (6.3) implies that

$$\int_P \frac{1}{(|\lambda|^2 + |\mu| + 1)(|d| + 1)} d\rho < +\infty.$$

Consequently, for any $t_0 > 0$, $0 < \delta < t_0$:

$$\int_P \frac{e^{-(t_0-\delta)d}}{|\lambda|^2 + |\mu| + 1} d\rho < +\infty.$$

Fix an arbitrary $t_0 > 0$. Now, just as in the proof of Lemma 5.1, one can conclude from the Lebesgue theorem that

$$\int_{\sigma(A,B)} |\psi_1(\lambda,\mu,t) - \psi_1(\lambda,\mu,t_0)|^2|\mu|^2 d(Ef_1,f_1)$$

$$= \int_P |\psi_1(\lambda,\mu,t) - \psi_1(\lambda,\mu,t_0)|^2 d\rho \to 0 \text{ as } t \to t_0,$$

i.e., $B\psi_1(A,B,t)f_1$ is continuous at $t = t_0$.

Taking into account Lemma 1.1, one can now prove that $f_1 \in D(A)$ and (6.3) imply $y_1''(t) = \psi_1''(A,B,t)f_1 \in C(R_+,H)$.

It remains to show that the following two conditions are equivalent:

1) $f_1 \in D(A)$ and (6.2) holds;

2) $f_1 \in D(A)$ and (6.3) holds.

To this end, we notice that $f_1 \in D(A)$ means: $\int_P (|\lambda|^2 + 1) d(Ef_1,f_1) < +\infty$. Further, set for any $x > 0$:

$$P_x' = P_x \cap \{(\lambda,\mu) \in C^2 |\ |\mu| \le |\lambda|^2 + 1\},$$

$$P_x'' = P_x \cap \{(\lambda,\mu) \in C^2 |\ |\mu| > |\lambda|^2 + 1\}.$$

Then

$$\int_{P_x} \frac{|\mu|^2}{|\lambda|^2 + |\mu| + 1} d(Ef_1,f_1) \le \int_{P_x} |\mu| d(Ef_1,f_1),$$

$$\int_{P_x'} |\mu| d(Ef_1,f_1) \le \int_{P_x'} (|\lambda|^2 + 1) d(Ef_1,f_1),$$

$$\int_{P_x''} |\mu| d(Ef_1,f_1) \le 2 \int_{P_x''} \frac{|\mu|^2}{|\lambda|^2 + |\mu| + 1} d(Ef_1,f_1).$$

Now it is clear that (6.2) implies (6.3), while $f_1 \in D(A)$ and (6.3) imply (6.2). The proof is over.

Corollary 1. *Fix an arbitrary* $0 < c < +\infty$. *Theorem* 6.1 *remains valid if one replaces* P_x *by* $P_x \cap \{(\lambda, \mu) \in \mathbf{C}^2 | \, |\mu| \geq c(|\lambda|^2 + 1)\}$.

Proof. Indeed, let $f_1 \in D(A)$, i.e., $\int_P (|\lambda|^2 + 1) d(Ef_1, f_1) < +\infty$. Then

$$\text{for any } x > 0: \int\limits_{P_x \cap \{(\lambda,\mu) \in \mathbf{C}^2 | \, |\mu| < c(|\lambda|^2+1)\}} |\mu| d(Ef_1, f_1) < +\infty,$$

$$\text{and} \int\limits_{P_x \cap \{(\lambda,\mu) \in \mathbf{C}^2 | \, |\mu| < c(|\lambda|^2+1)\}} |\mu| d(Ef_1, f_1)$$

$$\leq \int\limits_{P_x} c(|\lambda|^2 + 1) d(Ef_1, f_1) \to 0 \text{ as } x \to +\infty.$$

Thus, (6.2) holds if and only if

$$\int\limits_{P_x \cap \{(\lambda,\mu) \in \mathbf{C}^2 | \, |\mu| \geq c(|\lambda|^2+1)\}} |\mu| d(Ef_1, f_1) < +\infty \text{ for all } x > 0,$$

$$\text{and} \lim_{x \to +\infty} \int\limits_{P_x \cap \{(\lambda,\mu) \in \mathbf{C}^2 | \, |\mu| \geq c(|\lambda|^2+1)\}} |\mu| d(Ef_1, f_1) = 0.$$

2. Denote by F the set of all $f_1 \in H$ such that (6.1) (or, what is the same, (6.2)) holds. Obviously $D(|B|^{1/2}) \subseteq F \subseteq H$.

We have proved that $F_C = D(B) \times (D(A) \cap F)$.

Corollary 2. *Let A and B be c.n.o. in H such that both conditions* (3.3), (3.4) *hold, i.e., all of the equivalent conditions* 1), 2), 3), 4) *of Statement* 5.1 *hold. There exists a one-to-one mapping of the set of all solutions of equation* (1) *on* $\mathbf{R}_+$ *(we denote this set by Y) onto the set of their initial data* $F_C = D(B) \times (D(A) \cap F)$. *This mapping is given by the formula:*

$$F_C = D(B) \times (D(A) \cap F) \ni (f_0, f_1) \leftrightarrow \sum_{i=0}^{1} \psi_i(A, B, t) f_i \ (t \in \mathbf{R}_+) \in Y,$$

or, what is the same,

$$F_C = D(B) \times (D(A) \cap F) \ni (y(0), y'(0)) \leftrightarrow y(t) \ (t \in \mathbf{R}_+) \in Y.$$

3. Corollary 3. *Let A and B be c.n.o. in H such that both conditions* (3.3), (3.4) *hold, and moreover, let* $D(A^2) \subseteq D(B)$. *Then* $F_C = D(B) \times D(A)$.

Indeed, in this case $f_1 \in D(A) \Rightarrow f_1 \in D(|B|^{1/2}) \Rightarrow f_1 \in F$; so, $D(A) \cap F = D(A)$.

Corollary 4. *Let A and B be c.n.o. in H such that* (3.3) *holds, and moreover, let A be bounded. Then $F_C = D(B) \times D(|B|^{1/2})$.*

Proof. Set $\gamma_1 = \sup_{(\lambda,\mu)\in\sigma(A,B)} |\lambda|$; here $\gamma_1 = \| A \| < +\infty$. Now $\forall(\lambda,\mu) \in P = \sigma(A,B) : d = a - b_1 \leq a = \frac{1}{2}\operatorname{Re}\lambda \leq \gamma_1 < +\infty$; thus, $P = (P \cap \{(\lambda,\mu) \in C^2 | d < \gamma_1,\ a < \gamma_1,\ b_2 < \gamma_1\}) \cup P_{\gamma_1}$. By virtue of Corollary 2 of Lemma 5.7, $P \cap \{(\lambda,\mu) \in C^2 | d < \gamma_1,\ a < \gamma_1,\ b_2 < \gamma_1\}$ is bounded, and so

$$\int_{P\cap\{(\lambda,\mu)\in C^2|d<\gamma_1,a<\gamma_1,b_2<\gamma_1\}} |\mu| d(Ef_1, f_1) < +\infty, \quad \forall f_1 \in H.$$

Let $f_1 \in F$, i.e., (6.2) hold; then, in particular, $\int_{P_{\gamma_1}} |\mu| d(Ef_1, f_1) < +\infty$; consequently, $\int_P |\mu| d(Ef_1, f_1) < +\infty$, i.e., $f_1 \in D(|B|^{1/2})$. Thus, $F \subseteq D(|B|^{1/2})$. This means that $F = D(|B|^{1/2})$. Taking into account that here $D(A) = H$, we obtain: $D(A) \cap F = D(|B|^{1/2})$.

4. Let us clarify the statement of Theorem 6.1 in the special case that A and B are self-adjoint operators in H.

Corollary 5. *Let A and B be commuting self-adjoint operators in H such that*

$$\exists 0 < \gamma < +\infty : \sigma(A,B) \subseteq \{(\lambda,\mu) \in R^2 | \lambda \geq -\gamma,\ \mu \geq -\gamma\lambda - \gamma^2\},$$

i.e., all of conditions 1), 2), 3), 4) *of Statement* 5.1 *hold. Set $B_+ \overset{\text{def}}{=} BE_B([0,+\infty))$ (the nonnegative part of B). For each $x > 0$, denote: $Q_x = [-\gamma, x] \times [x^2, +\infty)$* (see Fig. 6.1). *The Cauchy problem for equation* (1) *on R_+ with initial conditions $y(0) = f_0$, $y'(0) = f_1$ admits a unique solution if and only if $f_0 \in D(B)$, $f_1 \in D(A)$, and*

$$\forall x > 0 : E(Q_x) f_1 \in D(B_+^{1/2}), \text{ and } \lim_{x\to+\infty} B_+^{1/2} E(Q_x) f_1 = 0. \tag{6.4}$$

Let's give some equivalent formulations for (6.4):

1)
$$\begin{gathered}\forall x > 0 : E_A([-\gamma, x]) f_1 \in D(B_+^{1/2}), \\ \text{and } B_+^{1/2} E_B([x^2, +\infty)) E_A([-\gamma, x]) f_1 \to 0 \text{ as } x \to +\infty;\end{gathered} \tag{6.5}$$

2) (6.4), with $B_+^{1/2}$ replaced by $|B|^{1/2}$;

3) (6.5), with $B_+^{1/2}$ replaced by $|B|^{1/2}$;

4) $\forall x > 0 : \int_{Q_x} \mu d(Ef_1, f_1) < +\infty$, and $\lim_{x\to+\infty} \int_{Q_x} \mu d(Ef_1, f_1) = 0$.

Corollary 5 follows from Theorem 6.1 and Corollary 1 of it.

Fix an arbitrary $0 < c < +\infty$. Corollary 5 remains valid if one replaces x^2 in its formulation by cx^2.

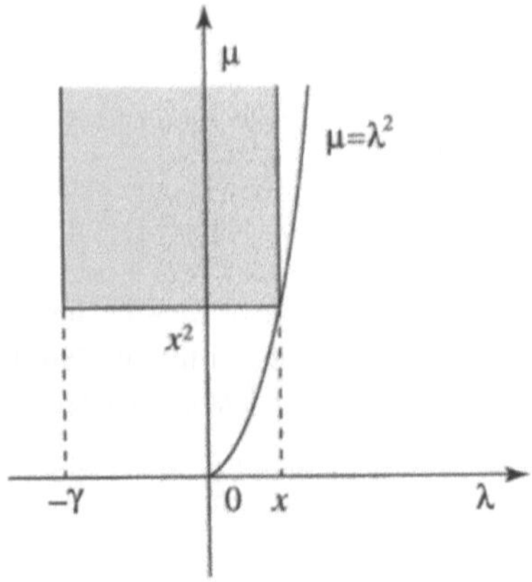

Fig. 6.1

2 When $F_C = D(B) \times (D(A) \cap D(|B|^{1/2}))$?

Let A and B be arbitrary commuting normal operators in H. In this case, we preserve the notation F_C for the set of all (f_0, f_1) such that the Cauchy problem for equation (1) on R_+ with initial conditions $y(0) = f_0$, $y'(0) = f_1$ admits a unique solution.

The well-posedness of the Cauchy problem for equation (1) on R_+ in $G = D(B) \times (D(A) \cap D(|B|^{1/2}))$ means, by definition, that

$$F_C \supseteq D(B) \times (D(A) \cap D(|B|^{1/2})).$$

Thus, the statement of Theorem 3.2 may be rewritten in the following form:

Let A and B be c.n.o. in H. The set F_C *contains* $D(B) \times (D(A) \cap D(|B|^{1/2}))$ if and only if $P = \sigma(A, B)$ satisfies (2).

The question now arises: what must be c.n.o. A and B in H for the set F_C *to coincide* with $D(B) \times (D(A) \cap D(|B|^{1/2}))$?

Theorem 6.2. *Let A and B be c.n.o. in H. The set F_C coincides with $D(B) \times (D(A) \cap D(|B|^{1/2}))$ if and only if $P = \sigma(A, B)$ satisfies condition (2) and the following condition:*

$$\begin{aligned} &\exists 0 < C_1,\ \gamma_1 < +\infty : \\ &P \subseteq \{(\lambda, \mu) \in \mathbf{C}^2 | d \leq \gamma_1\} \cup \{(\lambda, \mu) \in \mathbf{C}^2 |\ |\mu| \leq C_1(|\lambda|^2 + 1)\}. \end{aligned} \tag{6.6}$$

Remark 1. Recall that $P = \sigma(A, B)$ satisfies (2) if and only if both conditions (3.3), (3.4) hold. Further, $\{(\lambda, \mu) \in \mathbf{C}^2 | d \leq \gamma_1\} = \mathbf{C}^2 \setminus \{(\lambda, \mu) | \lambda_1 > 2\gamma_1, \mu_1 > \frac{1}{(\lambda_1 - 2\gamma_1)^2}(\mu_2^2 - \lambda_1\lambda_2 \cdot \mu_2 + \gamma_1(\lambda_1 - \gamma_1)(\lambda_1^2 + (\lambda_1 - 2\gamma_1)^2))\}$.

Remark 2. For

$$\exists 0 < \gamma_1 < +\infty : P = \sigma(A, B) \subseteq \{(\lambda, \mu) \in C^2 | d \leq \gamma_1\},$$

it suffices that Re A be bounded (see Corollary 1 of Theorem 4.1). For

$$\exists C_1 < +\infty : P \subseteq \{(\lambda, \mu) \in C^2 |\ |\mu| \leq C_1(|\lambda|^2 + 1)\},$$

it is necessary and sufficient that B be subordinate to A^2, i.e., $D(A^2) \subseteq D(B)$ (see Theorem 1.2 for c.n.o. A and B and the remark to it).

Proof of Theorem 6.2. First of all, $F_C = D(B) \times (D(A) \cap D(|B|^{1/2}))$ implies that $F_C \supseteq D(B) \times (D(A) \cap D(|B|^{1/2}))$, i.e., that $P = \sigma(A, B)$ satisfies (2).

In what follows, we assume that $P = \sigma(A, B)$ satisfies (2). In this case, $F_C = D(B) \times (D(A) \cap F)$. Thus, $F_C = D(B) \times (D(A) \cap D(|B|^{1/2}))$ if and only if $D(A) \cap D(|B|^{1/2}) = D(A) \cap F$.

We have shown previously that $D(|B|^{1/2}) \subseteq F$; so, $D(A) \cap D(|B|^{1/2}) \subseteq D(A) \cap F$. Further, $D(A) \cap F \subseteq D(A)$. We conclude that $F_C = D(B) \times (D(A) \cap D(|B|^{1/2}))$ if and only if

$$D(A) \cap F \subseteq D(|B|^{1/2}). \tag{6.7}$$

By the definitions, (6.7) may be rewritten in the following form: for $f_1 \in H$, if $\int_P (|\lambda|^2 + 1) d(Ef_1, f_1) < +\infty$ and (6.2) holds, then $\int_P |\mu| d(Ef_1, f_1) < +\infty$.

By virtue of Theorem 5.6 and Statement 5.5, this is the case if and only if (6.6) holds.

In order to prove this, we set $f(\lambda, \mu) = \frac{|\mu|}{|\lambda|^2+1}$ on P. Now condition (6.6) is equivalent to (5.25).

Let (6.6) hold. Fix an arbitrary $f_1 \in H$. Define a σ-finite measure ρ on P such that for each Borel set $\Delta \subseteq P : \rho_1(\Delta) = \int_\Delta (|\lambda|^2 + 1) d(Ef_1, f_1)$. Now $\int_P (|\lambda|^2 + 1) d(Ef_1, f_1) < +\infty$ means (5.16); (6.2) means (5.21); $\int_P |\mu| d(Ef_1, f_1) < +\infty$ means (5.24). By Theorem 5.6, if $\int_P (|\lambda|^2 + 1) d(Ef_1, f_1) < +\infty$ and (6.2) holds, then $\int_P |\mu| d(Ef_1, f_1) < +\infty$.

Conversely, let (6.6) be not valid. For each $g \in H$, define a σ-finite measure m_g on P by the formula: $m_g(\Delta) = (E(\Delta)g, g)$, $\forall \Delta \in \mathcal{B}(P)$. It is known (see, for instance, [34]) that there exists $h \in H$ such that $\operatorname{supp} m_h = \operatorname{supp} E = \sigma(A, B)$. Set $\rho_0 = m_h$. In view of Statement 5.5, there exists such a nonnegative continuous function φ on P that for a σ-finite measure ρ on P such that $\rho_1(\Delta) = \int_\Delta \varphi(\lambda, \mu) d\rho_0$, $\forall \Delta \in \mathcal{B}(P)$, — (5.16) and (5.21) hold but (5.24) does not.

Set $f_1 = (\varphi(\lambda, \mu)(|\lambda|^2 + 1)^{-1})^{1/2}(A, B)h$; here $f_1 \in H$ since $\| f_1 \|^2 = \int_P (|\lambda|^2 + 1)^{-1} \varphi(\lambda, \mu) d(Eh, h) \leq \rho_1(P) < +\infty$. Now $\rho_1(\Delta) = \int_\Delta (|\lambda|^2 + 1) d(Ef_1, f_1)$, $\forall \Delta \in \mathcal{B}(P)$. Thus, $\int_P (|\lambda|^2 + 1) d(Ef_1, f_1) < +\infty$ means (5.16); (6.2) means (5.21); $\int_P |\mu| d(Ef_1, f_1) < +\infty$ means (5.24). By Statement 5.5, for this $f_1 \in H : \int_P (|\lambda|^2 + 1) d(Ef_1, f_1) < +\infty$ and (6.2) holds, but $\int_P |\mu| d(Ef_1, f_1) = +\infty$. The proof is now completed.

Corollary 1. *Let A and B be commuting self-adjoint operators in H. The set F_C coincides with $D(B) \times (D(A) \cap D(|B|^{1/2}))$ if and only if one of the following equivalent conditions 1), 2) holds:*

1) $\exists 0 < \gamma, C < +\infty : \sigma(A,B) \subseteq \{(\lambda,\mu) | -\gamma \leq \lambda \leq \gamma,\ \mu \geq -\gamma\lambda - \gamma^2\} \cup \{(\lambda,\mu) | \lambda > \gamma,\ -\gamma\lambda - \gamma^2 \leq \mu \leq C(\lambda^2+1)\}$ (see Fig. 6.2);
2) *the space H can be represented as a direct orthogonal sum of Hilbert spaces $H_{(1)}$ and $H_{(2)}$ which are invariant with respect to both operators A and B (or, what is the same, with respect to all operators $E(\cdot)$) and such that:*

   a) for c.s.o. A_1 and B_1 in $H_{(1)}$ (the restrictions of A and B to $H_{(1)}$), A_1 is bounded, B_1 is semibounded below;

   b) for c.s.o. A_2 and B_2 in $H_{(2)}$ (the restrictions of A and B to $H_{(2)}$), A_2 is semibounded below, $(B_2)_-$ is subordinate to A_2, and $(B_2)_+$ is subordinate to A_2^2.

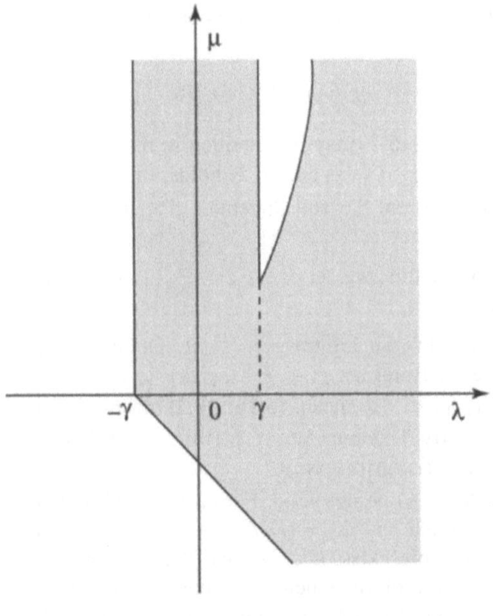

Fig. 6.2

Corollary 1 follows from Theorem 6.2, taking into account Theorem 1.2 and the remark to it and Lemma 1.2.

Example 6.1. Let A and B be the c.s.o. from Example 1.1. For equation (1), F_C contains $D(B) \times (D(A) \cap D(|B|^{1/2}))$ if and only if for each of $i = 1, 2$ there holds one of the following conditions:

1) $a_i > 0$;
2) $a_i = 0$, $b_i \geq 0$.

F_C coincides with $D(B) \times (D(A) \cap D(|B|^{1/2}))$ in all of these cases except when

$a_1 = 0$, $a_2 > 0$, and $b_1 > 0$,

or

$a_2 = 0$, $a_1 > 0$, and $b_2 > 0$.

3 When $F_C = D(B) \times D(A)$?

Let A and B be arbitrary commuting normal operators in H.

By definition, F_C is the set of all (f_0, f_1) such that there exists a unique solution $y(t)$ of (1) on R_+ with $y(0) = f_0$, $y'(0) = f_1$. In view of Definition 1.1, for every solution $y(t)$ of equation (1) on $R_+ : y(0) \in D(B)$, $y'(0) \in D(A)$. Thus,

$$F_C \subseteq D(B) \times D(A).$$

The question now arises: what must be c.n.o. A and B in H for F_C *to coincide* with $D(B) \times D(A)$?

Theorem 6.3. *Let A and B be c.n.o. in H. The set F_C coincides with $D(B) \times D(A)$ if and only if $P = \sigma(A, B)$ satisfies* (2) *and $D(A^2) \subseteq D(B)$.*

Proof. First of all, $F_C = D(B) \times D(A)$ implies that $F_C \supseteq D(B) \times (D(A) \cap D(|B|^{1/2}))$, i.e., that $P = \sigma(A, B)$ satisfies (2).

In what follows, we assume that $P = \sigma(A, B)$ satisfies (2). In this case, $F_C = D(B) \times (D(A) \cap F)$. Thus, $F_C = D(B) \times D(A)$ if and only if $D(A) \cap F = D(A)$, i.e.,

$$D(A) \subseteq F. \tag{6.8}$$

By the definitions, (6.8) may be rewritten in the following form: for $f_1 \in H$, $\int_P (|\lambda|^2 + 1) d(Ef_1, f_1) < +\infty$ implies (6.2).

By virtue of Theorem 5.5 and Statement 5.4, this is the case if and only if

$$\exists K < +\infty \forall (\lambda, \mu) \in P : \frac{|\mu|}{|\lambda|^2 + 1} \leq K. \tag{6.9}$$

By Theorem 1.2 for c.n.o. A and B and the remark to it, (6.9) holds if and only if $D(A^2) \subseteq D(B)$. The proof is over.

Remark 1. If $P = \sigma(A, B)$ satisfies (2) and $D(A^2) \subseteq D(B)$, then $F_C = D(B) \times (D(A) \cap D(|B|^{1/2})) = D(B) \times D(A)$.

Corollary 1. *Let A and B be arbitrary commuting self-adjoint operators in H. The set F_C coincides with $D(B) \times D(A)$ if and only if one of the following equivalent conditions* 1), 2) *holds:*

1) $\exists 0 < C, \gamma < +\infty : \sigma(A, B) \subseteq \{(\lambda, \mu) | \lambda \geq -\gamma, -\gamma\lambda - \gamma^2 \leq \mu \leq C(\lambda^2 + 1)\}$ (see Fig. 6.3)*;*
2) *A is semibounded below, B_- is subordinate to A, and B_+ is subordinate to A^2.*

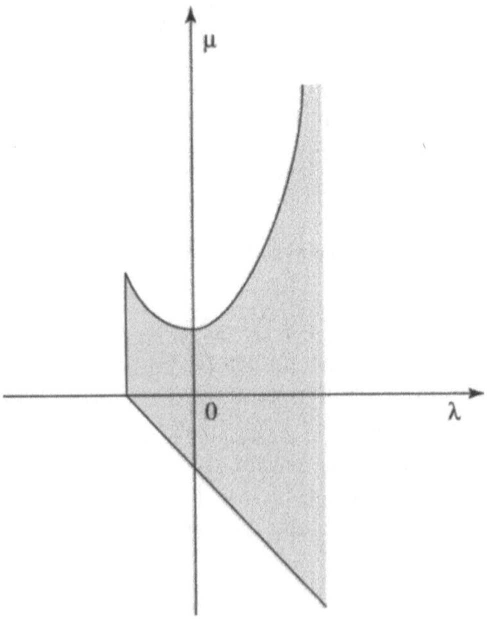

Fig. 6.3

Example 6.2. Let A and B be the c.s.o. from Example 1.1. For equation (1), the set F_C coincides with $D(B) \times D(A)$ if and only if for each of $i = 1, 2$ there holds one of the following conditions:

1) $a_i > 0$;
2) $a_i = 0$, $b_i = 0$.

4 $\mathbb{E}$-sequences of vectors and the general expression for weak solutions

In the following sections of Chapter 6, we go on to study similar questions for *weak* solutions. In order to answer these questions, we need to obtain the general expression for weak solutions of equation (1).

To this end, we introduce the concept of an $\mathbb{E}$-sequence of vectors.

1. Let A and B be c.n.o. in H.

Definition 6.1. We say that a sequence $\{g^{(k)}\}_{k=1}^{\infty}$ is an $\mathbb{E}$-sequence if:

1) $\forall k \geq 1 : g^{(k)} \in E_k H$;
2) $\forall m \geq k \geq 1 : E_k g^{(m)} = g^{(k)}$.

For instance, for any $g \in H : \{E_k g\}_{k=1}^{\infty}$ is an $\mathbb{E}$-sequence.

Fix an arbitrary $\mathbb{E}$-sequence $\{g^{(k)}\}_{k=1}^{\infty}$. For each Borel set $\Delta \subseteq C^2$, set: $(E(\Delta)g,g) \overset{\text{def}}{=} \lim_{k\to+\infty}(E(\Delta)g^{(k)},g^{(k)})$. Then $(E(\cdot)g,g)$ is a σ-finite measure on C^2.

Let $\varphi(\lambda,\mu)$ be a continuous real-valued function on C^2 such that

$$\exists n \geq 0,\ 1 \leq c < +\infty \forall(\lambda,\mu) \in \sigma(A,B) : 1 \leq \varphi(\lambda,\mu) \leq c(|\lambda|^n + |\mu|^n + 1).$$

By the definitions, $g^{(k)} \in H \subseteq H_-(\varphi(A,B))$, $\forall k \geq 1$. The sequence $\{g^{(k)}\}_{k=1}^{\infty}$ has a limit in $H_-(\varphi(A,B))$ as $k \to \infty$ if and only if $\int_{\sigma(A,B)} \frac{1}{\varphi^2(\lambda,\mu)} d(Eg,g) < +\infty$. If this is the case, denote by g the limit of $\{g^{(k)}\}_{k=1}^{\infty}$ in $H_-(\varphi(A,B))$ as $k \to \infty$; here $g \in H_-(\varphi(A,B))$. On the other hand, for every $k \geq 1 : E_k g = g^{(k)}$. See also Section 6 of Chapter 1. Cf. [6].

2. Let A and B be c.n.o. in H, $0 < T \leq +\infty$, and $y(t)$ be a weak solution of equation (1) on $(0,T)$.

Fix an arbitrary $k \geq 1$. $E_k y(t)$ is a weak solution on $(0,T)$ of the equation with bounded operator coefficients $u''(t) + Au'(t) + Bu(t) = 0$ in the space $E_k H$.

Fix an arbitrary $0 < T_1 < T$. Similarly to the proof of Lemma 2.1, it is easy to see that

$$E_k y(t) = \psi_0(A,B,t-T_1)(E_k y)(T_1) + \psi_1(A,B,t-T_1)(E_k y)'(T_1), \quad \forall t \in (0,T),$$

where $(E_k y)'(T_1) \in E_k H$ is the weak over $E_k H$ derivative of $E_k y(t)$ at $t = T_1$. Thus, $E_k y(t)$ can be extended up to a usual solution of the last equation on $(-\infty,+\infty)$. Denote: $f_0^{(k)} = (E_k y)(0)$, $f_1^{(k)} = (E_k y)'(0)$. Then $E_k y(t) = \psi_0(A,B,t)f_0^{(k)} + \psi_1(A,B,t)f_1^{(k)}$, $\forall t \in (0,T)$.

It is easy to see that $\{f_0^{(k)}\}_{k=1}^{\infty}$, $\{f_1^{(k)}\}_{k=1}^{\infty}$ are $\mathbb{E}$-sequences. We have proved the following lemma.

Lemma 6.1. *Let A and B be c.n.o. in H, $0 < T \leq +\infty$, and $y(t)$ be a weak solution of equation (1) on $(0,T)$. Denote for each $k \geq 1$: $f_0^{(k)} = (E_k y)(0)$, $f_1^{(k)} = (E_k y)'(0)$ (here $(E_k y)^{(i)}(0)$ in $E_k H$ are understood in the usual sense). Then $\{f_0^{(k)}\}_{k=1}^{\infty}$ and $\{f_1^{(k)}\}_{k=1}^{\infty}$ are $\mathbb{E}$-sequences, and*

$$y(t) = \lim_{k\to\infty} (\psi_0(A,B,t)f_0^{(k)} + \psi_1(A,B,t)f_1^{(k)}), \quad \forall t \in (0,T). \tag{6.10}$$

Remark 1. Let $y(t)$ be a weak solution of (1) on $[0,T)$ with i.d. (f_0, f_1). Then $\forall k \geq 1 : f_0^{(k)} = E_k f_0$; $f_1^{(k)} \in E_k H$, and $\forall h \in E_k H \subseteq \Phi : (f_1^{(k)}, h) = (f_1, h)$.

Corollary 1. *Let A and B be c.n.o. in H, $0 < T \leq +\infty$. The mapping $y(t) \mapsto (\{f_0^{(k)}\}_{k=1}^{\infty}, \{f_1^{(k)}\}_{k=1}^{\infty})$ is a one-to-one mapping of the set of all weak solutions of equation (1) on $(0,T)$ into the set of ordered pairs of $\mathbb{E}$-sequences.*

3. The previous statements were established for arbitrary c.n.o. A and B in H. Now we go on to consider the special case that conditions (3.3) and (3.4) hold.

Corollary 2. *Let A and B be c.n.o. in H such that both conditions (3.3) and (3.4) hold, $0 < T < +\infty$, and $y(t)$ be a weak solution of equation (1) on $(0,T]$. $y(t)$ can be extended up to a weak solution of (1) on $[0,T]$ if and only if:*

1) $\exists f_0 \in H : f_0^{(k)} = E_k f_0$, $\forall k \geq 1$;
2) $y_1(t) \stackrel{\text{def}}{=} \lim_{k\to\infty} \psi_1(A,B,t)f_1^{(k)}$ $(t \in [0,T])$ *is a weak solution of* (1) *on* $[0,T]$.

If this is the case then $y(0) \stackrel{\text{def}}{=} \lim_{t\to 0} y(t) = f_0$, *and*

$$y(t) = \psi_0(A,B,t)f_0 + \lim_{k\to\infty} \psi_1(A,B,t)f_1^{(k)}, \quad t \in [0,T].$$

By the definitions, $\| y_1(t) \|^2 = \int_{\sigma(A,B)} |\psi_1(\lambda,\mu,t)|^2 d(Ef_1, f_1)$, $t \in [0,T]$. Thus, if $y(t)$ is a weak solution of (1) on $[0,T]$ then, in particular,

$$\int_{\sigma(A,B)} |\psi_1(\lambda,\mu,t)|^2 d(Ef_1, f_1) \in C([0,T]).$$

Corollary 3. *Let A and B be c.n.o. in H such that both conditions* (3.3) *and* (3.4) *hold, $0 < T < +\infty$. A weak solution of equation* (1) *on* $(0, T]$ $y(t)$ *has a limit in H as $t \to 0$ if and only if:*

1) $\exists f_0 \in H : f_0^{(k)} = E_k f_0, \forall k \geq 1;$

2) $y_1(t) \overset{\text{def}}{=} \lim_{k\to\infty} \psi_1(A, B, t) f_1^{(k)}$ $(t \in (0, T])$ *is a weak solution of* (1) *on* $(0, T]$*;*

3) $y_1(t) \to 0$ *in the space H as $t \to 0$.*

If this is the case then $y(0) \overset{\text{def}}{=} \lim_{t\to 0} y(t) = f_0$, *and*

$$y(t) = \psi_0(A, B, t) f_0 + \lim_{k\to\infty} \psi_1(A, B, t) f_1^{(k)}, \quad t \in (0, T].$$

Here 2) implies that $\| y_1(t) \|^2 = \int_{\sigma(A,B)} |\psi_1(\lambda, \mu, t)|^2 d(Ef, f_1)$ is continuous on $(0, T]$; 3) holds if and only if $\int_{\sigma(A,B)} |\psi_1(\lambda, \mu, t)|^2 d(Ef_1, f_1)$ $(t \in [0, T])$ converges to 0 as $t \to 0$, i.e., is continuous at $t = 0$.

Corollary 4. *Let A and B be c.n.o. in H such that both conditions* (3.3) *and* (3.4) *hold, $0 < T < +\infty$. A weak solution of equation* (1) *on* $(0, T]$ $y(t)$ *is bounded (i.e., $\exists C_y < +\infty \forall t \in (0, T] : \| y(t) \| \leq C_y$) if and only if:*

1) $\exists f_0 \in H : f_0^{(k)} = E_k f_0, \forall k \geq 1;$

2) $y_1(t) \overset{\text{def}}{=} \lim_{k\to\infty} \psi_1(A, B, t) f_1^{(k)}$ $(t \in (0, T])$ *is a bounded weak solution of* (1) *on* $(0, T]$.

If this is the case then $y(t) = \psi_0(A, B, t) f_0 + \lim_{k\to\infty} \psi_1(A, B, t) f_1^{(k)}, t \in (0, T]$.

Here 2) implies that $\| y_1(t) \|^2 = \int_{\sigma(A,B)} |\psi_1(\lambda, \mu, t)|^2 d(Ef_1, f_1)$ is continuous and bounded on $(0, T]$.

We now prove Corollary 4; the previous two corollaries can be proved similarly.

Proof of Corollary 4. Indeed, let $y(t)$ be a weak solution of (1) on $(0, T]$ and $\exists C_y < +\infty \forall t \in (0, T] : \| y(t) \| \leq C_y$. Fix an arbitrary $k \geq 1$. By Lemma 6.1, $E_k y(t) \to f_0^{(k)}$ in H as $t \to 0$. But for any $t \in (0, T] : \| E_k y(t) \| \leq \| y(t) \| \leq C_y$; thus, $\| f_0^{(k)} \| \leq C_y$.

We have obtained that $\| f_0^{(k)} \| \leq C_y < +\infty, \forall k \geq 1$. Since $\{f_0^{(k)}\}_{k=1}^{\infty}$ is an $\mathbb{E}$-sequence, this implies that $\exists f_0 \in H : f_0^{(k)} \to f_0$ in H as $k \to +\infty$. For any $k \geq 1 : f_0^{(k)} = E_k f_0$.

By the assumption, c.n.o. A and B in H satisfy both conditions (3.3), (3.4). Therefore $y_0(t) \overset{\text{def}}{=} \psi_0(A, B, t) f_0$ is a weak solution of (1) on $[0, T]$ (see Theorem 3.1). Hence $\forall t \in (0, T] : \lim_{k\to\infty} \psi_0(A, B, t) f_0^{(k)} = \psi_0(A, B, t) f_0$.

We see that $y(t)$ is a bounded weak solution of (1) on $(0,T]$, $y_0(t) = \lim_{k\to\infty} \psi_0(A,B,t)f_0^{(k)}$ is a weak solution of (1) on $[0,T]$ and, in particular, a bounded weak solution of (1) on $(0,T]$. In view of (6.10), for any $t \in (0,T]$: $\psi_1(A,B,t)f_1^{(k)} \to (y(t)-y_0(t))$ in H as $k \to +\infty$, and so

$$y_1(t) = \lim_{k\to\infty} \psi_1(A,B,t)f_1^{(k)}$$

is a bounded weak solution of (1) on $(0,T]$. The proof is over.

5 The set of initial data of weak solutions

1. Let A and B be c.n.o. in H such that $P = \sigma(A,B)$ satisfies (2), i.e., the Cauchy problem for equation (1) on R_+ is weakly well-posed in $G = H \times H_{-1} = H \times H_-((|A|^2+|B|+1)^{1/2})$. This means that for any $(f_0,f_1) \in H \times H_{-1}$: the Cauchy problem for equation (1) on R_+ with initial data (f_0,f_1) admits a unique weak solution.

The question is: what is the set of *all* (f_0,f_1) such that the Cauchy problem for equation (1) on R_+ with initial data (f_0,f_1) admits a unique weak solution?

Denote this set by F'_C. By Definition 1.5 and Theorem 3.1, $H \times H_{-1} \subseteq F'_C \subseteq H \times \Phi'$. (Recall that Φ' is the space of all antilinear functionals on Φ).

In what follows, we also consider $H_-(\varphi(A,B))$ where $\varphi(\lambda,\mu) = (|\lambda|^2 + |\mu|+1)^{1/2}(|\lambda|+1)^{1/2}$ for each $(\lambda,\mu) \in C^2$ (see Section 6 of Chapter 1).

Theorem 6.4. *Let A and B be c.n.o. in H such that $P = \sigma(A,B)$ satisfies (2), i.e., the Cauchy problem for equation (1) on R_+ is weakly well-posed in $G = H \times H_{-1}$. Denote: $P = \sigma(A,B)$. The Cauchy problem for equation (1) on R_+ with initial data (f_0,f_1) admits a unique weak solution if and only if $f_0 \in H$, $f_1 \in H_-((|A|^2+|B|+1)^{1/2}(|A|+1)^{1/2})$, and*

$$\forall x > 0: \int_{P_x} \frac{d(Ef_1,f_1)}{|\lambda|^2+|\mu|+1} < +\infty, \text{ and } \lim_{x\to+\infty} \int_{P_x} \frac{d(Ef_1,f_1)}{|\lambda|^2+|\mu|+1} = 0. \quad (6.11)$$

If this is the case, then $f_1 \in H_-((|A|^2+|B|+1)^{1/2}(|d|+1)^{1/2}(A,B))$, and the weak solution of the Cauchy problem for (1) on R_+ with initial data (f_0,f_1) has the form:

$$y(t) = \psi_0(A,B,t)f_0 + \psi_1(A,B,t)f_1 \quad (t \in R_+).$$

Obviously (6.11) is a weaker condition than $f_1 \in H_{-1}$, i.e.,

$$\int\limits_P \frac{d(Ef_1, f_1)}{|\lambda|^2 + |\mu| + 1} < +\infty$$

(cf. Remark 2 to Theorem 6.1).

The proof of Theorem 6.4 is based on Corollary 2 of Lemma 6.1 and Theorems 5.1 and 5.3.

2. Denote by F_1 the set of all $f_1 \in H_-((|A|^2 + |B| + 1)^{1/2}(|A| + 1)^{1/2})$ such that (6.11) holds. We have shown that $F'_C = H \times F_1$.

Corollary 1. *Let A and B be c.n.o. in H such that $P = \sigma(A, B)$ satisfies (2). Then*

$$\begin{aligned} H \times H \subseteq H \times H_{-1} = H \times H_-((|A|^2 + |B| + 1)^{1/2}) \subseteq F'_C = H \times F_1 \\ \subseteq H \times H_-((|A|^2 + |B| + 1)^{1/2}(|d| + 1)^{1/2}(A, B)) \\ \subseteq H \times H_-((|A|^2 + |B| + 1)^{1/2}(|\operatorname{Re} A| + 1)^{1/2}) \\ \subseteq H \times H_-((|A|^2 + |B| + 1)^{1/2}(|A| + 1)^{1/2}) \subseteq H \times \Phi'. \end{aligned}$$

Corollary 2. *Let A and B be c.n.o. in H such that $P = \sigma(A, B)$ satisfies (2). There exists a one-to-one mapping of the set Y' of all weak solutions of equation (1) on R_+ onto the set of their initial data $F'_C = H \times F_1$. This mapping is given by the formula:*

$$F'_C = H \times F_1 \ni (f_0, f_1) \leftrightarrow \psi_0(A, B, t)f_0 + \psi_1(A, B, t)f_1 \quad (t \in R_+) \quad \in Y'.$$

3. Corollary 3. *Let A and B be c.n.o. in H such that $P = \sigma(A, B)$ satisfies (2), and moreover, $D(A) \subseteq D(B)$ (in other words: $\operatorname{Re} A$ is semibounded below, and $D(A) \subseteq D(B)$). Then $F'_C = H \times H_-(|A| + 1)$.*

Corollary 4. *Let A and B be c.n.o. in H such that $P = \sigma(A, B)$ satisfies (2), and moreover, A is bounded (in other words: A is bounded, and (3.3) holds). Then $F'_C = H \times H_-(|B|^{1/2} + 1)$.*

In order to prove Corollaries 3 and 4, we notice that in these special cases $P = \sigma(A, B)$ satisfies (5.20) (see Corollaries 1 and 2 of Theorem 4.1), and so $F_1 = H_{-1}$ (cf. the proof of Corollary 4 of Theorem 6.1).

4. These results on F'_C remain valid for an arbitrary Φ such that $\bigcup\limits_{k=1}^{\infty} E_k H \subseteq \Phi \subseteq D(A^2) \cap D(B)$. But in the general case F'_C may depend on Φ. For instance, let A

and B be c.n.o. in H such that $P = \sigma(A,B)$ satisfies (2) and $D(A) \subseteq D(B)$. If $\bigcup_{k=1}^{\infty} E_k H \subseteq \Phi \subseteq D(A^2) \cap D(B)$, then $F'_C = H \times F_1 = H \times H_-(|A|+1)$. But if one sets $\Phi = D(A) \cap D(B)$, then $F'_C = H \times H$. If A is unbounded, then this set is strictly narrower than $H \times F_1 = H \times H_-(|A|+1)$.

On the other hand, if A and B are c.n.o. in H such that $P = \sigma(A,B)$ satisfies (2) and A is bounded, then for any Φ such that $\bigcup_{k=1}^{\infty} E_k H \subseteq \Phi \subseteq D(A) \cap D(B)$: $F'_C = H \times F_1 = H \times H_-(|B|^{1/2}+1)$.

5. Corollary 5. *Let A and B be commuting self-adjoint operators in H such that*

$$\exists 0 < \gamma < +\infty : \sigma(A,B) \subseteq \{(\lambda,\mu) \in R^2 | \lambda \geq -\gamma, \quad \mu \geq -\gamma\lambda - \gamma^2\},$$

i.e., the Cauchy problem for equation (1) *on R_+ is weakly well-posed in $G = H \times H_{-1}$. For each $x > 0$, denote:* $Q'_x = (\{(\lambda,\mu) \in R^2 | -\gamma \leq \lambda \leq 4x,\ \mu \geq -\gamma\lambda - \gamma^2\} \cup \{(\lambda,\mu) \in R^2 | \lambda > 4x,\ -\gamma\lambda - \gamma^2 \leq \mu \leq x\lambda\}) \setminus \{(\lambda,\mu) \in R^2 | \lambda < x,\ \mu < \frac{x^2}{2}\} = \{(\lambda,\mu) \in R^2 | \lambda \geq -\gamma,\ \mu \geq -\gamma\lambda - \gamma^2\} \setminus (\{(\lambda,\mu) | \lambda > 4x,\ \mu > x\lambda\} \cup \{(\lambda,\mu) | \lambda < x,\ \mu < \frac{x^2}{2}\})$ (see Fig. 6.4). *The Cauchy problem for equation* (1) *on R_+ with i.d. (f_0, f_1) admits a unique weak solution if and only if $f_0 \in H$, $f_1 \in H_-((|A|^2 + |B| + 1)^{1/2}(|A|+1)^{1/2})$, and*

$$\forall x > 0 : \int_{Q'_x} \frac{d(Ef_1, f_1)}{|\lambda|^2 + |\mu| + 1} < +\infty, \text{ and } \lim_{x \to +\infty} \int_{Q'_x} \frac{d(Ef_1, f_1)}{|\lambda|^2 + |\mu| + 1} = 0.$$

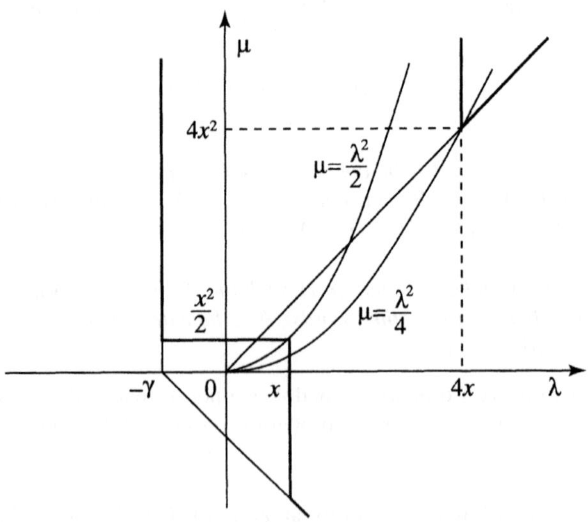

Fig. 6.4

6 When $F'_C = H \times H_{-1}$?

1. Let A and B be arbitrary commuting normal operators in H. In this case, we preserve the notation F'_C for the set of all (f_0, f_1) such that the Cauchy problem for equation (1) on R_+ with i.d. (f_0, f_1) admits a unique weak solution.

The weak well-posedness of the Cauchy problem for equation (1) on R_+ in $G = H \times H_{-1}$ means, by definition, that F'_C contains $H \times H_{-1}$.

Thus, the statement of Theorem 3.1 may be rewritten in the following form:

Let A and B be c.n.o. in H. The set F'_C *contains* $H \times H_{-1}$ if and only if both conditions (3.3), (3.4) hold.

The question now arises: what must be c.n.o. A and B in H for the set F'_C *to coincide* with $H \times H_{-1} \overset{\text{def}}{=} H \times H_-(|A| + |B|^{1/2} + 1)$?

Theorem 6.5. *Let A and B be c.n.o. in H. The set F'_C coincides with $H \times H_{-1} = H \times H_-(|A| + |B|^{1/2} + 1)$ if and only if $P = \sigma(A, B)$ satisfies* (2) *and*

$$\exists 0 < \gamma_1 < +\infty : P \subseteq \{(\lambda, \mu) \in \mathrm{C}^2 | d \leq \gamma_1\}. \tag{6.12}$$

In other words, $F'_C = H \times H_{-1}$ if and only if

$$\exists 0 < C,\ \gamma,\ \gamma_1 < +\infty :$$

$$P \subseteq \{(\lambda, \mu) \in \mathrm{C}^2 | -\gamma \leq d \leq \gamma_1, \quad |\operatorname{Im} \lambda|^2 \leq C(a^2 + d^2 + b_2^2 + 1)\}.$$

See Remark 1 to Theorem 6.2.

The proof is based on Theorem 6.4 and the remark to it, Theorem 5.4 and Statement 5.3, and Remark 1 to them.

2. Corollary 1. *Let A and B be c.n.o. in H such that $D(A) \subseteq D(B)$ and* (3.3) *holds (i.e., $D(A) \subseteq D(B)$ and* $\operatorname{Re} A$ *is semibounded below). Then $F'_C = H \times H_{-1} = H \times H_-(|A| + 1)$.*

See Corollary 3 of Theorem 3.1 and the remark after Corollaries 3 and 4 of Theorem 6.4.

Corollary 2. *Let A and B be c.n.o. in H such that A is bounded and* (3.3) *holds. Then $F'_C = H \times H_{-1} = H \times H_-(|B|^{1/2} + 1)$.*

See Theorem 3.1 and the remark after Corollaries 3 and 4 of Theorem 6.4.

Remark 1. Thus, if (3.3) holds and $A = 0$ (and even in the more general case of A bounded) or $B = 0$ (and even in the more general case of $D(A) \subseteq D(B)$) then $F'_C = H \times H_{-1}$.

3. Corollary 3. *Let A and B be commuting self-adjoint operators in H. The set F'_C coincides with $H \times H_{-1}$ if and only if one of the following equivalent conditions* 1), 2), 3) *holds:*

1) $\exists 0 < \gamma < +\infty \forall (\lambda, \mu) \in \sigma(A, B) : -\gamma \le \max_{i=1,2} \operatorname{Re} \omega_i(\lambda, \mu) \le \gamma;$

2) $\exists 0 < \gamma < +\infty : \sigma(A, B) \subseteq \{(\lambda, \mu) | -\gamma \le \lambda \le \gamma, \mu \ge -\gamma\lambda - \gamma^2\} \cup \{(\lambda, \mu) | \lambda > \gamma, -\gamma\lambda - \gamma^2 \le \mu \le \gamma\lambda + \gamma^2\} = \{(\lambda, \mu) \in R^2 | \lambda \ge -\gamma, \mu \ge -\gamma\lambda - \gamma^2\} \setminus \{(\lambda, \mu) | \lambda > \gamma, \mu > \gamma\lambda + \gamma^2\}$ (see Fig. 6.5);

3) *the space H can be represented as a direct orthogonal sum of Hilbert spaces $H_{(1)}$ and $H_{(2)}$ which are invariant with respect to both operators A and B (or, what is the same, with respect to all operators* $E(\cdot)$*) and such that:*
   a) *for c.s.o. A_1 and B_1 in $H_{(1)}$ (the restrictions of A and B to $H_{(1)}$), A_1 is bounded, B_1 is semibounded below;*
   b) *for c.s.o. A_2 and B_2 in $H_{(2)}$ (the restrictions of A and B to $H_{(2)}$), A_2 is semibounded below, B_2 is subordinate to A_2 (i.e., $D(A_2) \subseteq D(B_2)$).*

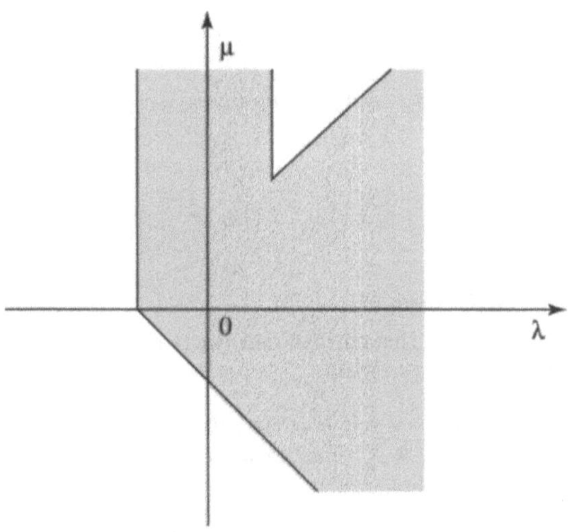

Fig. 6.5

Example 6.3. Let A and B be the c.s.o. from Example 1.1. For equation (1), F'_C contains $H \times H_{-1}$ if and only if for each of $i = 1, 2$ there holds one of the following conditions:

1) $a_i > 0$; 2) $a_i = 0$, $b_i \ge 0$.

F'_C coincides with $H \times H_{-1}$ in all of these cases except when

$$a_1 = 0, \quad a_2 > 0, \quad \text{and} \quad b_1 > 0,$$

or

$$a_2 = 0, \quad a_1 > 0, \quad \text{and} \quad b_2 > 0.$$

7 Fatou-Riesz property

Definition 6.2. Let A and B be c.n.o. in H. We say that equation (1) has Fatou-Riesz property if, for every $0 < T < +\infty$, each bounded weak solution of (1) on $(0, T]$ has a limit in the space H as $t \to 0$.

Originally, this property was established by P.J.L. Fatou and F. Riesz for the Laplace equation; weak solutions of this equation are harmonic functions. See [137] (references to Section 6 of Chapter 2), [81, 290, 279].

The monograph [137] contains a generalization of this result which may be written in the following form:

Let $A = 0$, B be a nonpositive self-adjoint operator in a separable Hilbert space H. Then equation (1) has Fatou-Riesz property.

We now consider the following questions.

Let A and B be c.n.o. in H such that $P = \sigma(A, B)$ satisfies (2). Does equation (1) necessarily have Fatou-Riesz property? If no, what must be operators A and B for equation (1) to have this property?

In order to answer these questions, we need the following lemma.

Lemma 6.2. *Let A and B be c.n.o. in H such that $P = \sigma(A, B)$ satisfies* (2). *Denote by F_1 the set of all $f_1 \in H_-((|A|^2 + |B| + 1)^{1/2}(|A| + 1)^{1/2})$ such that* (6.11) *holds; by F_2 the set of all $f_1 \in H_-((|A|^2 + |B| + 1)^{1/2}(|A| + 1)^{1/2})$ such that*

$$\exists K < +\infty \forall x > 0 : \int\limits_{P_x} \frac{d(Ef_1, f_1)}{|\lambda|^2 + |\mu| + 1} \leq K.$$

Further, fix an arbitrary $0 < T < +\infty$. Denote by Y_1 the set of all weak solutions of equation (1) *on $(0, T]$ such that*

$$\exists y_0 \in H : y(t) \to y_0 \ \text{ in } \ H \ \text{ as } \ t \to 0;$$

by Y_2 the set of all weak solutions of equation (1) *on $(0, T]$ such that*

$$\exists C_y < +\infty \forall t \in (0, T] : \| y(t) \| \leq C_y.$$

There exists a one-to-one mapping of $H \times F_1 \subseteq H \times H_-(((|A|^2 + |B| + 1)^{1/2}(|d| + 1)^{1/2}(A, B))$ onto Y_1 which is defined by the formula:

$$H \times F_1 \ni (f_0, f_1) \leftrightarrow \psi_0(A, B, t)f_0 + \psi_1(A, B, t)f_1 \quad (t \in (0, T]) \in Y_1;$$

there exists a one-to-one mapping of $H \times F_2 \subseteq H \times H_-(((|A|^2 + |B| + 1)^{1/2}(|d| + 1)^{1/2}(A, B))$ onto Y_2 which is defined by the formula:

$$H \times F_2 \ni (f_0, f_1) \leftrightarrow \psi_0(A, B, t)f_0 + \psi_1(A, B, t)f_1 (t \in (0, T]) \in Y_2.$$

The proof of Lemma 6.2 is based on Corollaries 3, 4 of Lemma 6.1 and Theorems 5.1, 5.2, 5.3.

Notice that

$$\begin{aligned} H\times H &\subseteq H\times H_{-1}\subseteq H\times F_1\subseteq H\times F_2\\ &\subseteq H\times H_{-}((|A|^2+|B|+1)^{1/2}(|d|+1)^{1/2}(A,B))\\ &\subseteq H\times H_{-}((|A|^2+|B|+1)^{1/2}(|\operatorname{Re}A|+1)^{1/2})\\ &\subseteq H\times H_{-}((|A|^2+|B|+1)^{1/2}(|A|+1)^{1/2}). \end{aligned}$$

By virtue of Lemma 6.2, $Y_1 = Y_2$ if and only if $F_1 = F_2$. Further, $F_1 = F_2$ if and only if $P = \sigma(A, B)$ satisfies (6.12). This follows from Theorem 5.4 and Statement 5.3 and Remark 1 to them.

We have proved the following theorem.

Theorem 6.6. *Let A and B be c.n.o. in H such that $P = \sigma(A, B)$ satisfies* (2). *Equation* (1) *has Fatou-Riesz property if and only if $P = \sigma(A, B)$ satisfies* (6.12). *Moreover, if $P = \sigma(A, B)$ does not satisfy* (6.12) *then for any $0 < T < +\infty$ there exists a bounded weak solution of* (1) *on* $(0, T]$ *which does not have a limit in H as $t \to 0$.*

See Remark 1 to Theorem 6.2.

Corollaries 1, 2 and Remark 1 of Section 6 of this chapter remain valid with «$F'_C = H \times H_{-1}$» replaced by «equation (1) has Fatou-Riesz property». Corollary 3 of Theorem 6.5 and the statement of Example 6.3 remain valid with «F'_C coincides with $H \times H_{-1}$» replaced by «$P = \sigma(A, B)$ satisfies (2), and equation (1) has Fatou-Riesz property».

Part III
Extension, Stability, and Stabilization of Weak Solutions

Introduction to Part III

Fix an arbitrary $0 < T < +\infty$.

In Section 7 of Chapter 6 we studied the following question:

1. What must be c.n.o. A and B in H for every bounded weak solution of equation (1) on $(0, T]$ to have a limit in H as $t \to 0$?

We have answered this question in the special case that $P = \sigma(A, B)$ satisfies (2).

This additional assumption is not essential for the resolution of the problem. In Chapter 8, we waive this assumption and answer this question in the general case where A and B are arbitrary c.n.o. in H.

To this end, we consider c.n.o. A and B in H in a more general case than in Section 7 of Chapter 6. Namely, we consider the case where (3.3) (or, what is the same, (2.2)) holds but (3.4) may not be valid. In this more general case, we obtain the general form of weak solutions of equation (1) on $(0, T]$ which have a limit in H as $t \to 0$, as well as the general form of bounded weak solutions of (1) on $(0, T]$. As well, we obtain the general form and the set of initial data of weak solutions of equation (1) on $[0, T]$.

Chapter 7 is devoted entirely to solution of these problems.

Further treatments are based on these results.

First, this enables us to answer the question 1. As well, we answer the following question:

What must be c.n.o. A and B in H for every bounded weak solution of equation (1) on $[0, T)$ to have a limit in H as $t \to T$?

These questions are answered in Chapter 8.

The last question may be considered as one of a series of questions relating to behaviour as $t \to T$ and extension of weak solutions of equation (1) on a finite interval $[0, T)$. Let us give some of these questions.

What must be c.n.o. A and B in H for every weak solution of (1) on $[0, T)$ to be bounded?

What must be c.n.o. A and B in H for every weak solution of (1) on $[0, T)$ to have a limit in H as $t \to T$?

What must be c.n.o. A and B in H for every weak solution of (1) on $[0, T)$ to be extendable up to a weak solution of (1) on the whole R_+?

These and similar questions are answered in Chapter 9.

In Chapter 10, we pose similar questions where $T = +\infty$. Let's give some of these questions.

What must be c.n.o. A and B in H for every weak solution of (1) on $R_+ = [0, +\infty)$ to be bounded?

What must be c.n.o. A and B in H for every weak solution of (1) on R_+ to have a limit in H as $t \to +\infty$?

These and other similar questions are answered in Chapter 10. In fact, of concern are problems of stability, stabilization of weak solutions, asymptotic stability, exponential stability of equation (1).

For ordinary linear differential equations in a finite-dimensional Hilbert space, these problems were studied beginning, apparently, from A.M. Lyapunov. For first order linear differential equations in Banach spaces

$$y'(t) + Ay(t) = 0,$$

these problems have been studied extensively during the last decades (see, for instance, [57, 138]).

In Chapter 10, we obtain criteria for stability, stabilization of weak solutions, asymptotic stability, exponential stability on $R_+ = [0, +\infty)$ for the equation

$$y''(t) + Ay'(t) + By(t) = 0$$

where A and B are arbitrary commuting normal operators in a Hilbert space H.

We consider all these questions in the «weak» setting, i.e., for weak solutions. Similarly, one can do this for usual solutions, with $\| y(t) \|$ replaced by $\max_{i=0,1,2} \| y^{(i)}(t) \|$. We omit the corresponding statements.

Chapter 7
The general form of weak solutions

Fix an arbitrary $0 < T \leq +\infty$.

In Section 5 of Chapter 6, we have established the general form of weak solutions of equation (1) with c.n.o. A and B on $[0,T]$ in the special case that both conditions (3.3) and (3.4) hold (the results for $R_+ = [0,+\infty)$ remain valid for an arbitrary $[0,T]$, $0 < T < +\infty$). In this chapter, we obtain the general form of weak solutions of (1) on $[0,T]$ in a more general case where A and B are c.n.o. in H and (3.3) holds, i.e.,

$$\sup_{(\lambda,\mu)\in\sigma(A,B),i=1,2} \operatorname{Re}\omega_i(\lambda,\mu) < +\infty. \tag{2.2}$$

Now, we do not assume that (3.4) (or, what is the same, (3.2)) holds.

1 Another general expression for weak solutions

1. In this case, Lemma 6.1, Remark 1 and Corollary 1 to it are valid as before. But $\psi_0(\lambda,\mu,t)$ may be unbounded on $\sigma(A,B)$ for all $t > 0$ (see Example 3.2).

We give a new expression for weak solutions of equation (1) on $(0,T)$, which is more appropriate for our further treatments than (6.10).

As previously, we set for each $(\lambda,\mu) \in C^2 : \operatorname{Re}\omega_1 \geq \operatorname{Re}\omega_2$; if $\operatorname{Re}\omega_1 = \operatorname{Re}\omega_2$ then $\operatorname{Im}\omega_1 \geq \operatorname{Im}\omega_2$.

Lemma 7.1. *Let A and B be c.n.o. in H, $0 < T \leq +\infty$, and $y(t)$ be a weak solution of equation* (1) *on* $(0,T)$. *Denote for each $k \geq 1 : f_0^{(k)} = (E_k y)(0)$, $f_1^{(k)} = (E_k y)'(0)$ (here $(E_k y)^{(i)}(0)$ in $E_k H$ are understood in the usual sense), and $g^{(k)} = -\omega_1(A,B)f_0^{(k)} + f_1^{(k)}$. Then $\{f_0^{(k)}\}_{k=1}^{\infty}$, $\{f_1^{(k)}\}_{k=1}^{\infty}$, $\{g^{(k)}\}_{k=1}^{\infty}$ are $\mathbb{E}$-sequences, and*

$$y(t) = \lim_{k\to\infty} (e^{\omega_1 t}(A,B)f_0^{(k)} + \psi_1(A,B,t)g^{(k)}), \quad \forall t \in (0,T). \tag{7.1}$$

In order to prove this, we notice that $\forall(\lambda,\mu,t) \in C^2 \times R_+ : \psi_0(\lambda,\mu,t) = e^{\omega_1 t} - \omega_1\psi_1(\lambda,\mu,t)$. Thus, for any $k \geq 1$ and $f_0^{(k)} \in E_k H : \psi_0(A,B,t)f_0^{(k)} = e^{\omega_1 t}(A,B)f_0^{(k)} + \psi_1(A,B,t)(-\omega_1(A,B)f_0^{(k)})$, $t \in [0,T)$. Then we make use of Lemma 6.1.

Corollary 1. *Let A and B be c.n.o. in H, $0 < T \leq +\infty$. The mapping $y(t) \mapsto (\{f_0^{(k)}\}_{k=1}^{\infty}, \{g^{(k)}\}_{k=1}^{\infty})$ is a one-to-one mapping of the set of all weak solutions of equation* (1) *on* $(0,T)$ *into the set of ordered pairs of $\mathbb{E}$-sequences.*

2. Lemma 7.1 and Corollary 1 to it are valid for arbitrary c.n.o. A and B in H. Let's go on to consider the special case that $\sigma(A,B)$ satisfies (2.2).

Statement 7.1. Let A and B be c.n.o. in H such that (2.2) holds. In this case, Corollaries 2, 3, 4 of Lemma 6.1 remain valid, with $\{f_1^{(k)}\}_{k=1}^{\infty}$ replaced by $\{g^{(k)}\}_{k=1}^{\infty}$, $\psi_0(A,B,t)f_0$ replaced by $e^{\omega_1 t}(A,B)f_0$.

Indeed, let A and B be c.n.o. in H such that (2.2) holds. It is easy to see that for any $f_0 \in H : y_0(t) = e^{\omega_1 t}(A,B)f_0$ is a weak solution of equation (1) on $[0,T]$. Now one can prove Statement 7.1 just as Corollaries 2, 3, 4 of Lemma 6.1, with $\psi_0(A,B,t)f_0$ replaced by $\lim_{k\to\infty} e^{\omega_1 t}(A,B)f_0^{(k)}$, $\lim_{k\to\infty} \psi_1(A,B,t)f_1^{(k)}$ replaced by $\lim_{k\to\infty} \psi_1(A,B,t)g^{(k)}$.

2 Continuity, boundedness, and integrability of $R(t)$ on $[0,T]$ in a more general case

Fix an arbitrary $0 < T \leq +\infty$. Let A and B be c.n.o. in H such that $P = \sigma(A,B)$ satisfies (2.2).

We have shown that if $y(t)$ is a weak solution of (1) on $[0,T]$ then

$$y(t) = e^{\omega_1 t}(A,B)f_0 + \lim_{k\to\infty} \psi_1(A,B,t)g^{(k)} \quad (t \in [0,T])$$

where $f_0 \in H$ and $\{g^{(k)}\}_{k=1}^{\infty}$ is an $\mathbb{E}$-sequence.

On the other hand, if $f_0 \in H$ then $y_0(t) = e^{\omega_1 t}(A,B)f_0$ is a weak solution of (1) on $[0,T]$. Now, in order to obtain the general form of weak solutions of (1) on $[0,T]$, we need to answer the following question: what must be an $\mathbb{E}$-sequence $\{g^{(k)}\}_{k=1}^{\infty}$ for $y_1(t) = \lim_{k\to\infty} \psi_1(A,B,t)g^{(k)}$ $(t \in [0,T])$ to be a weak solution of (1) on $[0,T]$?

To do this, we need to answer the following question: what must be a σ-finite measure ρ on $P = \sigma(A,B)$ for $R(t) = \int_{\sigma(A,B)} |\psi_1(\lambda,\mu,t)|^2 d\rho_{\lambda,\mu}$ to be continuous on $[0,T]$?

Up to this point, our treatment is just the same as in Chapter 6. But the last question is essentially more complicated than the same one in Chapter 5.

Indeed, now $P = \sigma(A,B)$ satisfies the condition

$$\exists 0 < \gamma < +\infty \forall (\lambda,\mu) \in P : \ d = -\max_{i=1,2} \operatorname{Re}\omega_i \geq -\gamma, \tag{7.2}$$

but may not satisfy the condition

$$\exists 0 < C < +\infty \forall (\lambda,\mu) \in P : \ |\operatorname{Im}\lambda|^2 \leq C(a^2 + d^2 + b_2^2 + 1). \tag{5.12}$$

Thus, the statement that

$$\exists 0 < C < +\infty \forall (\lambda,\mu) \in P: \; |\lambda|^2 + |\mu| + 1 \leq C(a^2 + d^2 + b_2^2 + 1) \qquad (7.3)$$

may not be valid.

Further, there exists a continuous mapping κ of P into $\Omega = \{(d,a,b_2) \in R^3 | d \geq -\gamma,\, a \geq d,\, b_2 \geq 0\}$ which is given by the formula: $P \ni (\lambda,\mu) \mapsto (d,a,b_2) \in \Omega$. Just as previously, to any σ-finite measure ρ on P there corresponds a measure ν on Ω which is defined by the following formula: for each Borel set $\Delta \subseteq \Omega$, $\nu(\Delta) \stackrel{\text{def}}{=} \rho(\kappa^{-1}(\Delta)) = \rho(\{(\lambda,\mu) \in P | (d,a,b_2) \in \Delta\})$.

For any $t \in R_+$:

$$R(t) \stackrel{\text{def}}{=} \int_P |\psi_1(\lambda,\mu,t)|^2 d\rho_{\lambda,\mu}$$

$$= \int_\Omega e^{-2at} \frac{\sin^2 b_2 t \cdot \cosh^2 b_1 t + \cos^2 b_2 t \cdot \sinh^2 b_1 t}{b_1^2 + b_2^2} d\nu_{d,a,b_2}.$$

But now the measure ν on Ω may not be σ-finite, in spite of the σ-finiteness of the measure ρ on P.

Example 7.1. Let P coincide with $\sigma(A,B)$ from Example 3.2: $P = \{(\lambda,\mu) \in C^2 | \operatorname{Re}\lambda = 0,\, \operatorname{Im}\lambda \in R^1,\, \mu = \frac{\lambda^2}{4}\}$. Then P is a closed subset of C^2, and for any $(\lambda,\mu) \in P : \operatorname{Re}\omega_1 = \operatorname{Re}\omega_2 = -\frac{1}{2}\operatorname{Re}\lambda = 0$. Thus, P satisfies our assumption. For each Borel set $\Delta \subseteq R^1$, we set $\rho(\{(\lambda,\mu) \in P | \operatorname{Im}\lambda \in \Delta\}) = m(\Delta)$ where m is the Lebesgue measure on R^1. Obviously ρ is a σ-finite measure on P. On the other hand, for any $(\lambda,\mu) \in P : a = d = b_1 = b_2 = 0$. Thus, κ maps $P \subseteq C^2$ onto the set $\{(0,0,0)\} \subset R^3$. Accordingly, $\nu(\{(0,0,0)\}) = \rho(P) = +\infty$.

In fact, we arrive at the following problem: to obtain a criterion for the continuity of $R(t)$ on $[0,T]$ in terms of the measure ν without the additional assumption that ν is σ-finite.

This question is answered by the following lemma. We make use of the notations from Lemma 5.8.

Lemma 7.2. *Let $0 < T \leq +\infty$, P be a closed subset of C^2 such that* (7.2) *holds, and ρ be a σ-finite measure on P. The integral transform $R(t) = \int_P |\psi_1(\lambda,\mu,t)|^2 d\rho$ is continuous on $[0,T]$ if and only if:*

1) *ν is a σ-finite measure on Ω;*
2) *for each $x > 0 : \nu_1(\Omega_x) < +\infty$;*
3) $\lim_{x \to +\infty} \nu_1(\Omega_x) = 0$.

Proof. Indeed, let $R(t) \in C((0,T])$. Set for each $(\lambda,\mu) \in C^2 : b = b_1 + ib_2$. Then $R(t) = t^2 \int_P e^{-2at} |\frac{\sinh bt}{bt}|^2 d\rho$, for any $t \in (0,T]$. Fix a $0 < \delta_1 < +\infty$ such that $\forall z \in C^1 : |z| \leq \delta_1 \Rightarrow |\frac{\sinh z}{z}| \geq \frac{1}{2}$; set $\delta = \frac{1}{\sqrt{2}}\delta_1$. Then $\forall(\lambda,\mu) \in P$, $t \in (0,T]$: if $a < \frac{\delta}{t}$, $0 \leq b_1 < \frac{\delta}{t}$, and $0 \leq b_2 < \frac{\delta}{t}$, then $e^{-2at}|\frac{\sinh bt}{bt}|^2 \geq \frac{1}{4}e^{-2\delta}$. Thus, $\forall t \in (0,T] : R(t) \geq \frac{t^2}{4}e^{-2\delta}\rho(\{(\lambda,\mu) \in P | a < \frac{\delta}{t}, b_1 < \frac{\delta}{t}, b_2 < \frac{\delta}{t}\}) = \frac{t^2}{4}e^{-2\delta}\nu(\{(d,a,b_2) \in \Omega | a < \frac{\delta}{t}, 0 \leq b_1 < \frac{\delta}{t}, 0 \leq b_2 < \frac{\delta}{t}\})$.

Let $\Delta \subseteq \Omega = \{(d,a,b_2) \in R^3 | d \geq -\gamma, b_1 = a-d \geq 0, b_2 \geq 0\}$ be a bounded Borel set in R^3. Then $\exists K < +\infty \forall(d,a,b_2) \in \Delta : -\gamma \leq d \leq K$, $d \leq a \leq K$, $0 \leq b_1 = a-d \leq K+\gamma$, $0 \leq b_2 \leq K$. Fix a $0 < t \leq T$ such that $t < \frac{\delta}{K+\gamma}$. Then $\nu(\Delta) \leq \nu(\{(d,a,b_2) \in \Omega | a < \frac{\delta}{t}, 0 \leq b_1 < \frac{\delta}{t}, 0 \leq b_2 < \frac{\delta}{t}\}) \leq \frac{4e^{2\delta}}{t^2}R(t) < +\infty$.

We have proved that if $R(t)$ is continuous on $(0,T]$ then ν is a σ-finite measure on Ω. Now Lemma 7.2 follows from Lemma 5.8.

Similarly, one can prove the following lemmas.

Lemma 7.3. *Let $0 < T < +\infty$, P be a closed subset of C^2 such that (7.2) holds, and ρ be a σ-finite measure on P. The integral transform $R(t)$ is continuous and bounded on $(0,T]$ if and only if:*

1) *ν is a σ-finite measure on Ω;*
2) *$\exists K < +\infty \forall x > 0 : \nu_1(\Omega_x) \leq K$.*

Lemma 7.4. *Let $0 < T < +\infty$, P be a closed subset of C^2 such that (7.2) holds, and ρ be a σ-finite measure on P. The integral transform $R(t)$ is integrable on $(0,T]$ (i.e., $R(t) \in L_1(0,T]$) if and only if:*

$$\int_\Omega \frac{d\nu_1}{|d|+1} = \int_\Omega \frac{d\nu}{(a^2+d^2+b_2^2+1)(|d|+1)} < +\infty. \tag{7.4}$$

Moreover, if (7.4) holds then $R(t) \in C((0,T])$.

Notice that (7.4) implies, in particular, that ν is a σ-finite measure on Ω.

Let's return to a σ-finite measure ρ on P. Notice that (7.4) holds if and only if

$$\int_P \frac{d\rho}{(a^2+d^2+b_2^2+1)(|d|+1)} < +\infty. \tag{7.5}$$

Thus, (7.5) implies that ν is a σ-finite measure on Ω. Now we can rewrite Lemmas 7.2, 7.3, 7.4 in the following form.

Theorem 7.1. *Let $P \subseteq C^2$ be a closed set such that (7.2) holds, and ρ be a σ-finite measure on P. For each Borel set $\Delta \subseteq P$, denote:*

$$\rho_2(\Delta) = \int_\Delta \frac{1}{a^2+d^2+b_2^2+1} d\rho.$$

Set for each $x > 0 : P_x = P \cap (\{(\lambda,\mu) | d \leq 2x\} \setminus \{(\lambda,\mu) | d < x, a < x, b_2 < x\})$.

1. *Let* $0 < T \leq +\infty$. *The integral transform* $R(t) = \int_P |\psi_1(\lambda,\mu,t)|^2 d\rho$ *is continuous on* $[0,T]$ *if and only if:*

1) (7.5) *holds;*

2) $\forall x > 0 : \rho_2(P_x) < +\infty$;

3) $\lim_{x\to+\infty} \rho_2(P_x) = 0$.

2. *Let* $0 < T < +\infty$. $R(t)$ *is continuous and bounded on* $(0,T]$ *if and only if:*

1) (7.5) *holds;*

2) $\exists K < +\infty \forall x > 0 : \rho_2(P_x) \leq K$.

3. *Let* $0 < T < +\infty$. $R(t)$ *is integrable on* $(0,T]$ *if and only if* (7.5) *holds. Moreover, if* (7.5) *holds then* $R(t)$ *is continuous on* $(0,T]$.

Remark 1. (7.5) holds if and only if $\int_P \frac{d\rho_2}{|d|+1} < +\infty$.

Remark 2. In points 1, 2 of Theorem 7.1, one can replace condition 1) by the weaker condition that ν is a σ-finite measure on Ω. Theorem 7.1 so reformulated may be considered as a generalization of Theorem 5.1 and Corollary 2 of it and Theorems 5.2, 5.3 to the case where (5.12) may not be valid. See Lemma 5.7 and Corollaries 1, 2 to it.

Remark 3. If (5.12) holds, then one can replace $(a^2 + d^2 + b_2^2 + 1)$ in Theorem 7.1 by $(|\lambda|^2 + |\mu| + 1)$ and, accordingly, ρ_2 by ρ_1 (see Theorems 5.1, 5.2, 5.3). In the general case, one cannot replace $(a^2 + d^2 + b_2^2 + 1)$ by $(|\lambda|^2 + |\mu| + 1)$, since (7.3) may not be valid.

Remark 4. If (5.12) holds, then ν is a σ-finite measure on Ω, and one can omit condition 1) in points 1, 2 of Theorem 7.1. In the general case, one cannot omit this condition or the weaker condition that ν is a σ-finite measure on Ω. Indeed, let P and ρ be from Example 7.1. Then $P \subseteq \{(\lambda,\mu) \in C^2 | d = a = b_2 = 0\}$, and for any $x > 0 : P_x \subseteq P \setminus \{(\lambda,\mu) | d < x, a < x, b_2 < x\} \Rightarrow P_x = \emptyset$. Thus, $\forall x > 0 : \rho_2(P_x) = 0$. But $\int_P \frac{d\rho}{(a^2+d^2+b_2^2+1)(|d|+1)} = \rho(P) = +\infty$, i.e., (7.5) does not hold. Moreover, $\nu(\{(0,0,0)\}) = +\infty$, and so ν is not a σ-finite measure on $\Omega \subset R^3$.

3 The general form of weak solutions where (2.2) holds

We can now prove the following theorems.

Theorem 7.2. *Let A and B be c.n.o. in H such that* (2.2) *(or, what is the same,* (3.3)*) holds. Denote: $P = \sigma(A, B)$. Let G_1 be the set of all $g \in H_-(((a^2+d^2+b_2^2+1)^{1/2}(|d|+1)^{1/2})(A, B))$ such that*

$$\forall x > 0 : \int\limits_{P_x} \frac{d(Eg, g)}{a^2+d^2+b_2^2+1} < +\infty, \text{ and } \lim_{x\to+\infty} \int\limits_{P_x} \frac{d(Eg, g)}{a^2+d^2+b_2^2+1} = 0. \tag{7.6}$$

Let $0 < T \leq +\infty$. There exists a one-to-one mapping of the set $H \times G_1 \subseteq H \times H_-(((a^2+d^2+b_2^2+1)^{1/2}(|d|+1)^{1/2})(A, B))$ onto the set Y' of all weak solutions of equation (1) *on* $[0, T]$ *which is given by the formula:*

$$H \times G_1 \ni (f_0, g) \leftrightarrow y(t) = e^{\omega_1 t}(A, B)f_0 + \psi_1(A, B, t)g \ (t \in [0, T]) \in Y'.$$

Remark 1. Obviously $G_1 \subseteq H_-(((a^2+d^2+b_2^2+1)^{1/2}(|d|+1)^{1/2})(A, B)) \subseteq H_-((|A|^2+|B|+1)^{1/2}(|d|+1)^{1/2}(A, B)) \subseteq H_-((|A|^2+|B|+1)^{1/2}(|\operatorname{Re} A|+1)^{1/2}) \subseteq H_-((|A|^2+|B|+1)^{1/2}(|A|+1)^{1/2})$.

Similarly, one can prove the following statement.

Lemma 7.5. *Let A and B be c.n.o. in H such that*

$$\exists 0 < \gamma < +\infty \forall (\lambda, \mu) \in \sigma(A, B) : \min_{i=1,2} \operatorname{Re} \omega_i(\lambda, \mu) \geq -\gamma.$$

Let $-\infty \leq T_1 < T < +\infty$. There exists a one-to-one mapping of the set $H \times G_1 \subseteq H \times H_-(((a^2+d^2+b_2^2+1)^{1/2}(|d|+1)^{1/2})(A, B))$ onto the set Y' of all weak solutions of (1) *on* $[T_1, T]$ *which is given by the formula:*

$$H \times G_1 \ni (f_0, g) \leftrightarrow y(t) = e^{\omega_2(t-T)}(A, B)f_0 + \psi_1(A, B, t-T)g \ (t \in [T_1, T]) \in Y'.$$

Theorem 7.3. *Let A and B be c.n.o. in H such that* (2.2) *(or, what is the same,* (3.3)*) holds. Denote: $P = \sigma(A, B)$.*

Let G_1 be the set of all $g \in H_-(((a^2+d^2+b_2^2+1)^{1/2}(|d|+1)^{1/2})(A, B))$ such that (7.6) *holds;*

G_2 be the set of all $g \in H_-(((a^2+d^2+b_2^2+1)^{1/2}(|d|+1)^{1/2})(A, B))$ such that

$$\exists K < +\infty \forall x > 0 : \int\limits_{P_x} \frac{d(Eg, g)}{a^2+d^2+b_2^2+1} \leq K. \tag{7.7}$$

Fix an arbitrary $0 < T < +\infty$. *Denote by* Y_1 *the set of all weak solutions of equation* (1) *on* $(0,T]$ *such that*

$$\exists y_0 \in H : y(t) \to y_0 \text{ in } H \text{ as } t \to 0;$$

by Y_2 *the set of all weak solutions of* (1) *on* $(0,T]$ *such that*

$$\exists C_y < +\infty \forall t \in (0,T] :\| y(t) \| \leq C_y.$$

There exists a one-to-one mapping of $H \times G_1 \subseteq H \times H_-(((a^2+d^2+b_2^2+1)^{1/2}(|d|+1)^{1/2})(A,B))$ *onto the set* Y_1 *which is given by the formula:*

$$H \times G_1 \ni (f_0, g) \leftrightarrow y(t) = e^{\omega_1 t}(A,B)f_0 + \psi_1(A,B,t)g \quad (t \in (0,T]) \in Y_1.$$

There exists a one-to-one mapping of $H \times G_2 \subseteq H \times H_-(((a^2+d^2+b_2^2+1)^{1/2}(|d|+1)^{1/2})(A,B))$ *onto the set* Y_2 *which is given by the formula:*

$$H \times G_2 \ni (f_0, g) \leftrightarrow y(t) = e^{\omega_1 t}(A,B)f_0 + \psi_1(A,B,t)g \quad (t \in (0,T]) \in Y_2.$$

Corollary 1. *Let* A *and* B *be c.n.o. in* H *such that* (2.2) *holds,* $0 < T \leq +\infty$, *and* $y(t)$ *be a weak solution of equation* (1) *on* $(0,T]$. *If* $y(t)$ *has a limit in* H *as* $t \to 0$, *then this* $y(t)$, *extended by continuity to* $t = 0$, *is a weak solution of equation* (1) *on* $[0,T]$.

Furthermore, we conclude from Theorem 7.3 that $Y_1 = Y_2$ if and only if $G_1 = G_2$.

Theorem 5.4 and Statement 5.3 and Remark 1 to them are based on the corresponding statements for a σ-finite measure ν_1 on $\kappa(P)$. Therefore, Theorem 5.4 and Statement 5.3 and Remark 1 to them remain valid if one waives the assumption (5.12), considers σ-finite measures ρ on P such that the corresponding measure ν on Ω is σ-finite as well, and replaces ρ_1 by ρ_2 from the previous section. From this, we conclude that $G_1 = G_2$ if and only if (6.12) holds. We have proved the following theorem.

Theorem 7.4. *Let* A *and* B *be c.n.o. in* H *such that* (2.2) *holds. Equation* (1) *has Fatou-Riesz property if and only if* $P = \sigma(A,B)$ *satisfies* (6.12). *Moreover, if* $P = \sigma(A,B)$ *does not satisfy* (6.12) *then for any* $0 < T < +\infty$ *there exists a bounded weak solution of equation* (1) *on* $(0,T]$ *which does not have a limit in* H *as* $t \to 0$.

4 Initial data of weak solutions where (2.2) holds

Theorem 7.5. *Let A and B be c.n.o. in H such that* (2.2) *holds,* $0 < T \leq +\infty$. *The Cauchy problem for equation* (1) *on* $[0,T]$ *with initial data* (f_0, f_1) *has a weak solution if and only if* $f_0 \in H$, $f_1 \in H'_- = H_-((|A|^2+|B|+1)^{1/2}(|A|+1)^{1/2})$, *and* $h = \frac{A}{2}f_0 + f_1 \in G_1$. *(Here we understand* A, $\omega_i(A,B)$ $(i=1,2)$ *as linear bounded operators acting from* H *into* $H_{-1} = H_-((|A|^2+|B|+1)^{1/2}) \subseteq H'_- = H_-((|A|^2+|B|+1)^{1/2}(|A|+1)^{1/2})$ *which are extended by continuity to the whole* H; G_1 *is defined in Theorem 7.2.) If this is the case then* $g = -\omega_1(A,B)f_0 + f_1 \in G_1 \subseteq H_-(((a^2+d^2+b_2^2+1)^{1/2}(|d|+1)^{1/2})(A,B))$, *and a weak solution of* (1) *on* $[0,T]$ *with i.d.* (f_0, f_1) *is unique and has the form:*

$$\begin{aligned} y(t) &= e^{\omega_1 t}(A,B)f_0 + \psi_1(A,B,t)g \\ &= e^{\omega_1 t}(A,B)f_0 + \lim_{k\to\infty} \psi_1(A,B,t)(-\omega_1(A,B)E_k f_0 + E_k f_1) \quad (t \in [0,T]). \end{aligned}$$

Remark 1. In Theorem 7.5, one can replace $H'_- = H_-((|A|^2+|B|+1)^{1/2}(|A|+1)^{1/2})$ by $H_-((|A|^2+|B|+1)^{1/2}(|d|+1)^{1/2}(A,B))$.

Proof. Taking into account Lemma 6.1 and Remark 1 to it and Lemma 7.1, we can conclude from Theorem 7.2 that if $y(t)$ is a weak solution of (1) on $[0,T]$ with i.d. (f_0, f_1) then

$$y(t) = e^{\omega_1 t}(A,B)f_0 + \lim_{k\to\infty} \psi_1(A,B,t)g^{(k)} \quad (t \in [0,T]),$$

where $f_0 = y(0) \in H$, $\forall k \geq 1 : g^{(k)} = -\omega_1(A,B)f_0^{(k)} + f_1^{(k)}$, and $\exists g \in G_1 :$ $g^{(k)} = E_k g$, $\forall k \geq 1$.

Notice that the existence of a $g \in G_1$ such that $g^{(k)} \stackrel{\text{def}}{=} -\omega_1(A,B)f_0^{(k)} + f_1^{(k)} = E_k g$, $\forall k \geq 1$, is equivalent to the existence of an $h \in G_1$ such that $h^{(k)} \stackrel{\text{def}}{=} \frac{A}{2}f_0^{(k)} + f_1^{(k)} = E_k h$, $\forall k \geq 1$. Indeed, $\forall (\lambda, \mu) \in (A,B) : -\omega_1 - \frac{\lambda}{2} = \frac{1}{2}(\omega_2 - \omega_1)$; here $|\omega_2 - \omega_1|^2 = 4(b_1^2 + b_2^2)$. Thus, for $f_0 \in H : f'_0 = \frac{1}{2}(\omega_2 - \omega_1)(A,B)f_0 \in H_-((d^2+b_1^2+b_2^2+1)^{1/2}(A,B)) = H_-((a^2+d^2+b_2^2+1)^{1/2}(A,B)) \subseteq G_1 \subseteq H'_-$. For any $k \geq 1 : g^{(k)} - h^{(k)} = E_k f'_0$. We have proved the statement; moreover, $g = h + f'_0$.

Further, $f_0 \in H \Rightarrow \frac{A}{2}f_0 \in H_{-1} = H_-((|A|^2+|B|+1)^{1/2}) \subseteq H'_-$; for any $k \geq 1 : \frac{A}{2}f_0^{(k)} = E_k(\frac{A}{2}f_0)$. We obtain that for $f = h - \frac{A}{2}f_0 \in H'_- : f_1^{(k)} = E_k f$, $\forall k \geq 1$.

We have proved that

$$y(t) = e^{\omega_1 t}(A,B)f_0 + \psi_1(A,B,t)(-\omega_1(A,B)f_0 + f) \quad (t \in [0,T]) \tag{7.8}$$

where $f_0 \in H$, $f \in H'_-$. It is easy to see that for any $g_1 \in \Phi : (y(t), g_1)'|_{t=0} = (f, g_1)$. In other words, $f_1 = f$ as antilinear functionals on Φ.

Summarizing, for a weak solution of (1) on $[0,T]$ $y(t)$ with i.d. (f_0, f_1) : $f_0 \in H$, $f_1 \in H'_-$, $h = \frac{A}{2} f_0 + f_1 \in G_1$, and $y(t)$ has the form (7.8) with f replaced by f_1.

Conversely, if this is the case then $g = -\omega_1(A,B) f_0 + f_1 = h + f'_0 \in G_1$, and so $y(t) = e^{\omega_1 t}(A,B) f_0 + \psi_1(A,B,t) g$ $(t \in [0,T])$ is a weak solution of (1) on $[0,T]$, by virtue of Theorem 7.2. Furthermore, $y(0) = f_0$, and $\forall g_1 \in \Phi$: $(y(t), g_1)'|_{t=0} = (f_1, g_1)$. It is shown before that if a weak solution of (1) on $[0,T]$ with i.d. (f_0, f_1) exists then it is unique.

The proof is over.

Let's analyze this result. We consider c.n.o. A and B in H such that (2.2) holds. It is shown that for the Cauchy problem for (1) on $[0,T]$ with i.d. (f_0, f_1) to have a unique weak solution, not only f_0 and f_1 separately must satisfy certain conditions, but also $\frac{A}{2} f_0 + f_1$ must satisfy a certain condition.

Example 7.2. Let A and B be the operators from Example 3.2. Then for any $(\lambda, \mu) \in \sigma(A,B) : a = d = b_1 = b_2 = 0$. It follows from Theorem 7.5 and Remark 1 to it that the Cauchy problem for equation (1) on $[0,T]$ with i.d. (f_0, f_1) has a unique weak solution if and only if $f_0 \in H$, $f_1 \in H_{-1}$, and $\frac{A}{2} f_0 + f_1 \in H$.

In the special case that (3.2) (or, what is the same, (3.4)) holds, our general result on initial data of weak solutions of (1) on $[0,T]$ turns into the previous result on this problem which was obtained in Section 5 of Chapter 6 for this special case.

Indeed, let A and B be c.n.o. in H such that (2.2) and (3.2) hold. In this case, $G_1 = F_1$ (see Lemma 5.7). Taking into account that $H_{-1} \subseteq F_1 \subseteq H'_-$ and $\forall f_0 \in H : \frac{A}{2} f_0 \in H_{-1} \subseteq F_1$, we conclude that $f_0 \in H$, $f_1 \in H'_-$ and $h = \frac{A}{2} f_0 + f_1 \in F_1$ if and only if $f_0 \in H$ and $f_1 \in F_1$.

5 Weak well-posedness of the Cauchy problem in a special space of initial data

Let A and B be c.n.o. in H, $0 < T \leq +\infty$. In Chapter 3, we have shown that condition (2.2) is neither necessary nor sufficient for the weak well-posedness of the Cauchy problem for (1) on $[0,T]$ in $G = H \times H$. For the weak well-posedness in a wider space G such that $H \times H_-(|A| + 1) \subseteq G \subseteq H \times H_{-1} = H \times H_-((|A|^2 + |B| + 1)^{1/2})$, (2.2) is necessary, but it is not sufficient.

Now we can indicate such a space of initial data G_C, which is a linear subspace of $H \times H_{-1}$, that (2.2) is necessary and sufficient for the weak well-posedness of the Cauchy problem for (1) on $[0,T]$ in G_C.

Theorem 7.6. *Let A and B be c.n.o. in H, $0 < T \leq +\infty$, and G_C be a linear subspace of $H \times H_{-1}$ which consists of all $(f_0, f_1) \in H \times H_{-1}$ such that $\frac{A}{2} f_0 + f_1 \in H_-((a^2 + d^2 + b_2^2 + 1)^{1/2}(A, B))$. The Cauchy problem for equation (1) on $[0, T]$ is weakly well-posed in G_C if and only if (2.2) (or, what is the same, (3.3)) holds.*

Proof. Let A and B be c.n.o. in H such that (2.2) holds. Then for any $(f_0, f_1) \in G_C$: $f_0 \in H$, $f_1 \in H_{-1} \subseteq H'_-$, and $h = \frac{A}{2} f_0 + f_1 \in H_-((a^2 + d^2 + b_2^2 + 1)^{1/2}(A, B)) \subseteq G_1$; by Theorem 7.5, there exists a unique weak solution of (1) on $[0, T]$ with i.d. (f_0, f_1).

Conversely, let the Cauchy problem for (1) on $[0, T]$ be weakly well-posed in G_C. Fix an arbitrary $f_0 \in H$ and set $f_1 = \omega_1(A, B) f_0$; $f_1 \in H_{-1}$. Then $h = \frac{A}{2} f_0 + f_1 = \frac{1}{2}(\omega_1 - \omega_2)(A, B) f_0 \in H_-((a^2 + d^2 + b_2^2 + 1)^{1/2}(A, B))$. By the assumption, there exists a unique weak solution of (1) on $[0, T]$ with i.d. (f_0, f_1); denote this weak solution by $y_{f_0}(t)$. Notice that $\forall k \geq 1 : f_0^{(k)} = E_k f_0$, $f_1^{(k)} = E_k f_1$, $g^{(k)} = -\omega_1(A, B) f_0^{(k)} + f_1^{(k)} = 0$. By Lemma 7.1 and Statement 7.1, $y_{f_0}(t) = \lim_{k \to \infty} e^{\omega_1 t}(A, B) E_k f_0$ $(t \in [0, T])$.

Fix an arbitrary $0 < t_0 < T$. We have shown that for every $f_0 \in H$ there exists the limit of $e^{\omega_1 t_0}(A, B) E_k f_0$ in H as $k \to \infty$. In other words, the normal operator $e^{\omega_1 t_0}(A, B)$ is defined on the whole H. By virtue of the Banach closed graph theorem, $e^{\omega_1 t_0}(A, B)$ is a bounded linear operator on H, i.e., $\sup_{(\lambda,\mu) \in \sigma(A,B)} |e^{\omega_1 t_0}| = \| e^{\omega_1 t_0}(A, B) \| < +\infty$. Finally, $\sup_{(\lambda,\mu) \in \sigma(A,B)} |e^{\omega_1 t_0}| < +\infty \Rightarrow \sup_{(\lambda,\mu) \in \sigma(A,B)} \operatorname{Re} \omega_1 < +\infty$, i.e., (2.2) is valid.

Example 7.3. Let A and B be the operators from Example 3.2. Then (2.2) holds, and so the Cauchy problem for (1) on $[0, T]$ is weakly well-posed in G_C. Here $G_C = \{(f_0, f_1) \in H \times H_{-1} | \frac{A}{2} f_0 + f_1 \in H\}$. G_C does not contain $H \times H$; moreover, it does not contain even $H \times \{0\}$. We have shown in Example 3.2 that the weak well-posedness of the Cauchy problem for (1) on $[0, T]$ in $G = H \times H$ does not hold.

Notice that G_C does not contain $H \times H$ but does contain, for instance, all (f_0, f_1) such that $f_0 \in H$, $f_1 = -\frac{A}{2} f_0$; some of these vectors do not belong to $H \times H$.

Example 7.4. Let A and B be the operators from Example 3.1. Then (2.2) does not hold, and so the Cauchy problem for (1) on $[0, T]$ is not weakly well-posed in G_C. Here $G_C = H \times H_{-1}$. Now G_C is a strictly wider space than $H \times H$. We have shown in Example 3.1 that the weak well-posedness of the Cauchy problem for (1) on $[0, T]$ in $G = H \times H$ holds.

Chapter 8
Fatou-Riesz property

1 Fatou-Riesz property

Definition 8.1. Let A and B be c.n.o. in H. Equation (1) is said to have Fatou-Riesz property if for every $0 < T < +\infty$: each bounded weak solution of equation (1) on $(0, T]$ has a limit in H as $t \to 0$.

Theorem 8.1. *Let A and B be c.n.o. in H. Equation* (1) *has Fatou-Riesz property if and only if one of the following equivalent conditions holds:*

1) $\exists 0 < \gamma < +\infty \forall (\lambda, \mu) \in \sigma(A, B) : \max\limits_{i=1,2} \operatorname{Re} \omega_i(\lambda, \mu) \geq -\gamma;$ (8.1)

2) $\exists 0 < \gamma < +\infty : \sigma(A, B) \subseteq C^2 \setminus \{(\lambda, \mu) | \lambda_1 > 2\gamma,\ \mu_1 > \frac{1}{(\lambda_1 - 2\gamma)^2}(\mu_2^2 - \lambda_1 \lambda_2 \cdot \mu_2 + \gamma(\lambda_1 - \gamma)(\lambda_2^2 + (\lambda_1 - 2\gamma)^2))\}$.

Moreover, if (8.1) *does not hold then for any $0 < T < +\infty$: there exists a bounded weak solution of* (1) *on* $(0, T]$ *which does not have a limit in H as $t \to 0$.*

Proof. Let A and B be c.n.o. in H, $0 < T < +\infty$, and $y(t)$ be a bounded weak solution of equation (1) on $(0, T]$.

Fix an arbitrary $0 < \gamma_0 < +\infty$. Denote:

$$\Delta_1 = \{(\lambda, \mu) \in C^2 | \min_{i=1,2} \operatorname{Re} \omega_i(\lambda, \mu) > -\gamma_0\};$$

$$\Delta_2 = \{(\lambda, \mu) \in C^2 | \min_{i=1,2} \operatorname{Re} \omega_i(\lambda, \mu) \leq -\gamma_0,\ \max_{i=1,2} \operatorname{Re} \omega_i(\lambda, \mu) > \gamma_0\};$$

$$\Delta_3 = \{(\lambda, \mu) \in C^2 | \max_{i=1,2} \operatorname{Re} \omega_i(\lambda, \mu) \leq \gamma_0\} \setminus \Delta_1.$$

It is clear that $\Delta_1 \cup \Delta_2 \cup \Delta_3 = C^2$. For each of $i = 1, 2, 3$, let $H_i = E(\Delta_i)H$, A_i and B_i be the restrictions of A and B respectively to H_i. Obviously, for each of $i = 1, 2, 3 :$ A_i and B_i are c.n.o. in H_i, and $\sigma(A_i, B_i) \subseteq \sigma(A, B) \cap \overline{\Delta_i}$ (here $\overline{}$ denotes the closure of a set).

Now one can write:

$$y(t) = y_1(t) + y_2(t) + y_3(t) \quad (t \in (0, T])$$

where $y_i(t) = E(\Delta_i)y(t)$ $(t \in (0, T])$, $i = 1, 2, 3$. Obviously, $y(t)$ is bounded in H if and only if each of $y_i(t)$ $(i = 1, 2, 3)$ is bounded in H_i. Furthermore, $y(t)$

has a limit in H as $t \to 0$ if and only if each of $y_i(t)$ $(i = 1,2,3)$ has a limit in H_i as $t \to 0$.

1. $y_1(t)$ is a weak solution of the equation

$$y_1''(t) + A_1 y_1'(t) + B_1 y_1(t) = 0$$

on $(0,T]$ where A_1 and B_1 are c.n.o. in H_1 such that

$$\sigma(A_1,B_1) \subseteq \{(\lambda,\mu) \in C^2 | \min_{i=1,2} \operatorname{Re} \omega_i(\lambda,\mu) \geq -\gamma_0\}.$$

For this equation, every weak solution on $(0,T]$ has a limit in H_1 as $t \to 0$. Let's show this. For the sake of simplicity, we replace H_1, A_1, B_1 by H, A, B respectively.

By Lemma 7.5: $y_1(t) = e^{\omega_2(t-T)}(A,B)f_0 + \psi_1(A,B,t-T)g$ $(t \in (0,T])$ where $f_0 \in H$, $g \in G_1$. Setting $T_1 = -\infty$ in Lemma 7.5, we obtain that for these $f_0 \in H$ and $g \in G_1$: $e^{\omega_2(t-T)}(A,B)f_0 + \psi_1(A,B,t-T)g$ $(t \in (-\infty,T])$ is a weak solution of (1) on the whole $(-\infty,T]$. In particular, $y_1(t) = e^{\omega_2(t-T)}(A,B)f_0 + \psi_1(A,B,t-T)g$ $(t \in (0,T])$ has a limit in H as $t \to 0$.

2. $y_2(t)$ is a bounded weak solution of the equation

$$y_2''(t) + A_2 y_2'(t) + B_2 y_2(t) = 0$$

on $(0,T]$ where A_2 and B_2 are c.n.o. in H_2 such that

$$\sigma(A_2,B_2) \subseteq \{(\lambda,\mu) \in C^2 | \min_{i=1,2} \operatorname{Re} \omega_i(\lambda,\mu) \leq -\gamma_0, \max_{i=1,2} \operatorname{Re} \omega_i(\lambda,\mu) \geq \gamma_0\}.$$

For this equation every bounded weak solution on $(0,T]$ has a limit in H_2 as $t \to 0$. Let's show this. For simplicity, we replace H_2, A_2, B_2 by H, A, B respectively.

First, notice that $\forall(\lambda,\mu) \in \sigma(A,B)$:

$$\min_{i=1,2} \operatorname{Re} \omega_i \leq -\gamma_0, \max_{i=1,2} \operatorname{Re} \omega_i \geq \gamma_0, \left|(\omega_1 - \omega_2) - \frac{2\pi i l}{T}\right| \geq |\operatorname{Re}(\omega_1 - \omega_2)| \geq 2\gamma_0 > 0, \ \forall l \in \mathbb{Z}.$$

By Theorem 4.1, the Dirichlet problem for this equation on $[0,T]$ is weakly well-posed in $G = H \times H$, and for any $(f_0, f_1) \in H \times H : \varphi_0(A,B,t)f_0 + \varphi_1(A,B,t)f_1$ is a weak solution of (1) on $[0,T]$ with boundary values f_0, f_1.

By the assumption, $\exists C_y < +\infty \forall t \in (0,T] : \| y_2(t) \| \leq C_y$.

Fix an arbitrary $k \geq 1$. The Dirichlet problem for equation (1) on $[0,T]$ is weakly well-posed in $E_k H \times E_k H$, and for any solution $u(t)$ of (1) in $E_k H$ on $[0,T]$:

$$u(t) = \varphi_0(A,B,t)u(0) + \varphi_1(A,B,t)u(T) \quad (t \in [0,T]).$$

We noted before Lemma 6.1 that $E_k y_2(t)$ $(t \in (0,T])$ can be extended up to a usual solution of (1) in $E_k H$ on the whole $(-\infty,+\infty)$. Denote: $f_0^{(k)} = \lim_{t\to 0} E_k y_2(t)$. Then

$$\begin{aligned} E_k y_2(t) &= \varphi_0(A,B,t)f_0^{(k)} + \varphi_1(A,B,t)(E_k y_2)(T) \\ &= \varphi_0(A,B,t)f_0^{(k)} + \varphi_1(A,B,t)E_k y_2(T), \quad t \in (0,T]. \end{aligned}$$

Moreover, $\forall t \in (0,T] : \| E_k y_2(t) \| \le \| y_2(t) \| \le C_y < +\infty$; thus, $\| f_0^{(k)} \| = \lim_{t\to 0} \| E_k y_2(t) \| \le C_y$.

Obviously $\{f_0^{(k)}\}_{k=1}^{\infty}$ is an $\mathbb{E}$-sequence; we have shown that $\exists C_y < +\infty \forall k \ge 1 : \| f_0^{(k)} \| \le C_y$; thus, $\exists f_0 \in H \forall k \ge 1 : f_0^{(k)} = E_k f_0$.

It was shown before that for $f_0 \in H$, $y_2(T) \in H$:

$$\varphi_0(A,B,t)f_0, \quad \varphi_1(A,B,t)y_2(T) \in C([0,T],H).$$

For any $k \ge 1$:

$$\begin{aligned} E_k y_2(t) &= \varphi_0(A,B,t)E_k f_0 + \varphi_1(A,B,t)E_k y_2(T) \\ &= E_k(\varphi_0(A,B,t)f_0 + \varphi_1(A,B,t)y_2(T)), \quad t \in (0,T]; \end{aligned}$$

thus, $y_2(t) = \varphi_0(A,B,t)f_0 + \varphi_1(A,B,t)y_2(T)$, $t \in (0,T]$.

We conclude that $y_2(t)$ can be extended up to a weak solution of (1) on $[0,T]$ with boundary values f_0, $y_2(T)$; in particular, $y_2(t)$ has a limit in H as $t \to 0$.

3. So, a bounded weak solution of (1) on $(0,T]$ $y(t)$ has a limit in H as $t \to 0$ if and only if $y_3(t) = E(\Delta_3)y(t)$ has a limit in H_3 as $t \to 0$.

We have proved the following fact: equation (1) with c.n.o. A and B in H has Fatou-Riesz property if and only if the equation

$$y_3''(t) + A_3 y_3'(t) + B_3 y_3(t) = 0 \tag{8.2}$$

with c.n.o. A_3 and B_3 in H_3 has Fatou-Riesz property. But $\sigma(A_3,B_3) \subseteq \{(\lambda,\mu) \in \mathbb{C}^2 | \max_{i=1,2} \operatorname{Re}\omega_i \le \gamma_0\}$. By Theorem 7.4, equation (8.2) has Fatou-Riesz property if and only if

$$\exists 0 < \gamma < +\infty : \sigma(A_3,B_3) \subseteq \{(\lambda,\mu) \in \mathbb{C}^2 | \max_{i=1,2} \operatorname{Re}\omega_i \ge -\gamma\}.$$

By the definition, for any $(\lambda,\mu) \in \overline{\Delta_1}$: $\max_{i=1,2} \operatorname{Re}\omega_i \ge \min_{i=1,2} \operatorname{Re}\omega_i \ge -\gamma_0$; for any $(\lambda,\mu) \in \overline{\Delta_2}$: $\max_{i=1,2} \operatorname{Re}\omega_i \ge -\gamma_0$; $\sigma(A,B) \subseteq \sigma(A_1,B_1) \cup \sigma(A_2,B_2) \cup \sigma(A_3,B_3) \subseteq \overline{\Delta_1} \cup \overline{\Delta_2} \cup \sigma(A_3,B_3)$. Thus,

$$\exists 0 < \gamma < +\infty \forall (\lambda,\mu) \in \sigma(A_3,B_3) \subseteq \sigma(A,B) \cap \Delta_3 : \max_{i=1,2} \operatorname{Re}\omega_i \ge -\gamma$$

if and only if (8.1) holds. The proof is over.

In passing, we have obtained the following result.

Theorem 8.2. *Let A and B be c.n.o. in H, $0 < T < +\infty$, and $y(t)$ be a weak solution of equation* (1) *on* $(0, T]$. *If $y(t)$ has a limit in H as $t \to 0$, then $y(t)$* $(t \in (0, T])$, *extended by continuity to $t = 0$, is a weak solution of equation* (1) *on* $[0, T]$.

The similar results are valid for $[0, T)$ and $(0, T)$ as well.

2 Two-sided Fatou-Riesz property

Just as previously, we can prove the following theorem.

Theorem 8.3. *Let A and B be c.n.o. in H, $0 < T < +\infty$. The following conditions are equivalent:*

1) *each bounded weak solution of equation* (1) *on* $[0, T)$ *has a limit in H as $t \to T$;*
2) $\exists 0 < \gamma < +\infty \forall (\lambda, \mu) \in \sigma(A, B) : \min_{i=1,2} \operatorname{Re} \omega_i(\lambda, \mu) \leq \gamma;$
3) $\exists 0 < \gamma < +\infty : \sigma(A, B) \subseteq \mathbb{C}^2 \setminus \{(\lambda, \mu) | \lambda_1 < -2\gamma,\ \mu_1 > \frac{1}{(\lambda_1 + 2\gamma)^2}(\mu_2^2 - \lambda_1\lambda_2 \cdot \mu_2 - \gamma(\lambda_1 + \gamma)(\lambda_2^2 + (\lambda_1 + 2\gamma)^2))\}$.

Definition 8.2. Let A and B be c.n.o. in H, and for every $-\infty < a < b < +\infty$: each bounded weak solution of equation (1) on (a, b) has limits in H as $t \to a$ and $t \to b$. Then equation (1) is said to have two-sided Fatou-Riesz property.

Obviously, if equation (1) has two-sided Fatou-Riesz property then it has Fatou-Riesz property. The converse statement is not valid.

Theorems 8.1 and 8.3 imply the following statement.

Theorem 8.4. *Let A and B be c.n.o. in H, $-\infty < a < b < +\infty$. The following conditions are equivalent:*

1) *each bounded weak solution of equation* (1) *on* (a, b) *has limits in H as $t \to a$ and $t \to b$;*
2) $\exists 0 < \gamma < +\infty \forall (\lambda, \mu) \in \sigma(A, B) :$
$$\max_{i=1,2} \operatorname{Re} \omega_i(\lambda, \mu) \geq -\gamma,\ \min_{i=1,2} \operatorname{Re} \omega_i(\lambda, \mu) \leq \gamma; \tag{2.8}$$
3) $\exists 0 < \gamma < +\infty : \sigma(A, B) \subseteq \mathbb{C}^2 \setminus (\{(\lambda, \mu) | \lambda_1 > 2\gamma,$
$\mu_1 > \frac{1}{(\lambda_1 - 2\gamma)^2}(\mu_2^2 - \lambda_1\lambda_2 \cdot \mu_2 + \gamma(\lambda_1 - \gamma)(\lambda_2^2 + (\lambda_1 - 2\gamma)^2))\} \cup \{(\lambda, \mu) |$
$\lambda_1 < -2\gamma,\ \mu_1 > \frac{1}{(\lambda_1 + 2\gamma)^2}(\mu_2^2 - \lambda_1\lambda_2 \cdot \mu_2 - \gamma(\lambda_1 + \gamma)(\lambda_2^2 + (\lambda_1 + 2\gamma)^2))\})$.

3 First order equation and incomplete second order equations

The fact that equation (1) with c.n.o. A and B in H may not have Fatou-Riesz property is surprising.

1. Statement 8.1. The first order equation

$$y'(t) + Ay(t) = 0, \tag{8.3}$$

where A is a normal operator in H, does have Fatou-Riesz property.

Proof. Indeed, let A be a normal operator in H, $0 < T < +\infty$. For each $\lambda \in \mathbb{C}^1$, denote by ω the root of the characteristic polynomial $\omega + \lambda$; obviously $\omega = -\lambda$. Denote: $\Delta_1 = \{\lambda \in \mathbb{C}^1 | \operatorname{Re}\omega \geq 0\}$, $\Delta_2 = \{\lambda \in \mathbb{C}^1 | \operatorname{Re}\omega < 0\}$. For each of $i = 1, 2$, let $H_i = E_A(\Delta_i)H$, A_i be the restriction of A to H_i; then A_i is a normal operator in H_i, and $\sigma(A_i) \subseteq \sigma(A) \cap \overline{\Delta_i}$.

Let $y(t)$ be a bounded weak solution of (8.3) on $(0, T]$; then $y(t) = y_1(t) + y_2(t)$ $(t \in (0, T])$ where $y_i(t) = E_A(\Delta_i)y(t)$ $(t \in (0, T])$, $i = 1, 2$. Obviously, $y(t)$ has a limit in H as $t \to 0$ if and only if for each of $i = 1, 2$: $y_i(t)$ has a limit in H_i as $t \to 0$.

$y_1(t)$ is a weak solution of the equation

$$y_1'(t) + A_1 y_1(t) = 0 \tag{8.4}$$

on $(0, T]$ where A_1 is a normal operator in H_1 such that $\sigma(A_1) \subseteq \{\lambda \in \mathbb{C}^1 | \operatorname{Re}\omega \geq 0\}$. One can show that $y_1(t) = e^{\omega(t-T)}(A_1)y_1(T)$ $(t \in (0, T])$. On the other hand, $e^{\omega(t-T)}(A_1)y_1(T)$ is a weak solution of (8.4) on the whole $(-\infty, T]$. In particular, $y_1(t) = e^{\omega(t-T)}(A_1)y_1(T)$ $(t \in (0, T])$ has a limit in H_1 as $t \to 0$.

$y_2(t)$ is a bounded weak solution of the equation

$$y_2'(t) + A_2 y_2(t) = 0 \tag{8.5}$$

on $(0, T]$ where A_2 is a normal operator in H_2 such that $\sigma(A_2) \subseteq \{\lambda \in \mathbb{C}^1 | \operatorname{Re}\omega \leq 0\}$. One can show that there exists an $\mathbb{E}$-sequence $\{f_0^{(k)}\}_{k=1}^{\infty}$ in H_2 such that $y_2(t) = \lim_{k\to\infty} e^{\omega t}(A_2)f_0^{(k)}$, $t \in (0, T]$. By the assumption, $\exists C_y < +\infty \forall t \in (0, T]$: $\| y_2(t) \| \leq C_y$. For any $k \geq 1$: $\| E_k y_2(t) \| \leq \| y_2(t) \| \leq C_y$, $t \in (0, T]$, and $f_0^{(k)} = \lim_{t\to 0} E_k y_2(t)$; so, $\| f_0^{(k)} \| \leq C_y$. Thus, $\{f_0^{(k)}\}_{k=1}^{\infty}$ is an $\mathbb{E}$-sequence in H_2 such that $\| f_0^{(k)} \| \leq C_y < +\infty$, $\forall k \geq 1$; this implies that $\exists f_0 \in H_2 : f_0^{(k)} = E_k f_0$, $\forall k \geq 1$. Thus, $\exists f_0 \in H_2 : y_2(t) = e^{\omega t}(A_2)f_0$, $t \in (0, T]$. On the other hand, taking into account that $\sigma(A_2) \subseteq \{\lambda \in \mathbb{C}^1 | \operatorname{Re}\omega \leq 0\}$, we conclude that $e^{\omega t}(A_2)f_0$ is a weak solution of (8.5) in H_2 on $[0, +\infty)$. In particular, $y_2(t) = e^{\omega t}(A_2)f_0$ $(t \in (0, T])$ has a limit in H_2 as $t \to 0$.

The statement is proved.

In the special case that A is a self-adjoint nonnegative operator in H, this result was obtained in [137].

Statement 8.2. The first order equation $y'(t) + Ay(t) = 0$ with a normal operator A in H has two-sided Fatou-Riesz property.

2. We now give two corollaries to Theorem 8.4.

Corollary 1. *Let A and B be c.n.o. in H such that* $\operatorname{Re} A$ *is bounded. Then equation* (1) *has two-sided Fatou-Riesz property.*

See Corollary 1 to Theorem 4.1.

Corollary 2. *Let A and B be c.n.o. in H such that B is subordinate to A (i.e., $D(A) \subseteq D(B)$). Then equation* (1) *has two-sided Fatou-Riesz property.*

See Corollary 2 to Theorem 4.1.

3. Inference. *First order* equation (8.3) with a normal A in H does have two-sided Fatou-Riesz property. *Incomplete second order* equation (1) with commuting normal operators A and B in H, i.e., equation (1) with $A = 0$ (and even in the more general case of A bounded) or $B = 0$ (and even in the more general case of B subordinate to A), also does have two-sided Fatou-Riesz property. Nevertheless, *complete second order* equation (1) with commuting normal operators A and B in H where both operators A and B are unbounded, and moreover, $\operatorname{Re} A$ is unbounded and B is not subordinate to A, — may not have Fatou-Riesz property.

4 The case of self-adjoint A and B

Let's clarify our results in the special case that both operators A and B are self-adjoint.

Corollary 3. *Let A and B be c.s.o. in H, $-\infty < a < b < +\infty$. The following conditions are equivalent:*

1) *each bounded weak solution of equation* (1) *on* $(a, b]$ *has a limit in H as $t \to a$;*
2) $\exists 0 < \gamma < +\infty : \sigma(A, B) \subseteq R^2 \setminus \{(\lambda, \mu) | \lambda > 2\gamma, \mu > \gamma\lambda - \gamma^2\}$ (see Fig. 8.1).

Corollary 4. *Let A and B be c.s.o. in H, $-\infty < a < b < +\infty$. The following conditions are equivalent:*

1) *each bounded weak solution of equation* (1) *on* $[a, b)$ *has a limit in H as $t \to b$;*
2) $\exists 0 < \gamma < +\infty : \sigma(A, B) \subseteq R^2 \setminus \{(\lambda, \mu) | \lambda < -2\gamma, \mu > -\gamma\lambda - \gamma^2\}$ (see Fig. 8.2).

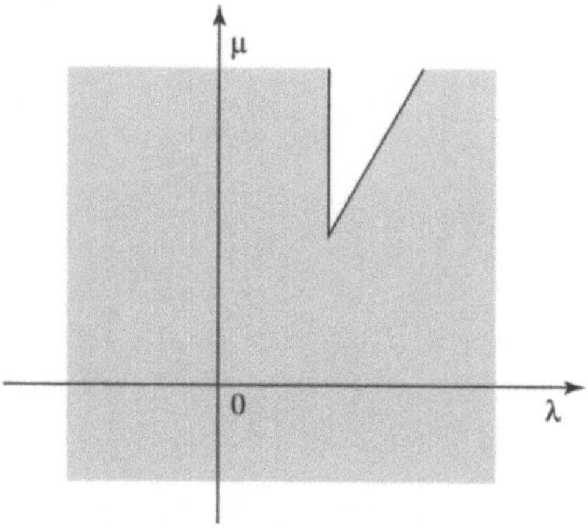

Fig. 8.1

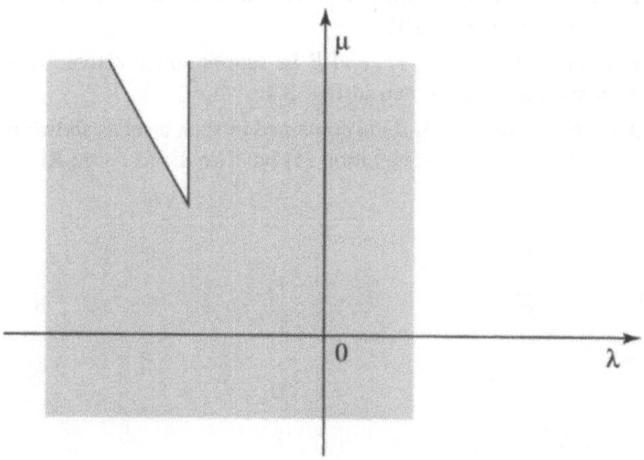

Fig. 8.2

Corollary 5. *Let A and B be c.s.o. in H, $-\infty < a < b < +\infty$. The following conditions are equivalent:*

1) *each bounded weak solution of equation* (1) *on* (a, b) *has limits in* H *as* $t \to a$ *and* $t \to b$;
2) $\exists 0 < \gamma < +\infty : \sigma(A, B) \subseteq R^2 \setminus \{(\lambda, \mu) | \, |\lambda| > 2\gamma, \, \mu > \gamma|\lambda| - \gamma^2\}$ (8.6) (see Fig. 8.3).

Example 8.1. 1) Let A and B be c.s.o. in H, and $D(A) \subseteq D(B)$. Then $\sigma(A, B)$ is contained in such a set as shown in Fig. 8.4a (see Theorem 1.2).

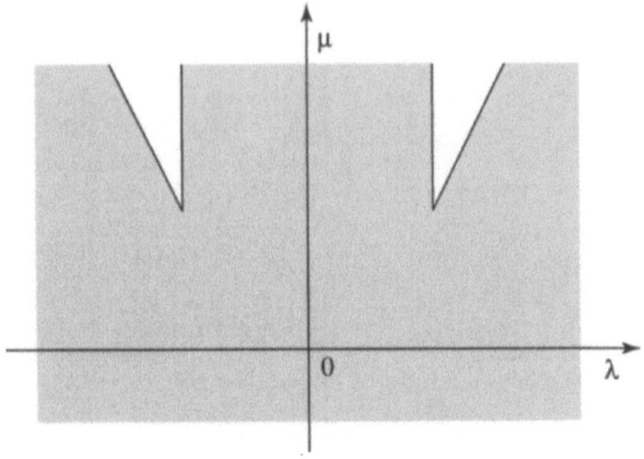

Fig. 8.3

2) Let A and B be c.s.o. in H, and A be bounded. Then $\sigma(A, B)$ is contained in such a set as shown in Fig. 8.4b.

3) Let A and B be c.s.o. in H, and B be semibounded above. Then $\sigma(A, B)$ is contained in such a set as shown in Fig. 8.4c.

In each of these cases, $\sigma(A, B)$ is contained in such a set as shown in Fig. 8.3, i.e., (8.6) holds; by Corollary 5, equation (1) has two-sided Fatou-Riesz property.

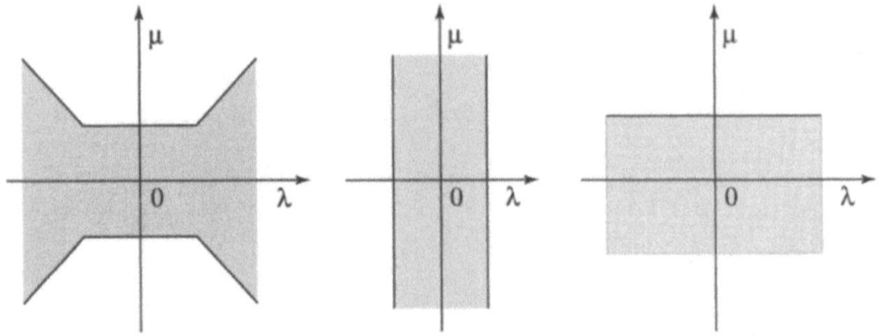

Fig. 8.4a, 8.4b, 8.4c

Example 8.2. Let A and B be the operators from Example 1.1.

Equation (1) has Fatou-Riesz property in all cases except the following four cases:

1) $a_1 = 0$, $a_2 > 0$, $b_1 > 0$;

2) $a_2 = 0$, $a_1 > 0$, $b_2 > 0$;

3) $a_1 < 0$, $a_2 > 0$, $\frac{b_1}{a_1} < \frac{b_2}{a_2}$;

4) $a_2 < 0$, $a_1 > 0$, $\frac{b_2}{a_2} < \frac{b_1}{a_1}$.

Let $0 < T < +\infty$. Any bounded weak solution of equation (1) on $[0, T)$ has a limit in H as $t \to T$ in all cases except the following four cases:

1) $a_1 = 0$, $a_2 < 0$, $b_1 > 0$;

2) $a_2 = 0$, $a_1 < 0$, $b_2 > 0$;

3) $a_1 > 0$, $a_2 < 0$, $\frac{b_2}{a_2} < \frac{b_1}{a_1}$;

4) $a_2 > 0$, $a_1 < 0$, $\frac{b_1}{a_1} < \frac{b_2}{a_2}$.

Equation (1) has two-sided Fatou-Riesz property in all cases except the following four cases:

1) $a_1 = 0$, $a_2 \neq 0$, $b_1 > 0$;

2) $a_2 = 0$, $a_1 \neq 0$, $b_2 > 0$;

3) $a_1 > 0$, $a_2 < 0$, $\frac{b_2}{a_2} < \frac{b_1}{a_1}$;

4) $a_2 > 0$, $a_1 < 0$, $\frac{b_1}{a_1} < \frac{b_2}{a_2}$.

5 Spectrum of the associated operator pencil

Let A and B be c.n.o. in H. Recall (see Section 6 of Chapter 3) that the pencil of linear operators in H $P(z) = z^2 + Az + B$, $z \in C^1$, defined on $D(A) \cap D(B)$, is called the operator pencil associated with equation (1). The spectrum of this operator pencil is denoted by $\sigma(P)$. By definition, $\sigma(P) \subseteq C^1$, and $C^1 \setminus \sigma(P)$ is the set of all $z \in C^1$ such that the operator $P(z) = z^2 + Az + B$ in H has an inverse one in H which is densely defined in H and bounded.

Theorem 8.1 gives a necessary and sufficient condition for equation (1) with c.n.o. A and B in H to have Fatou-Riesz property; this condition is expressed in terms of the position of the joint spectrum of A and B $\sigma(A, B)$ in C^2.

Statement 8.3. It is impossible to express a necessary and sufficient condition for equation (1) with c.n.o. A and B in H to have Fatou-Riesz property in terms of the position in C^1 of the spectrum $\sigma(P)$ of the associated operator pencil $z^2 + Az + B$.

Proof. In order to prove Statement 8.3, we give the following two examples. In both cases, A and B are c.s.o. in H. Spectrum of an associated operator pencil is one and the same in both cases, but equation (1) has two-sided Fatou-Riesz property in the first case and does not have Fatou-Riesz property in the second one.

Denote: $S = \{\omega_i(\lambda, \mu) | (\lambda, \mu) \in \sigma(A, B),\ i = 1, 2\}$. By Lemma 3.8, $S \subseteq \sigma(P) \subseteq \overline{S}$.

Example 8.3. Let $H = l_2(C^1)$, $A = 0$, $B = (b_{kl})_{k,l=1}^{\infty}$, where $b_{kl} = 0$ for $k \neq l$, $b_{kl} = -k^2$ for $k = l$.

Then $\sigma(A, B) = \{(\lambda_n, \mu_n) | n \in \mathbb{N},\ \lambda_n = 0,\ \mu_n = -n^2\}$. For any $n \in \mathbb{N}$, we have for $(\lambda_n, \mu_n) \in \sigma(A, B) : \{\omega_1, \omega_2\} = \{-n, n\}$. Thus, for any $(\lambda, \mu) \in \sigma(A, B) : \max\limits_{i=1,2} \operatorname{Re} \omega_i = n \geq 0$, $\min\limits_{i=1,2} \operatorname{Re} \omega_i = -n \leq 0$. By Theorem 8.4, equation (1) has two-sided Fatou-Riesz property.

Here $S = \mathbb{Z} \setminus \{0\}$; $\overline{S} = S$; so, by Lemma 3.8: $\sigma(P) = \mathbb{Z} \setminus \{0\}$.

Example 8.4. Let $H = l_2(C^1)$, $A = (a_{kl})_{k,l=1}^{\infty}$ and $B = (b_{kl})_{k,l=1}^{\infty}$ be linear operators in H defined by diagonal matrices, where $a_{kl} = b_{kl} = 0$ for $k \neq l$; $a_{kl} = -(2k + 1)$, $b_{kl} = k(k + 1)$ for $k = l$, if k is odd; $a_{kl} = 2k - 1$, $b_{kl} = k(k - 1)$ for $k = l$, if k is even.

Then $\sigma(A, B) = \{(\lambda_n, \mu_n) | n \in \mathbb{N},\ \lambda_n = a_{nn},\ \mu_n = b_{nn}\}$. For an arbitrary odd $n \in \mathbb{N}$, we have for $(\lambda_n, \mu_n) \in \sigma(A, B) : \{\omega_1, \omega_2\} = \{n, n + 1\}$, and so $\min\limits_{i=1,2} \operatorname{Re} \omega_i = n$. For an arbitrary even $n \in \mathbb{N}$, we have for $(\lambda_n, \mu_n) \in \sigma(A, B)$: $\{\omega_1, \omega_2\} = \{-(n - 1), -n\}$, and so $\max\limits_{i=1,2} \operatorname{Re} \omega_i = -(n - 1)$.

In particular, for every odd $n \geq 1 : \max\limits_{i=1,2} \operatorname{Re} \omega_i(\lambda, \mu) = -n$ for $(\lambda, \mu) = (\lambda_{n+1}, \mu_{n+1}) \in \sigma(A, B)$. By Theorem 8.1, equation (1) does not have Fatou-Riesz property.

Here $S = \mathbb{Z} \setminus \{0\}$; $\overline{S} = S$; so, by Lemma 3.8: $\sigma(P) = \mathbb{Z} \setminus \{0\}$.

Remark 1. In fact, we have proved that Statement 8.3 remains valid if one replaces «c.n.o.» by «c.s.o.». This distinguishes Statement 8.3 from Statement 3.2 which is not valid if one replaces «c.n.o.» by «c.s.o.» (see Statement 3.3).

Chapter 9
Extension of weak solutions

Let $0 < T < +\infty$, A and B be c.n.o. in H. In the previous chapter, we have answered the following question: what must be A and B for *each bounded* weak solution of equation (1) on $[0,T)$ to have a limit in H as $t \to T$?

In the present chapter, we answer the following question: what must be A and B for *each* weak solution of equation (1) on $[0,T)$ to have a limit in H as $t \to T$?

1 Extension of weak solutions on a finite interval

Theorem 9.1. *Let $0 < T < +\infty$, A and B be c.n.o. in H. The following conditions are equivalent:*

1) *each weak solution of equation* (1) *on* $[0,T)$ *has a limit in* H *as* $t \to T$;

2) $\exists 0 < \gamma < +\infty \forall (\lambda,\mu) \in \sigma(A,B) : \max_{i=1,2} \operatorname{Re} \omega_i(\lambda,\mu) \leq \gamma;$ (2.2)

3) $\exists 0 < \gamma < +\infty : \sigma(A,B) \subseteq \{(\lambda,\mu) \in \mathbf{C}^2 | \lambda_1 > -2\gamma,$

$$\mu_1 > \frac{1}{(\lambda_1+2\gamma)^2}(\mu_2^2 - \lambda_1\lambda_2 \cdot \mu_2 - \gamma(\lambda_1+\gamma)(\lambda_2^2 + (\lambda_1+2\gamma)^2))\}. \tag{3.3}$$

If (2.2) *holds then any weak solution of equation* (1) *on* $[0,T)$ *can be extended up to a weak solution of* (1) *on the whole* $R_+ = [0,+\infty)$.

Proof. The sufficiency of (2.2) follows immediately from Theorem 7.2. Indeed, let $y(t)$ be a weak solution of (1) on $[0,T)$; then by Theorem 7.2: $y(t) = e^{\omega_1 t}(A,B)f_0 + \psi_1(A,B,t)g$ $(t \in [0,T))$ where $f_0 \in H$, $g \in G_1$. Furthermore, by Theorem 7.2: $e^{\omega_1 t}(A,B)f_0 + \psi_1(A,B,t)g$ $(t \in [0,+\infty))$ is a weak solution of (1) on $R_+ = [0,+\infty)$; so, $y(t)$ $(t \in [0,T))$ can be extended up to a weak solution of (1) on $[0,+\infty)$. In particular, it has a limit in H as $t \to T$.

Let's prove the necessity.

Denote: $\Delta_1 = \{(\lambda,\mu) \in \mathbf{C}^2 | \operatorname{Re}\omega_1 \geq 0\}$, $\Delta_2 = \{(\lambda,\mu) \in \mathbf{C}^2 | \operatorname{Re}\omega_1 < 0\}$. Let for each of $i = 1,2 : H_i = E(\Delta_i)H$; A_i and B_i be the restrictions of A and B to H_i. Obviously, A_i and B_i are c.n.o. in H_i, and $\sigma(A_i,B_i) \subseteq \sigma(A,B) \cap \overline{\Delta_i}$.

It suffices to prove (2.2) with $\sigma(A,B)$ replaced by $\sigma(A_1,B_1)$; indeed,

$$\sigma(A,B) \subseteq \sigma(A_1,B_1) \cup \sigma(A_2,B_2),$$

and $\forall (\lambda,\mu) \in \sigma(A_2,B_2) : \operatorname{Re}\omega_1 \leq 0$.

Thus, it suffices to prove the necessity for the equation $y''(t) + A_1 y'(t) + B_1 y(t) = 0$ in H_1. For the sake of simplicity, we replace the notations A_1, B_1, H_1 by A, B, H respectively; now $\sigma(A,B) \subseteq \{(\lambda,\mu) \in C^2 | \operatorname{Re}\omega_1 \geq 0\}$.

Fix an arbitrary $f_0 \in H$ and consider $y_{f_0}(t) = ((\operatorname{Re}\omega_1)e^{\omega_1(t-T)})(A,B)f_0$, $t \in [0,T)$. It is easy to verify that $y_{f_0}(t)$ is a weak solution of (1) on $[0,T)$ (cf. the proof of Theorem 9.6). By the assumption, it has a limit in H as $t \to T$. It is easy to see that if this is the case then this limit is $(\operatorname{Re}\omega_1)(A,B)f_0$. Thus, $(\operatorname{Re}\omega_1)(A,B)f_0 \in H$.

We have obtained that the normal operator $(\operatorname{Re}\omega_1)(A,B)$ is defined on the whole H. By the Banach closed graph theorem, $(\operatorname{Re}\omega_1)(A,B)$ is a bounded operator on H, i.e., $\sup_{(\lambda,\mu)\in\sigma(A,B)} |(\operatorname{Re}\omega_1)(\lambda,\mu)| = \| (\operatorname{Re}\omega_1)(A,B) \| < +\infty$.

Corollary 1. *Let $0 < T_1 < T_2 \leq +\infty$. Each weak solution of equation (1) on $[0,T_1]$ can be extended up to a weak solution of (1) on $[0,T_2]$ if and only if (2.2) (or, what is the same, (3.3)) holds.*

2 Boundedness of weak solutions on a finite interval

Let $0 < T < +\infty$, A and B be c.n.o. in H. We have proved that (2.2) is sufficient for each weak solution of (1) on $[0,T)$ to be extendable up to a weak solution of (1) on R_+.

On the other hand, (2.2) is necessary even for each weak solution of (1) on $[0,T)$ to have a limit in H as $t \to T$.

We now prove the stronger assertion: (2.2) is necessary even for each weak solution of (1) on $[0,T)$ to be bounded on $[0,T)$.

Theorem 9.2. *Let $0 < T < +\infty$, A and B be c.n.o. in H. Each weak solution of equation (1) on $[0,T)$ is bounded if and only if (2.2) (or, what is the same, (3.3)) holds. If this is the case then any weak solution of (1) on $[0,T)$ can be extended up to a weak solution of (1) on the whole $R_+ = [0,+\infty)$.*

Proof. The proof of Theorem 9.2 coincides with the proof of Theorem 9.1 until the words «By the assumption». By the assumption, $y_{f_0}(t) = ((\operatorname{Re}\omega_1)e^{\omega_1(t-T)})(A,B)f_0$ is bounded on $[0,T)$.

But for any $t \in [0,T)$:

$$\| ((\operatorname{Re}\omega_1)e^{\omega_1(t-T)})(A,B) \| = \sup_{(\lambda,\mu)\in\sigma(A,B)} |(\operatorname{Re}\omega_1)e^{\omega_1(t-T)}|$$
$$\leq \sup_{x\geq 0} |xe^{x(t-T)}| < +\infty.$$

Thus, we have the family of bounded linear operators $((\operatorname{Re}\omega_1)e^{\omega_1(t-T)})(A,B)$ $(t \in [0,T))$ on H, and for each $f_0 \in H$: the actions of these operators on f_0

are uniformly bounded. By virtue of the Banach-Steinhaus theorem on uniform boundedness, these operators are uniformly bounded, i.e.,

$$\sup_{t\in[0,T)} \| ((\mathrm{Re}\,\omega_1)e^{\omega_1(t-T)})(A,B) \| < +\infty.$$

But

$$\sup_{t\in[0,T)} \| ((\mathrm{Re}\,\omega_1)e^{\omega_1(t-T)})(A,B) \| = \sup_{t\in[0,T)} \sup_{(\lambda,\mu)\in\sigma(A,B)} |(\mathrm{Re}\,\omega_1)e^{\omega_1(t-T)}|$$

$$= \sup_{(\lambda,\mu)\in\sigma(A,B)} \sup_{t\in[0,T)} |(\mathrm{Re}\,\omega_1)e^{\omega_1(t-T)}| = \sup_{(\lambda,\mu)\in\sigma(A,B)} |\mathrm{Re}\,\omega_1|.$$

We have proved that $\sup_{(\lambda,\mu)\in\sigma(A,B)} \mathrm{Re}\,\omega_1 < +\infty$, i.e., (2.2) holds.

Remark 1. In Theorems 9.4, 9.5 and Statement 9.1 (see below), the words «has a limit in H» («has limits in H») can be replaced by the words «is bounded» as well.

3 Exponential growth of weak solutions

Theorem 9.3. *Let A and B be c.n.o. in H. The following conditions are equivalent:*

1) *there exists $0 < \gamma_1 < +\infty$ such that for every weak solution $y(t)$ of equation (1) on R_+ there exists $0 \le C_y < +\infty$ such that*

$$\| y(t) \| \le C_y e^{\gamma_1 t}, \quad \forall t \in R_+;$$

2) *(2.2) holds;*
3) *(3.3) holds.*

Proof. The sufficiency of (2.2) follows immediately from Theorem 7.2. Indeed, let (2.2) hold. By Theorem 7.2, any weak solution of (1) on R_+ has the form: $y(t) = e^{\omega_1 t}(A,B)f_0 + \psi_1(A,B,t)g$ where $f_0 \in H$, $g \in H_-(((d^2+b_1^2+b_2^2+1)^{1/2}(|d|+1)^{1/2})(A,B))$. For any $(\lambda,\mu) \in \sigma(A,B) : d = -\max_{i=1,2} \mathrm{Re}\,\omega_i = -\mathrm{Re}\,\omega_1 \ge -\gamma$. Hence $\| e^{\omega_1 t}(A,B)f_0 \| \le e^{\gamma t} \| f_0 \|$, $\forall t \in R_+$. Furthermore, $\forall(\lambda,\mu) \in \sigma(A,B)$, $t \in R_+$:

$$|\psi_1(\lambda,\mu,t)|^2(d^2+b_1^2+b_2^2+1)(|d|+1)$$

$$= e^{-2dt}\left|\frac{e^{(\omega_2-\omega_1)t}-1}{\omega_2-\omega_1}\right|^2 (d^2+b_1^2+b_2^2+1)(|d|+1).$$

Taking into account that $|\omega_1 - \omega_2|^2 = 4(b_1^2 + b_2^2)$, and $\max_{z \in C^1, |z| \leq 1} |\frac{e^z - 1}{z}| < 2$ (see the proof of Lemma 2.2), we obtain:

$$|\psi_1(\lambda, \mu, t)|^2 (d^2 + b_1^2 + b_2^2 + 1)(|d| + 1)$$
$$\leq (1 + 4t^2(d^2 + 1))(|d| + 1)e^{-2dt}, \; \forall (\lambda, \mu) \in \sigma(A, B), \quad t \in R_+.$$

Hence $\forall \delta > 0 \exists C_\delta < +\infty \forall g \in H_-(((d^2 + b_1^2 + b_2^2 + 1)^{1/2}(|d| + 1)^{1/2})(A, B))$:

$$\| \psi_1(A, B, t)g \| \leq C_\delta e^{(\gamma+\delta)t} \| g \|_{H_-(((d^2+b_1^2+b_2^2+1)^{1/2}(|d|+1)^{1/2})(A,B))}, \; \forall t \geq 1$$

(cf. the proof of the sufficiency of Theorem 10.1).

Let's prove the necessity of (2.2) by contradiction. Let $\exists (\lambda_0, \mu_0) \in \sigma(A, B)$: $\operatorname{Re} \omega_1(\lambda_0, \mu_0) > \gamma_1$. Then $\exists 0 < \delta < +\infty$: $\operatorname{Re} \omega_1(\lambda_0, \mu_0) - \delta > \gamma_1$. Set $\Delta = \{(\lambda, \mu) \in C^2 | \; |\lambda - \lambda_0|^2 + |\mu - \mu_0|^2 < 1\} \cap \{(\lambda, \mu) \in C^2 | \; |\operatorname{Re} \omega_1(\lambda, \mu) - \operatorname{Re} \omega_1(\lambda_0, \mu_0)| < \delta\}$. Δ is an open bounded set in C^2. Since $\Delta \ni (\lambda_0, \mu_0)$, and $(\lambda_0, \mu_0) \in \sigma(A, B)$, we conclude that $E(\Delta) \neq 0$. Fix an arbitrary $f_0 \in E(\Delta)H$ such that $\| f_0 \| = 1$; this is possible since $E(\Delta) \neq 0$. Then $y(t) = e^{\omega_1 t}(A, B) f_0$ is a solution of (1) on R_+ and consequently a weak solution of (1) on R_+. Set $\gamma_2 = \operatorname{Re} \omega_1(\lambda_0, \mu_0) - \delta$; then $\gamma_2 > \gamma_1$, and $\forall t \in R_+$:

$$\| y(t) \|^2 = \int_\Delta e^{2 \operatorname{Re} \omega_1 t} d(E f_0, f_0) \geq e^{2\gamma_2 t} \| f_0 \|^2 = e^{2\gamma_2 t} \Rightarrow \| y(t) \| \geq e^{\gamma_2 t}.$$

Thus, there is no $C_y < +\infty$ such that $\forall t \in R_+ : \| y(t) \| \leq C_y e^{\gamma_1 t}$; so 1) does not hold.

We have proved that if 1) holds then $\forall (\lambda, \mu) \in \sigma(A, B) : \operatorname{Re} \omega_1 \leq \gamma_1$.

Corollary 1. *Let A and B be c.n.o. in H such that $P = \sigma(A, B)$ satisfies* (2) *(i.e.,* (3.3) *and* (3.4) *holds). There exist $0 < C < +\infty$, $0 < \gamma_1 < +\infty$ such that for any $(f_0, f_1) \in H \times H_{-1}$: there exists a unique weak solution $y(t)$ of equation* (1) *on R_+ with i.d. (f_0, f_1), and*

$$\| y(t) \| \leq C(\| f_0 \|_H + \| f_1 \|_{H_{-1}}) e^{\gamma_1 t}, \quad t \in R_+.$$

Proof. In view of Theorem 3.1, it remains to prove the estimate for $\| y(t) \|$. It follows from Theorem 9.3 similarly to the proof of the sufficiency of Theorem 10.2 below, with $\psi_i(A, B, t)$ replaced by $e^{-\gamma_1 t} \psi_i(A, B, t)$.

4 Two-sided extension of weak solutions

1. Just as previously, one can obtain the following result.

Theorem 9.4. *Let* $0 < T < +\infty$, *A and B be c.n.o. in H. The following conditions are equivalent:*

1) *each weak solution of equation* (1) *on* $(0, T]$ *has a limit in H as* $t \to 0$;

2) $\exists 0 < \gamma < +\infty \forall (\lambda, \mu) \in \sigma(A, B): \min_{i=1,2} \operatorname{Re} \omega_i(\lambda, \mu) \geq -\gamma$; (9.1)

3) $\exists 0 < \gamma < +\infty : \sigma(A, B) \subseteq \{(\lambda, \mu) \in C^2 | \lambda_1 < 2\gamma,$

$$\mu_1 > \frac{1}{(\lambda_1 - 2\gamma)^2}(\mu_2^2 - \lambda_1\lambda_2 \cdot \mu_2 + \gamma(\lambda_1 - \gamma)(\lambda_2^2 + (\lambda_1 - 2\gamma)^2))\}.$$

If (9.1) *holds then any weak solution of* (1) *on* $(0, T]$ *can be extended up to a weak solution of* (1) *on* $(-\infty, T]$.

Theorems 9.1 and 9.4 imply the following theorem.

Theorem 9.5. *Let* $-\infty < a < b < +\infty$, *A and B be c.n.o. in H. The following conditions are equivalent:*

1) *each weak solution of equation* (1) *on* (a, b) *has limits in H as* $t \to a$ *and* $t \to b$;

2) $\exists 0 < \gamma < +\infty \forall (\lambda, \mu) \in \sigma(A, B), i = 1, 2 : -\gamma \leq \operatorname{Re} \omega_i \leq \gamma$; (9.2)

3) $\exists 0 < \gamma < +\infty : \sigma(A, B) \subseteq \{(\lambda, \mu) \in C^2 | \, |\lambda_1| < 2\gamma,$

$$\mu_1 > \frac{1}{(|\lambda_1| - 2\gamma)^2}\left(\mu_2 - \frac{\lambda_1\lambda_2}{2}\right)^2 - \frac{\lambda_2^2}{4} + |\lambda_1|\gamma - \gamma^2\}.$$

If (9.2) *holds then any weak solution of* (1) *on* (a, b) *can be extended up to a weak solution of* (1) *on the whole* $R^1 = (-\infty, +\infty)$.

2. Corollary 1. *Let* $-\infty < a < b < +\infty$, *A and B be c.n.o. in H, and moreover, B be subordinate to A (i.e.,* $D(A) \subseteq D(B)$*). Then:*

1) *each weak solution of* (1) *on* $[a, b)$ *has a limit in H as* $t \to b$ *if and only if* Re *A is semibounded below;*

2) *each weak solution of* (1) *on* $(a, b]$ *has a limit in H as* $t \to a$ *if and only if* Re *A is semibounded above;*

3) *each weak solution of* (1) *on* (a, b) *has limits in H as* $t \to a$ *and* $t \to b$ *if and only if* Re *A is bounded.*

The proof is analogous to the proof of Corollary 3 of Theorem 3.1.

3. Let A and B be commuting self-adjoint operators in H. For this special case, conditions (2.2) and (9.1) are clarified in Chapter 2 (see Theorems 2.1 and 2.7). See also Examples 2.1 and 2.4. Condition (9.2) holds if and only if A is bounded, B is semibounded below. If A and B are the operators from Example 1.1, then (9.2) holds if and only if $a_1 = a_2 = 0$, $b_1 \geq 0$, $b_2 \geq 0$.

5 Spectrum of the associated operator pencil

Statement 9.1. Let $-\infty < a < b < +\infty$, A and B be c.n.o. in H. Then:

1) each weak solution of equation (1) on $[a, b)$ has a limit in H as $t \to b$ if and only if

$$\exists 0 < \gamma < +\infty : \sigma(P) \subseteq \{z \in C^1 | \operatorname{Re} z \leq \gamma\};$$

2) each weak solution of (1) on $(a, b]$ has a limit in H as $t \to a$ if and only if

$$\exists 0 < \gamma < +\infty : \sigma(P) \subseteq \{z \in C^1 | \operatorname{Re} z \geq -\gamma\};$$

3) each weak solution of (1) on (a, b) has limits in H as $t \to a$ and $t \to b$ if and only if

$$\exists 0 < \gamma < +\infty : \sigma(P) \subseteq \{z \in C^1 | -\gamma \leq \operatorname{Re} z \leq \gamma\}.$$

Here 1) follows from Theorem 9.1 and Lemma 3.9; 2) and 3) can be proved analogously.

Remark 1. Let $0 < T < +\infty$. Consider the following condition on equation (1) with c.n.o. A and B in H: each weak solution of (1) on $[0, T)$ has a limit in H as $t \to T$. We have shown that this condition can be expressed in terms of the position in C^1 of the spectrum $\sigma(P)$ of the associated operator pencil $P(z) = z^2 + Az + B$. This distinguishes the last condition from the weaker condition that each *bounded* weak solution of (1) on $[0, T)$ has a limit in H as $t \to T$, as well as from the stronger condition that the Cauchy problem for equation (1) on $[0, T)$ is weakly well-posed in $G = H \times H_{-1}$.

6 Comparison of the results on extension of weak solutions and bounded weak solutions

Let's compare the results obtained in Chapter 8 for the set of bounded weak solutions of equation (1) on an interval, on one hand, and the results obtained in Chapter 9 for the set of all weak solutions of (1) on this interval, on the other hand. In order to illustrate these results, we give a figure for each condition. In this figure, we show in what set must $\sigma(A, B)$ be contained in the special case of A and B self-adjoint (and, consequently, $\sigma(A, B) \subseteq R^2$) for the condition to hold.

Let $-\infty < a < b < +\infty$, A and B be c.n.o. in H.

Each bounded weak solution of equation (1) on $(a, b]$ has a limit in H as $t \to a$ if and only if

$$\exists 0 < \gamma < +\infty \forall (\lambda, \mu) \in \sigma(A, B) : \max_{i=1,2} \operatorname{Re} \omega_i \geq -\gamma$$

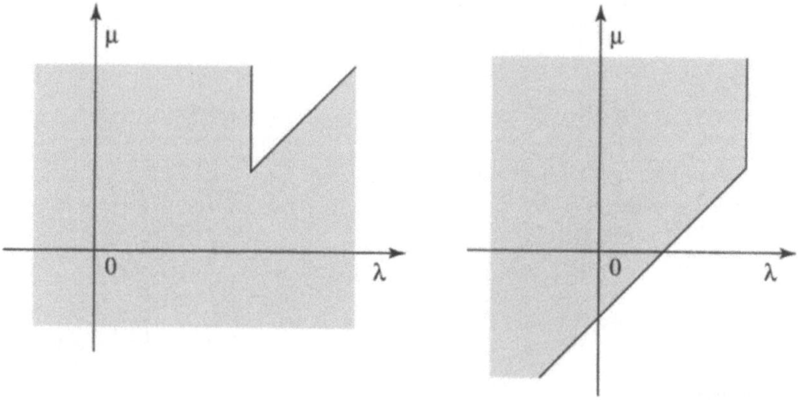

Fig. 9.1a, 9.1b

(see Fig. 9.1a); each weak solution of (1) on $(a, b]$ has a limit in H as $t \to a$ if and only if

$$\exists 0 < \gamma < +\infty \forall (\lambda, \mu) \in \sigma(A, B) : \min_{i=1,2} \operatorname{Re} \omega_i \geq -\gamma$$

(see Fig. 9.1b).

Each bounded weak solution of (1) on $[a, b)$ has a limit in H as $t \to b$ if and only if

$$\exists 0 < \gamma < +\infty \forall (\lambda, \mu) \in \sigma(A, B) : \min_{i=1,2} \operatorname{Re} \omega_i \leq \gamma$$

(see Fig. 9.2a); each weak solution of (1) on $[a, b)$ has a limit in H as $t \to b$ if and only if

$$\exists 0 < \gamma < +\infty \forall (\lambda, \mu) \in \sigma(A, B) : \max_{i=1,2} \operatorname{Re} \omega_i \leq \gamma$$

(see Fig. 9.2b).

Each bounded weak solution of (1) on (a, b) has limits in H as $t \to a$ and $t \to b$ if and only if

$$\exists 0 < \gamma < +\infty \forall (\lambda, \mu) \in \sigma(A, B) : \min_{i=1,2} \operatorname{Re} \omega_i \leq \gamma, \ \max_{i=1,2} \operatorname{Re} \omega_i \geq -\gamma$$

(see Fig. 9.3a); each weak solution of (1) on (a, b) has limits in H as $t \to a$ and $t \to b$ if and only if

$$\exists 0 < \gamma < +\infty \forall (\lambda, \mu) \in \sigma(A, B) : -\gamma \leq \operatorname{Re} \omega_i \leq \gamma, \quad i = 1, 2$$

(see Fig. 9.3b).

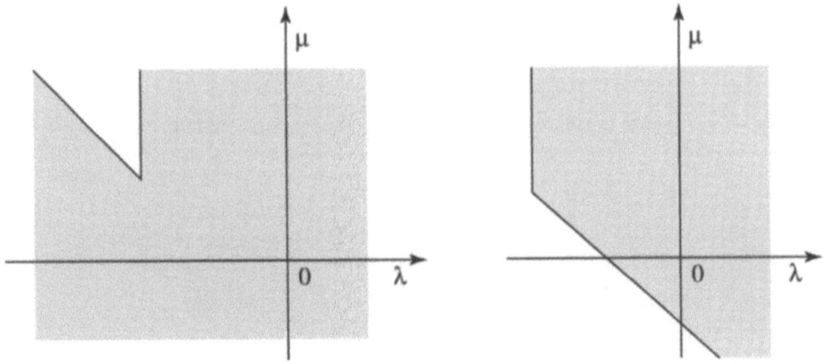

Fig. 9.2a, 9.2b

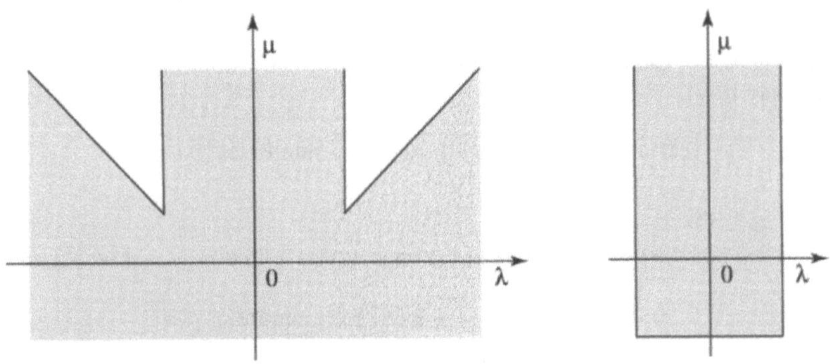

Fig. 9.3a, 9.3b

7 Intermediate classes of weak solutions

Let $-\infty < a < b < +\infty$. Consider the interval $[a,b)$ (the intervals $(a,b]$ and (a,b) can be considered analogously).

Previously, we studied questions of the following form: what must be c.n.o. A and B in H for each weak solution of equation (1) on $[a,b)$ from a certain class to have a limit in H as $t \to b$?

We considered two classes of weak solutions, namely, the class of *all bounded weak* solutions of (1) of $[a,b)$ and the class of *all weak* solutions of (1) on $[a,b)$.

There are many intermediate classes between these two classes. For each of these intermediate classes, one can pose the same question. In this section, we establish that even for classes closely related to the class of all bounded weak solutions, the answer to this question is the same as for the class of all weak solutions.

Let $\alpha \in R^1 = (-\infty, +\infty)$. The question is: what must be c.n.o. A and B in H for each weak solution of equation (1) on $[a, b)$ $y(t)$ such that

$$\exists C_y < +\infty : \| y(t) \| \leq C_y (b-t)^{-\alpha} \quad (t \in [a, b))$$

to have a limit in H as $t \to b$?

If $\alpha < 0$, then this is the case for arbitrary c.n.o. A and B in H. Indeed, if $\| y(t) \| \leq C_y (b-t)^{-\alpha}$ $(t \in [a, b))$ with $\alpha < 0$, then $y(t) \to 0$ in H as $t \to b$.

If $\alpha = 0$, then this is the case if and only if

$$\exists 0 < \gamma < +\infty \forall (\lambda, \mu) \in \sigma(A, B) : \min_{i=1,2} \operatorname{Re} \omega_i \leq \gamma.$$

This was shown in Theorem 8.3.

Theorem 9.6. *Let* $0 < \alpha < +\infty$, *A and B be c.n.o. in H. Each weak solution of equation* (1) *on* $[a, b)$ $y(t)$ *such that*

$$\exists C_y < +\infty : \| y(t) \| \leq C_y (b-t)^{-\alpha} \quad (t \in [a, b))$$

has a limit in H as t → b if and only if

$$\exists 0 < \gamma < +\infty \forall (\lambda, \mu) \in \sigma(A, B) : \max_{i=1,2} \operatorname{Re} \omega_i \leq \gamma \tag{2.2}$$

(i.e., (3.3) *is valid). On the other hand, if* (2.2) *holds then each weak solution of* (1) *on* $[a, b)$ *has a limit in H as t → b.*

Proof. For $\alpha = 1$, the proof coincides with the proof of Theorem 9.1, with 0 and T replaced by a and b respectively. In order to show this, we must verify that if $\sigma(A, B) \subseteq \{(\lambda, \mu) \in C^2 | \operatorname{Re} \omega_1 \geq 0\}$, then for any $f_0 \in H$ and the weak solution of (1) on $[a, b)$

$$y_{f_0}(t) = ((\operatorname{Re} \omega_1) e^{\omega_1 (t-b)})(A, B) f_0 \quad (t \in [a, b))$$

there exists $C_{f_0} < +\infty$ such that $\| y_{f_0}(t) \| \leq C_{f_0} (b-t)^{-1}$ $(t \in [a, b))$.

Indeed, let $\sigma(A, B) \subseteq \{(\lambda, \mu) \in C^2 | \operatorname{Re} \omega_1 \geq 0\}$, and $a \leq t < b$; then

$$\| ((\operatorname{Re} \omega_1)(b-t) e^{\omega_1 (t-b)})(A, B) \| = \sup_{(\lambda, \mu) \in \sigma(A, B)} | \operatorname{Re} \omega_1 (b-t) e^{\omega_1 (t-b)} |$$

$$= \sup_{(\lambda, \mu) \in \sigma(A, B)} | \operatorname{Re} \omega_1 (b-t) e^{\operatorname{Re} \omega_1 (t-b)} | \leq \sup_{x \geq 0} (x e^{-x}) = e^{-1} < 1.$$

Thus, $\forall f_0 \in H$, $t \in [a, b) : \| y_{f_0}(t) \| \leq \| ((\operatorname{Re} \omega_1) e^{\omega_1 (t-b)})(A, B) \| \cdot \| f_0 \| \leq (b-t)^{-1} \cdot \| f_0 \|$.

For an arbitrary $0 < \alpha < +\infty$, the proof is similar, with

$$y_{f_0}(t) = ((\operatorname{Re} \omega_1) e^{\omega_1 (t-b)})(A, B) f_0$$

replaced by

$$y_{f_0}(t) = ((\operatorname{Re} \omega_1)^{\alpha} e^{\omega_1 (t-b)})(A, B) f_0.$$

Remark 1. Let's consider the first order equation $y'(t)+Ay(t)=0$ with a normal operator A in H, and pose the same question: what must be A for each weak solution of this equation on $[a,b)$ $y(t)$ such that

$$\exists C_y<+\infty : \| y(t) \| \leq C_y(b-t)^{-\alpha} \quad (t\in[a,b))$$

to have a limit in H as $t\to b$?

If $\alpha<0$, then this is the case for an arbitrary A, similarly to second order equations.

If $\alpha>0$, then this is the case if and only if

$$\exists 0<\gamma<+\infty \forall \lambda\in\sigma(A) : \ \mathrm{Re}\,\omega\leq\gamma,$$

similarly to second order equations.

Difference between first and second order equations exists in the case of $\alpha=0$. What must be an equation for each its weak solution on $[a,b)$ $y(t)$ such that $\exists C_y<+\infty : \| y(t) \| \leq C_y$ $(t\in[a,b))$ to have a limit in H as $t\to b$? For first order equations, this is the case for an arbitrary normal operator A, just as in the case of $\alpha<0$. For second order equation (1) with c.n.o. A and B in H, this is the case if and only if

$$\exists 0<\gamma<+\infty \forall(\lambda,\mu)\in\sigma(A,B) : \min_{i=1,2}\mathrm{Re}\,\omega_i\leq\gamma.$$

Theorem 9.7. *Let $0<\alpha<+\infty$, A and B be c.n.o. in H. Each weak solution of equation (1) on $[a,b)$ $y(t)$ such that $y(t)\in L_\alpha([a,b),H)$ (i.e., $\int_a^b \| y(t) \|^\alpha dt<+\infty$) has a limit in H as $t\to b$ if and only if (2.2) (or, what is the same, (3.3)) holds.*

The proof is similar to the proof of Theorem 9.6.

8 Extension of weak solutions and weak well-posedness of boundary-value problems

Taking into account Theorems 3.1, 4.1, 4.8, we obtain the following statement.

Statement 9.2. Let $-\infty<a<b<+\infty$, A and B be c.n.o. in H. Then:

1) if the Cauchy problem for equation (1) on $[a,b]$ is weakly well-posed in $G=H\times H_{-1}$, then each weak solution of (1) on $[a,b)$ can be extended up to a weak solution of (1) on $[a,+\infty)$;

2) if the Dirichlet problem for (1) on $[a,b]$ is weakly well-posed in $G=H\times H$, then each bounded weak solution of equation (1) on (a,b) can be extended up to a weak solution of (1) on $[a,b]$;

3) if the inverse Cauchy problem for (1) on $[a,b]$ is weakly well-posed in $G=H\times H_{-1}$, then each weak solution of (1) on $(a,b]$ can be extended up to a weak solution of (1) on $(-\infty,b]$.

Notice that for each of these three statements the converse implication is not valid (i.e., the first condition is strictly stronger than the second one).

Chapter 10
Stability and stabilization of weak solutions

1 Stability and uniform stability of an equation

Definition 10.1. Let A and B be c.n.o. in H. Equation (1) is said to be stable on R_+ if for every weak solution $y(t)$ of (1) on R_+:

$$\sup_{t\in R_+} \| y(t) \| < +\infty$$

(i.e., $\exists C_y < +\infty : \| y(t) \| \leq C_y, \forall t \in R_+$).

Theorem 10.1. *Let A and B be c.n.o. in H. Equation (1) is stable on R_+ if and only if one of the following equivalent conditions 1), 2), 3) holds:*

1) *a)* $\forall(\lambda,\mu) \in \sigma(A,B) : \max_{i=1,2} \operatorname{Re}\omega_i \leq 0,$ (10.1)

b) $\exists 0 < \delta < +\infty : \sigma(A,B) \subseteq \{(\lambda,\mu) \in C^2 |\, |\omega_1 - \omega_2| \geq \delta\} \cup \{(\lambda,\mu) \in C^2 | \max_{i=1,2} \operatorname{Re}\omega_i \leq -\delta\};$ (10.2)

2) $\exists 0 < \delta < +\infty : \sigma(A,B) \subseteq \{(\lambda,\mu) \in C^2 | d \geq 0,\ d^2 + b_1^2 + b_2^2 \geq \delta^2\};$ (10.3)

3) *a)* $\sigma(A,B) \subseteq \{(\lambda,\mu) \in C^2 | \lambda_1 > 0,\ \mu_1 \geq \frac{\mu_2}{\lambda_1^2}(\mu_2 - \lambda_1\lambda_2)\} \cup \{(\lambda,\mu) \in C^2 | \lambda_1 = 0,\ \mu_2 = 0,\ \mu_1 \geq -\frac{\lambda_2^2}{4}\},$ (10.4)

b) $\exists 0 < \delta < +\infty : \sigma(A,B) \subseteq \{(\lambda,\mu) \in C^2 |\, |\mu - \frac{\lambda^2}{4}| \geq \delta^2\} \cup \{(\lambda,\mu) \in C^2 | \lambda_1 \geq \delta\}.$ (10.5)

Proof. It is clear that 2) is equivalent to 1), while 3b) is equivalent to 1b). Further, for any $\gamma \in R^1$:

$$\{(\lambda,\mu) \in C^2 | \max_{i=1,2} \operatorname{Re}\omega_i < \gamma\} = \{(\lambda,\mu) \in C^2 | \lambda_1 > -2\gamma,$$

$$\mu_1 > (\lambda_1 + 2\gamma)^{-2}(\mu_2^2 - \lambda_1\lambda_2 \cdot \mu_2 - \gamma(\lambda_1+\gamma)(\lambda_2^2 + (\lambda_1 + 2\gamma)^2))\};$$

taking into account that

$$\{(\lambda,\mu) \in C^2 | \max_{i=1,2} \operatorname{Re}\omega_i \leq 0\} = \bigcap_{\gamma>0} \{(\lambda,\mu) \in C^2 | \max_{i=1,2} \operatorname{Re}\omega_i < \gamma\},$$

we conclude that 1a) is equivalent to 3a).

Set $\gamma_1 = 0$. Let equation (1) be stable on R_+. By the definition, for every weak solution $y(t)$ of (1) on R_+ there exists $C_y < +\infty$ such that $\| y(t) \| \leq C_y e^{\gamma_1 t}$

$(t \in R_+)$. Just as in the proof of the necessity of Theorem 9.3, hence we obtain that $\forall(\lambda,\mu) \in \sigma(A,B) : \max_{i=1,2} \operatorname{Re}\omega_i \le \gamma_1 = 0$, i.e., 1a) is valid.

Then, by virtue of Theorem 7.2, for every $g \in H \subseteq H_-((a^2+d^2+b_2^2+1)^{1/2}(A,B)) : y(t) = \psi_1(A,B,t)g$ $(t \in R_+)$ is a weak solution of (1) on R_+. Since equation (1) is stable on R_+, we conclude that, in particular, for each $g \in H$: $y(t) = \psi_1(A,B,t)g$ $(t \in R_+)$ is bounded on R_+.

Thus, for any $t \in R_+$: the normal operator $\psi_1(A,B,t)$ is defined on the whole H; by the Banach closed graph theorem, $\psi_1(A,B,t)$ is a bounded linear operator. For each $g \in H$, $\exists C_g < +\infty :\| \psi_1(A,B,t)g \| \le C_g$, $\forall t \in R_+$. By the Banach-Steinhaus theorem on uniform boundedness, the linear bounded operators $\psi_1(A,B,t)$ $(t \in R_+)$ are uniformly bounded.

Thus, $\exists C < +\infty \forall t \in R_+ : \| \psi_1(A,B,t) \| = \sup_{(\lambda,\mu)\in\sigma(A,B)} |\psi_1(\lambda,\mu,t)| \le C.$

So,

$$\exists C < +\infty \forall (\lambda,\mu) \in \sigma(A,B),\ t \in R_+ : \\ |\psi_1(\lambda,\mu,t)| = e^{-at}\left|\frac{\sinh bt}{b}\right| \le C, \tag{10.6}$$

where $\forall(\lambda,\mu) \in \sigma(A,B) : a = -\frac{1}{2}\operatorname{Re}(\omega_1+\omega_2) \ge 0$, $b = \frac{1}{2}(\omega_1-\omega_2)$. We now show that $\exists 0 < \delta < +\infty \forall(\lambda,\mu) \in \sigma(A,B) : a^2+|b|^2 \ge \delta^2$.

Set $C_1 = \min_{z\in C^1,|z|\le 1} |\frac{\sinh z}{z}|$; then $C_1 > 0$.

1. For each $(\lambda,\mu) \in \sigma(A,B) : a^2+|b|^2 > 0$. Indeed, if $a = b = 0$ then $\forall t \in R_+ : |\psi_1(\lambda,\mu,t)| = t$, which contradicts (10.6).

2. $\exists \delta_1 > 0 \forall(\lambda,\mu) \in \sigma(A,B) \cap \{(\lambda,\mu) \in C^2 | a \ge |b|\} : a \ge \delta_1$.

Indeed, otherwise there exists a sequence of $(\lambda_n,\mu_n) \in \sigma(A,B)$ such that $a_n \ge |b_n|$ and $a_n \to 0$ as $n \to +\infty$. For each $n \ge 1 : a_n > 0$. (Indeed, if $a_n = 0$ then $|b_n| \le a_n \Rightarrow |b_n| = 0$; but we have shown that $a_n = b_n = 0$ is impossible for $(\lambda_n,\mu_n) \in \sigma(A,B)$.)

For each $n \ge 1$, set $t_n = a_n^{-1}$. Then $t_n \to +\infty$ as $n \to +\infty$. We have:

$$|\psi_1(\lambda_n,\mu_n,t_n)| = e^{-a_nt_n}\left|\frac{\sinh(b_n/a_n)}{(b_n/a_n)}\right| t_n \ge e^{-1}C_1t_n \to +\infty \quad \text{as} \quad n \to +\infty.$$

This contradicts (10.6).

3. $\exists \delta_2 > 0 \forall(\lambda,\mu) \in \sigma(A,B) \cap \{(\lambda,\mu) \in C^2 |\ |b| \ge a\} : |b| \ge \delta_2$.

Indeed, otherwise there exists a sequence of $(\lambda_n,\mu_n) \in \sigma(A,B)$ such that $|b_n| \ge a_n$ and $|b_n| \to 0$ as $n \to +\infty$. For each $n \ge 1 : |b_n| > 0$. (Indeed, if $|b_n| = 0$ then $a_n \le |b_n| \Rightarrow a_n = 0$; but we have shown that $a_n = b_n = 0$ is impossible for $(\lambda_n,\mu_n) \in \sigma(A,B)$.)

For each $n \ge 1$, set $t_n = |b_n|^{-1}$. Then $t_n \to +\infty$ as $n \to +\infty$. We have:

$$|\psi_1(\lambda_n,\mu_n,t_n)| = e^{-a_n|b_n|^{-1}}\left|\frac{\sinh(b_nt_n)}{b_nt_n}\right| t_n \ge e^{-1}C_1t_n \to +\infty \text{ as } n \to +\infty.$$

This contradicts (10.6).

So, $\exists 0 < \delta < +\infty \forall (\lambda, \mu) \in \sigma(A, B) : a^2 + |b|^2 \geq \delta^2$. By the definitions, $\forall (\lambda, \mu) \in \sigma(A, B) : |b|^2 = b_1^2 + b_2^2$; $a = d + b_1$. Thus, $\exists 0 < \delta < +\infty \forall (\lambda, \mu) \in \sigma(A, B) : d^2 + b_1^2 + b_2^2 \geq \delta^2$. Taking into account that $\forall (\lambda, \mu) \in \sigma(A, B) : |\omega_1 - \omega_2|^2 = 4(b_1^2 + b_2^2)$ and $\max_{i=1,2} \operatorname{Re} \omega_i = -d \leq 0$, we conclude that 1b) is valid.

Conversely, let 1) hold. Then, in particular, $\sigma(A, B) \subseteq \{(\lambda, \mu) \in C^2 | d \geq 0\}$. By Theorem 7.2, for every weak solution of (1) on R_+ $y(t)$ there exist $f_0 \in H$, $g \in G_1 \subseteq H_-(((d^2 + b_1^2 + b_2^2 + 1)^{1/2}(|d| + 1)^{1/2})(A, B))$ such that $y(t) = e^{\omega_1 t}(A, B) f_0 + \psi_1(A, B, t) g$ $(t \in R_+)$.

For any $(\lambda, \mu) \in \sigma(A, B)$, $t \in R_+ : |e^{\omega_1 t}| = e^{\operatorname{Re} \omega_1 t} \leq 1$, and thus, $\forall f_0 \in H :$ $\| e^{\omega_1 t}(A, B) f_0 \| \leq \| f_0 \|$, $\forall t \in R_+$.

Further, notice that $\forall (\lambda, \mu) \in \sigma(A, B) : |\omega_1 - \omega_2|^2 = 4(b_1^2 + b_2^2)$. Set $C_2 = \max_{x \geq 0}((x^2 + 1)(x + 1)e^{-x}) < +\infty$. Now we have for any $t \geq 1$ and any $(\lambda, \mu) \in \sigma(A, B)$ such that $|\omega_1 - \omega_2| \geq \delta : |\psi_1(\lambda, \mu, t)|^2(d^2 + b_1^2 + b_2^2 + 1)(|d| + 1) = e^{-2dt} \left| \frac{e^{(\omega_2 - \omega_1)t} - 1}{\omega_1 - \omega_2} \right|^2 (d^2 + b_1^2 + b_2^2 + 1)(d + 1) \leq (1 + 4\delta^{-2}(d^2 + 1))(d + 1)e^{-2dt} \leq (1 + 4\delta^{-2})((dt)^2 + 1)(dt + 1)e^{-2dt} \leq (1 + 4\delta^{-2})C_2 < +\infty$.

For any $t \geq 1$ and any $(\lambda, \mu) \in \sigma(A, B)$ such that $\operatorname{Re} \omega_1 \leq -\delta$, we have by virtue of the estimate obtained in the first paragraph of the proof of Lemma 2.2:

$$
\begin{aligned}
&|\psi_1(\lambda, \mu, t)|^2(d^2 + b_1^2 + b_2^2 + 1)(d + 1) \\
&= e^{-2dt} \left| \frac{e^{(\omega_2 - \omega_1)t} - 1}{\omega_1 - \omega_2} \right|^2 (d^2 + b_1^2 + b_2^2 + 1)(d + 1) \\
&\leq (1 + 4t^2(d^2 + 1))(d + 1)e^{-2dt} \leq 5C_2 t^2 e^{-\delta t} \leq 5C_2 \cdot 4\delta^{-2} e^{-2} < +\infty.
\end{aligned}
$$

Summarizing, $\exists C_3 = 5(1 + 4\delta^{-2})C_2 < +\infty$ such that $\forall (\lambda, \mu) \in \sigma(A, B)$, $t \geq 1 : |\psi_1(\lambda, \mu, t)|^2 \leq C_3(d^2 + b_1^2 + b_2^2 + 1)^{-1}(d + 1)^{-1}$. Hence $\forall g \in H_-(((d^2 + b_1^2 + b_2^2 + 1)^{1/2}(d + 1)^{1/2})(A, B))$:

$$
\| \psi_1(A, B, t) g \| \leq \sqrt{C_3} \, \| g \|_{H_-(((d^2 + b_1^2 + b_2^2 + 1)^{1/2}(d+1)^{1/2})(A, B))}, \quad \forall t \geq 1.
$$

The proof is now completed.

Remark 1. Consider the first order equation $y'(t) + Ay(t) = 0$ with a normal operator A in H. For each $\lambda \in C^1$, denote by ω the root of the characteristic polynomial $\omega + \lambda$. The equation $y'(t) + Ay(t) = 0$ is stable on R_+ if and only if $\forall \lambda \in \sigma(A) : \operatorname{Re} \omega \leq 0$, or, what is the same, $\sigma(A) \subseteq \{\lambda \in C^1 | \operatorname{Re} \lambda \geq 0\}$.

Remark 2. Notice that condition 1) of Theorem 10.1 is a direct generalization of the Lyapunov condition for stability of the scalar ordinary differential equation

$$
u''(t) + au'(t) + bu(t) = 0
$$

on R_+ where $a, b \in C^1$.

Corollary 1. *Let A and B be c.n.o. in H, and let* $\operatorname{Re} A = 0$. *Equation* (1) *is stable on R_+ if and only if* $\operatorname{Im} B = 0$ *and*

$$\exists 0 < \delta < +\infty : \sigma(A,B) \subseteq \left\{ (\lambda,\mu) \in C^2 | \lambda_1 = 0,\ \mu_2 = 0,\ \mu_1 \geq -\frac{\lambda_2^2}{4} + \delta \right\}.$$

In particular, the equation $y''(t) + By(t) = 0$ with a normal operator B in H is stable on R_+ if and only if B is a self-adjoint operator in H such that $\inf \sigma(B) > 0$.

Corollary 2. *The equation $y''(t) + Ay'(t) = 0$ with a normal operator A in H is stable on R_+ if and only if* $\inf \sigma(\operatorname{Re} A) \geq 0$ *and* $0 \notin \sigma(A)$.

Definition 10.2. Let A and B be c.n.o. in H, and let $G \subseteq H \times \Phi'$ be a Banach space. Equation (1) is said to be uniformly stable on R_+ in the space of initial data G if there exists $0 < C < +\infty$ such that for each $(f_0, f_1) \in G$: there exists a unique weak solution $y(t)$ of equation (1) on R_+ with i.d. (f_0, f_1), and

$$\| y(t) \| \leq C \| (f_0, f_1) \|_G, \quad \forall t \in R_+.$$

Theorem 10.2. *Let A and B be c.n.o. in H, and let $G = H \times H_{-1} = H \times H_-((|A|^2 + |B| + 1)^{1/2})$. Equation* (1) *is uniformly stable on R_+ in the space of initial data G if and only if* (3.2) *(i.e.,* (3.4)*) and* (10.3) *hold, or, what is the same, one of the following equivalent conditions* 1), 2), 3) *holds:*

1) *a)* (10.1) *holds,*
   b) $\exists 0 < \delta < +\infty : \sigma(A,B) \subseteq \{(\lambda,\mu) \in C^2 |\ |\omega_1 - \omega_2| \geq \delta |\omega_1 + \omega_2|\} \cup \{(\lambda,\mu) \in C^2 | \operatorname{Re}(\omega_1 + \omega_2) \leq -\delta |\omega_1 + \omega_2|\}$, (10.7)
   c) $(0,0) \notin \sigma(A,B)$;
2) $\exists 0 < C < +\infty : \sigma(A,B) \subseteq \{(\lambda,\mu) \in C^2 | d \geq 0, |\operatorname{Im} \lambda|^2 < C^2(d^2 + b_1^2 + b_2^2)\}$; (10.8)
3) *a)* (10.4) *holds,*
   b) $\exists 0 < \delta < +\infty : \sigma(A,B) \subseteq \{(\lambda,\mu) \in C^2 |\ |\mu - \frac{\lambda^2}{4}| \geq \delta^2 |\lambda|^2\} \cup \{(\lambda,\mu) \in C^2 | \lambda_1 \geq \delta |\lambda|\}$, (10.9)
   c) $(0,0) \notin \sigma(A,B)$.

Proof. At first, let's prove by contradiction that if 3) holds then $\exists 0 < \delta < +\infty \forall (\lambda,\mu) \in \sigma(A,B) : d^2 + b_1^2 + b_2^2 \geq \delta^2$. Indeed, if this is not the case then there exists a sequence of $(\lambda_n, \mu_n) \in \sigma(A,B)$ such that $d_n^2 + (b_1)_n^2 + (b_2)_n^2 \to 0$ as $n \to +\infty$. By virtue of 3b), $|\operatorname{Im} \lambda_n| \to 0$ as $n \to +\infty$ as well. Thus, $|\lambda_n|^2 + |\mu_n|^2 \to 0$ as $n \to +\infty$. Since $\sigma(A,B)$ is a closed subset of C^2, we obtain that $(0,0) \in \sigma(A,B)$. This contradicts 3c).

Now, it is easy to see that 3) implies 2). Obviously 2) implies 3a) and 3c). Considering separately $\{(\lambda, \mu) \in C^2 | \, |4\mu - \lambda^2| \geq |\operatorname{Re} \lambda|^2\}$ and $\{(\lambda, \mu) \in C^2 | \, |4\mu - \lambda^2| < |\operatorname{Re} \lambda|^2\}$, one can show that 2) implies 3b). The equivalence of conditions 1) and 3) is obvious.

Furthermore, it is shown that 2) holds if and only if $P = \sigma(A, B)$ satisfies (2) and (10.3) holds.

Let equation (1) be uniformly stable in G on R_+. In particular, the Cauchy problem for this equation is weakly well-posed in $G = H \times H_{-1}$ on R_+. By virtue of Theorem 3.1, $P = \sigma(A, B)$ satisfies (2).

Furthermore, for every solution $y(t)$ of (1) on R_+ (which is, consequently, a weak solution of (1) on R_+ with i.d. from $H \times H \subseteq H \times H_{-1}$): $\sup_{t \in R_+} \| y(t) \| < +\infty$; for every $f_1 \in H$ and $y(t) = \psi_1(A, B, t) f_1$ (which is a weak solution of (1) on R_+ with i.d. $(0, f_1) \in H \times H \subseteq H \times H_{-1}$): $\sup_{t \in R_+} \| y(t) \| < +\infty$. In the proof of the necessity of Theorem 10.1, we have shown that if this is the case then (10.3) holds.

Conversely, let $P = \sigma(A, B)$ satisfy (2), and let (10.3) hold. By Theorem 3.1, for every $(f_0, f_1) \in G = H \times H_{-1}$ there exists a unique weak solution $y(t)$ of (1) on R_+ with i.d. (f_0, f_1), and $y(t) = \psi_0(A, B, t) f_0 + \psi_1(A, B, t) f_1$ $(t \in R_+)$.

In particular, the normal operator $\psi_0(A, B, t)$ is defined on the whole H, for each $t \geq 0$. By the Banach closed graph theorem, $\psi_0(A, B, t)$ is a bounded linear operator on H, for each $t \geq 0$. For every $f_0 \in H$, $y(t) = \psi_0(A, B, t) f_0$ is a weak solution of (1) on R_+ with i.d. $(f_0, 0)$; by Theorem 10.1, (10.3) implies that $\exists C_y < +\infty \forall t \in R_+ : \| y(t) \| \leq C_y$. Thus, $\forall f_0 \in H \exists C_{f_0} < +\infty :$ $\| \psi_0(A, B, t) f_0 \| \leq C_{f_0}$, $\forall t \in R_+$. By the Banach-Steinhaus theorem on uniform boundedness, the operators $\psi_0(A, B, t)$ $(t \geq 0)$ are uniformly bounded. So, $\exists C_0 < +\infty \forall t \geq 0 : \| \psi_0(A, B, t) \| \leq C_0$, whence $\forall f_0 \in H : \| \psi_0(A, B, t) f_0 \| \leq C_0 \| f_0 \|_H$.

Similarly, one can prove that

$$\exists C_1 < +\infty \forall t \geq 0, \ \forall f_1 \in H_{-1} : \| \psi_1(A, B, t) f_1 \| \leq C_1 \| f_1 \|_{H_{-1}} .$$

Thus, for every $(f_0, f_1) \in H \times H_{-1}$ and the weak solution $y(t)$ of (1) on R_+ with i.d. (f_0, f_1):

$$\| y(t) \| = \| \psi_0(A, B, t) f_0 + \psi_1(A, B, t) f_1 \| \leq C_0 \| f_0 \|_H + C_1 \| f_1 \|_{H_{-1}}, \ \forall t \in R_+.$$

The proof is completed.

Remark 1. Notice that the stability of (1) on R_+ itself does not imply the uniform stability of this equation in $G = H \times H_{-1}$ on R_+, as well as (10.1) does not imply that the Cauchy problem for equation (1) on R_+ is weakly well-posed in $G = H \times H_{-1}$. For instance, if $\sigma(A, B) = \{(\lambda, \mu) \in C^2 | \lambda_1 = 1, \lambda_2 \in R^1, \mu = \frac{\lambda^2}{4}\}$ then (10.3) holds while (10.8) does not.

2 Stabilization of the Cesaro means for weak solutions

If the first order equation

$$y'(t) + Ay(t) = 0, \tag{8.3}$$

where A is a self-adjoint operator in H, is stable on R_+ then every weak solution of (8.3) on R_+ has a limit in H as $t \to +\infty$. Namely, if (8.3) is stable on R_+ then $\sigma(A) \subseteq \{\lambda \in R^1 | \omega \leq 0\}$; for every weak solution of (8.3) on R_+ $y(t)$ there exists $f_0 \in H$ such that $y(t) = e^{\omega t}(A)f_0$ (see [137]); then $y(t) \to E_A\{0\}f_0$ in H as $t \to +\infty$.

If complete second order equation (1) is stable on R_+ then a weak solution of (1) on R_+ may not have a limit in H as $t \to +\infty$, even in the case of commuting *self-adjoint* operators A and B. For instance, let $H = C^1$, and consider the equation $u''(t) + u(t) = 0$ in $H = C^1$; $u(t) = \cos t$ is a solution of this equation on R_+. The equation is stable on R_+, but $\cos t$ does not have a limit in C^1 as $t \to +\infty$.

Definition 10.3. Let $x(t) \in C(R_+, H)$. The Cesaro mean $\tilde{x}(t)$ for $x(t)$ is defined as follows: $\tilde{x}(t) = t^{-1} \int_0^t x(\tau)d\tau$ for $t > 0$, $\tilde{x}(0) = x(0)$.

Obviously $\tilde{x}(t) \in C(R_+, H)$. If $x(t) \to x_0 \in H$ in H as $t \to +\infty$, then $\tilde{x}(t) \to x_0$ in H as $t \to +\infty$ as well. Cf. [139].

Theorem 10.3. *Let A and B be c.n.o. in H, and let equation* (1) *be stable on R_+. Let $y(t)$ be a weak solution of* (1) *on R_+ with initial data (f_0, f_1). The Cesaro mean for $y(t)$ has a limit in H as $t \to +\infty$, and this limit is equal to $E_B\{0\}f_0 + A^{-1}E_B\{0\}f_1$.*

In order to prove this theorem, let's prove the following lemma.

Lemma 10.1. *Let A and B be c.n.o. in H such that* (10.3) *holds and $E\{(\lambda,\mu) \in C^2 | d = 0\} = 0$. Then any weak solution of equation* (1) *on R_+ converges to 0 in H as $t \to +\infty$.*

Proof. Indeed, for any weak solution of (1) on R_+ $u(t)$ with i.d. $(f_0, f_1) : u(t) = e^{\omega_1 t}(A,B)f_0 + \psi_1(A,B,t)g$ $(t \in R_+)$ where $f_0 \in H$, $g \in H_-(((d^2 + b_1^2 + b_2^2 + 1)^{1/2}(d+1)^{1/2})(A,B))$ (see Theorem 7.2). By the assumption, $\sigma(A,B)$ satisfies (10.3). In the proof of the sufficiency of Theorem 10.1, we have shown that in this case $\exists 0 < C_3 < +\infty \forall t \geq 1$:

$$|e^{\omega_1 t}|^2 \leq 1, \ |\psi_1(\lambda,\mu,t)|^2 \leq C_3((d^2+b_1^2+b_2^2+1)(|d|+1))^{-1}, \ \forall (\lambda,\mu) \in \sigma(A,B);$$

the majorizing functions are integrable with respect to $d(Ef_0, f_0)$ and $d(Eg, g)$ respectively. Taking into account that $E\{(\lambda,\mu) \in C^2 | d = 0\} = 0$, we can write:

$$\| e^{\omega_1 t}(A,B)f_0 \|^2 = \int\limits_{\sigma(A,B)\cap\{(\lambda,\mu)\in C^2 | d>0\}} |e^{\omega_1 t}|^2 d(Ef_0, f_0);$$

but for any $(\lambda, \mu) \in \sigma(A, B)$ such that $d > 0$: $|e^{\omega_1 t}|^2 = e^{-2dt} \to 0$ as $t \to +\infty$; by the Lebesgue theorem on majorized convergence,

$$\| e^{\omega_1 t}(A, B) f_0 \|^2 = \int_{\sigma(A,B)\cap\{(\lambda,\mu)\in C^2 | d>0\}} |e^{\omega_1 t}|^2 d(E f_0, f_0) \to 0 \text{ as } t \to +\infty.$$

Similarly, $\| \psi_1(A, B, t) g \|^2 \to 0$ as $t \to +\infty$.

Proof of Theorem 10.3. By Theorem 10.1, (10.3) holds. In particular, H is a direct orthogonal sum of Hilbert spaces H_+ and H_0, where $H_+ = E\{(\lambda, \mu) \in C^2 | d > 0\}H$ and $H_0 = E\{(\lambda, \mu) \in C^2 | d = 0\}H$.

Accordingly, the weak solution $y(t)$ of (1) on R_+ can be written in the form $y(t) = y_+(t) + y_0(t)$ $(t \in R_+)$ where $y_+(t) = E\{d > 0\}y(t)$ is a weak solution of (1) on R_+ in $H_+ = E\{d > 0\}H$, while $y_0(t) = E\{d = 0\}y(t)$ is a weak solution of (1) on R_+ in $H_0 = E\{d = 0\}H$ with i.d. $(E\{d = 0\}f_0, E\{d = 0\}f_1)$.

1. $H_+ = E\{d > 0\}H$ satisfies the conditions of Lemma 10.1 on a space H. By Lemma 10.1, $y_+(t) \to 0$ as $t \to +\infty$ in the space H_+ and, consequently, in H. Hence $\tilde{y}_+(t) \to 0$ in H as $t \to +\infty$.

2. It remains to consider $y_0(t)$ in H_0. For the sake of simplicity, we replace the notations H_0, $y_0(t)$, $E\{d = 0\}f_0$, $E\{d = 0\}f_1$, and the restrictions of A and B to H_0 by H, $y(t)$, f_0, f_1, A, B respectively. For each Borel set $\Delta \subseteq C^2$, $E(\Delta)$ stands in place of $E(\Delta \cap \{(\lambda, \mu) \in C^2 | d = 0\})$.

Now $\exists 0 < \delta < +\infty$:

$$\sigma(A, B) \subseteq \{(\lambda, \mu) \in C^2 | \max_{i=1,2} \operatorname{Re} \omega_i = -d = 0, \ |\omega_1 - \omega_2|^2 = 4(b_1^2 + b_2^2) \geq \delta^2\};$$

$H_-(((d^2 + b_1^2 + b_2^2 + 1)^{1/2}(|d| + 1)^{1/2})(A, B)) = H_-((b_1^2 + b_2^2 + 1)^{1/2}(A, B))$, and $y(t)$ is a weak solution of (1) on R_+ with initial data (f_0, f_1).

By Theorem 7.5 and Remark 1 to it,

$$y(t) = e^{\omega_1 t}(A, B) f_0 + (e^{\omega_1 t} - e^{\omega_2 t})(A, B) g_1$$

where $f_0 \in H$,

$$g = -\omega_1(A, B) f_0 + f_1 \in H_-((b_1^2 + b_2^2 + 1)^{1/2}(A, B)),$$
$$g_1 = (\omega_1 - \omega_2)^{-1}(A, B) g \in H.$$

Now

$$\tilde{y}(t) = \frac{e^{\omega_1 t} - 1}{\omega_1 t}(A, B)(f_0 + g_1) - \frac{e^{\omega_2 t} - 1}{\omega_2 t}(A, B) g_1 \ (t \in R_+).$$

For any $(\lambda, \mu) \in \sigma(A, B)$, $t \in R_+$, $i = 1, 2$: $\left|\frac{e^{\omega_i t} - 1}{\omega_i t}\right| \leq 2$; if $\omega_i \neq 0$, then $\frac{e^{\omega_i t} - 1}{\omega_i t} \to 0$ as $t \to +\infty$; if $\omega_i = 0$, then $\frac{e^{\omega_i t} - 1}{\omega_i t} \equiv 1 \to 1$ as $t \to +\infty$.

For any $(\lambda,\mu) \in \sigma(A,B)$: if $\mu \neq 0$, then $\omega_1 \neq 0$ and $\omega_2 \neq 0$. Similarly to the proof of Lemma 10.1, one can conclude that $E\{\mu \neq 0\}\tilde{y}(t) \to 0$ in H as $t \to +\infty$.

Further,

$$\sigma(A,B) \cap \{(\lambda,\mu) \in C^2 | \mu = 0\} \subseteq$$
$$\{(\lambda,\mu)|\omega_1 = 0,\ \omega_2 = -\lambda,\ |\omega_2| \geq \delta\} \cup \{(\lambda,\mu)|\omega_2 = 0,\ \omega_1 = -\lambda,\ |\omega_1| \geq \delta\}. \tag{10.10}$$

Hence $\forall(\lambda,\mu) \in \sigma(A,B) \cap \{\mu = 0\} : |\lambda| \geq \delta$. Taking into account that $\sigma(A,B) \subseteq \{(\lambda,\mu)|d = 0\}$, we conclude from Theorem 7.5 and Remark 1 to it: $E\{\mu = 0\}f_1 \in E\{\mu = 0\}H_-((|\lambda|^2 + |\mu| + 1)^{1/2}(A,B)) = E\{\mu = 0\}H_-(|A| + 1)$. Thus, $A^{-1}E\{\mu = 0\}f_1 \in H$. By virtue of (10.10), $E\{\mu = 0\}\tilde{y}(t) \to E\{\mu = 0\}f_0 + A^{-1}E\{\mu = 0\}f_1$ in H as $t \to +\infty$.

Let's return to the original notations. In fact, we have shown that for the weak solution $y(t)$ of (1) on R_+ with i.d. (f_0, f_1) :

$$\tilde{y}(t) \to E\{d = 0,\ \mu = 0\}f_0 + A^{-1}E\{d = 0,\ \mu = 0\}f_1 \text{ in } H \text{ as } t \to +\infty.$$

But $\forall(\lambda,\mu) \in \sigma(A,B) : \mu = 0 \Rightarrow \omega_1 = 0$ or $\omega_2 = 0 \Rightarrow d = -\max_{i=1,2} \operatorname{Re}\omega_i \leq 0$; taking into account that $d \geq 0$ by virtue of (10.3), we obtain that $d = 0$. So, $\sigma(A,B) \cap \{\mu = 0\} \subseteq \{d = 0\}$, whence $E\{d = 0, \mu = 0\} = E\{\mu = 0\}$. Finally, notice that $E\{(\lambda,\mu) \in C^2 | \mu = 0\} = E_B\{0\}$. This remark completes the proof of Theorem 10.3.

Corollary 1. *Let A and B be c.n.o. in H, and let equation (1) be stable on R_+. For each weak solution of equation (1) on R_+ its Cesaro mean converges to 0 in H as $t \to +\infty$ if and only if* $\operatorname{Ker} B = \{0\}$.

Corollary 2. *Let A and B be c.n.o. in H, and let equation (1) be stable on R_+. If a weak solution $y(t)$ of (1) on R_+ with i.d. (f_0, f_1) has a limit in H as $t \to +\infty$, then this limit is equal to $E_B\{0\}f_0 + A^{-1}E_B\{0\}f_1$.*

In what follows, we say that $x(t) \in C(R_+, H)$ stabilizes if $x(t)$ has a limit in H as $t \to +\infty$. It turns out that in the general case of c.n.o. A and B stabilization of the Cesaro means of all weak solutions holds not only for stable equations (1) but for a strictly wider class of equations (1). Namely, one can prove the following statement.

Corollary 3. *Let A and B be arbitrary c.n.o. in H. Stabilization of the Cesaro means of all weak solutions of equation (1) on R_+ is the case if and only if one of the following equivalent conditions 1), 2), 3) holds:*

1) *a)* (10.1) *holds,*

   b) $E\{(\lambda,\mu) \in C^2 | \operatorname{Re}\omega_1 = 0,\ \omega_1 = \omega_2\} = 0$,

   c) $(0,0) \notin \sigma(A,B)$;

2) $\exists 0 < \delta < +\infty : \sigma(A,B) \subseteq \{(\lambda,\mu) \in C^2 | d \geq 0,\ d^2 + b_1^2 + b_2^2 + |\operatorname{Im}\lambda|^2 \geq \delta^2\}$ *and* $E\{(\lambda,\mu) \in C^2 | d^2 + b_1^2 + b_2^2 = 0,\ |\operatorname{Im}\lambda| \geq \delta\} = 0$;

3) *a)* (10.4) *holds,*

b) $E\{(\lambda,\mu) \in C^2 | \lambda_1 = 0,\ \mu = -\frac{1}{4}\lambda_2^2\} = 0$,

c) $(0,0) \notin \sigma(A,B)$.

If this is the case and $y(t)$ *is a weak solution of* (1) *on* R_+ *with i.d.* (f_0, f_1), *then* $\lim_{t\to+\infty} \tilde{y}(t) = E_B\{0\}f_0 + A^{-1}E_B\{0\}f_1$.

3 Stabilization of a weak solution

Let A and B be c.n.o. in H such that equation (1) is stable on R_+. We have shown in the previous section that for any weak solution $y(t)$ of (1) on R_+, its Cesaro mean $\tilde{y}(t)$ has a limit in H as $t \to +\infty$.

Let $y(t)$ be a weak solution of (1) on R_+ with i.d. (f_0, f_1). What must be (f_0, f_1) for $y(t)$ itself to have a limit in H as $t \to +\infty$?

Denote:

$$\begin{aligned}
\Delta_1 &= \{(\lambda,\mu) \in C^2 | \operatorname{Re}\omega_1 = 0,\ \operatorname{Re}\omega_2 < 0,\ \omega_1 \neq 0\} \\
&= \{(\lambda,\mu) \in C^2 | d = 0,\ \lambda_1 > 0,\ \mu \neq 0\} \\
&= \{(\lambda,\mu) | \lambda_1 > 0,\ \mu_1 = \lambda_1^{-2}\mu_2(\mu_2 - \lambda_1\lambda_2),\ \mu_2 \neq 0\};
\end{aligned}$$

$$\begin{aligned}
\Delta_2 &= \{(\lambda,\mu) \in C^2 | \operatorname{Re}\omega_1 = 0,\ \operatorname{Re}\omega_2 < 0,\ \omega_1 = 0\} \\
&= \{(\lambda,\mu) \in C^2 | d = 0,\ \lambda_1 > 0,\ \mu = 0\} \\
&= \{(\lambda,\mu) | \lambda_1 > 0,\ \mu = 0\};
\end{aligned}$$

$$\begin{aligned}
\Delta_3 &= \{(\lambda,\mu) \in C^2 | \operatorname{Re}\omega_1 = \operatorname{Re}\omega_2 = 0,\ \omega_1 \neq 0,\ \omega_2 \neq 0\} \\
&= \{(\lambda,\mu) \in C^2 | d = 0,\ \lambda_1 = 0,\ \mu \neq 0\} \\
&= \{(\lambda,\mu) | \lambda_1 = 0,\ \mu_1 \geq -\frac{\lambda_2^2}{4},\ \mu_1 \neq 0,\ \mu_2 = 0\};
\end{aligned}$$

$$\begin{aligned}
\Delta_4 &= \{(\lambda,\mu) \in C^2 | \operatorname{Re}\omega_1 = \operatorname{Re}\omega_2 = 0,\ \omega_1\omega_2 = 0,\ \omega_1 + \omega_2 \neq 0\} \\
&= \{(\lambda,\mu) \in C^2 | d = 0,\ \lambda_1 = 0,\ \lambda_2 \neq 0,\ \mu = 0\} \\
&= \{(\lambda,\mu) | \lambda_1 = 0,\ \lambda_2 \neq 0,\ \mu = 0\}.
\end{aligned}$$

Notice that $\{(\lambda,\mu) \in C^2 | d = 0\} = \Delta_1 \cup \Delta_2 \cup \Delta_3 \cup \Delta_4 \cup \{(0,0)\}$. By virtue of (10.3), $(0,0) \notin \sigma(A,B)$; hence $\sigma(A,B) \cap \{d = 0\} \subseteq \Delta_1 \cup \Delta_2 \cup \Delta_3 \cup \Delta_4$. Further, $\Delta_i \cap \Delta_k = \emptyset$ for $i \neq k$.

Theorem 10.4. *Let A and B be c.n.o. in H such that* (10.3) *holds, $y(t)$ be a weak solution of equation* (1) *on R_+ with i.d. (f_0, f_1). $y(t)$ has a limit in H as $t \to +\infty$ if and only if the following conditions hold:*

1) $-\omega_2(A,B)E(\Delta_1)f_0 + E(\Delta_1)f_1 = (\frac{\lambda}{2} + \sqrt{\frac{\lambda^2}{4} - \mu})(A,B)E(\Delta_1)f_0 + E(\Delta_1)f_1 = 0$ *(choosing $\sqrt{z}$ such that* Re $\sqrt{z} \geq 0$, *and if* Re $\sqrt{z} = 0$ *then* Im $\sqrt{z} \geq 0$, *for each $z \in C^1$);*
2) $E(\Delta_3)f_0 = E(\Delta_3)f_1 = 0$;
3) $E(\Delta_4)f_1 = 0$.

If this is the case then $\lim_{t \to +\infty} y(t) = E_B\{0\}f_0 + A^{-1}E_B\{0\}f_1$.

In order to prove Theorem 10.4, we need the following lemma.

Lemma 10.2. *Let X be a self-adjoint operator in H such that $0 \notin \sigma(X)$, and let $g \in H$. The function $y(t) = (e^{ixt})(X)g$ ($t \in R_+$) has a limit in H as $t \to +\infty$ if and only if $g = 0$.*

Proof. Let $y(t) \to h \in H$ as $t \to +\infty$. Then $\tilde{y}(t) \to h$ as $t \to +\infty$ as well. But $\tilde{y}(t) = \frac{e^{ixt}-1}{ixt}(X)g \to 0$ as $t \to +\infty$; this can be shown just as in the proof of Theorem 10.3. Thus, $y(t) = (e^{ixt})(X)g \to 0$ in H as $t \to +\infty$, i.e., $\| y(t) \| \to 0$ as $t \to +\infty$. But $\forall t \in R_+ : (e^{ixt})(X)$ is a unitary operator, and $\| y(t) \| = \| e^{ixt}(X)g \| = \| g \|$. Thus, $\| y(t) \| \to 0$ as $t \to +\infty$ if and only if $g = 0$.

Proof of Theorem 10.4. Just as in the proof of Theorem 10.3, we conclude from Lemma 10.1 that $E\{d > 0\}y(t) \to 0$ in H as $t \to +\infty$. Now we need to consider only $E\{d = 0\}y(t)$.

For each of $i = 1,2,3,4$, denote: $H_i = E(\Delta_i)H$; let A_i and B_i be the restrictions of A and B respectively to H_i. Then A_i and B_i are c.n.o. in H_i; $\sigma(A_i, B_i) \subseteq \sigma(A,B) \cap \overline{\Delta_i}$. Further, $y_i(t) = E(\Delta_i)y(t)$ is a weak solution of the equation $y_i''(t) + A_i y_i'(t) + B_i y_i(t) = 0$ in H_i on R_+.

$E\{d = 0\}y(t) = E(\Delta_1)y(t) + E(\Delta_2)y(t) + E(\Delta_3)y(t) + E(\Delta_4)y(t)$ has a limit in H as $t \to +\infty$ if and only if for each of $i = 1,2,3,4 : y_i(t) = E(\Delta_i)y(t)$ has a limit in H_i as $t \to +\infty$.

Recall that, by virtue of Theorem 7.5 and Remark 1 to it,

$$y(t) = e^{\omega_1 t}(A,B)f_0 + \psi_1(A,B,t)(-\omega_1(A,B)f_0 + f_1) \quad (t \in R_+)$$

where $f_0 \in H, f_1 \in H_-(((|\lambda|^2 + |\mu| + 1)^{1/2}(|d| + 1)^{1/2})(A,B))$, and $-\omega_1(A,B)f_0 + f_1 \in H_-(((d^2 + b_1^2 + b_2^2 + 1)^{1/2}(|d| + 1)^{1/2})(A,B))$. Hence

$$E\{d = 0\}y(t)$$
$$= e^{\omega_1 t}(A,B)E\{d = 0\}f_0 + \psi_1(A,B,t)(-\omega_1(A,B)E\{d = 0\}f_0 + E\{d = 0\}f_1)$$

($t \in R_+$) where $E\{d = 0\}f_0 \in H$, $E\{d = 0\}f_1 \in H_-((|A|^2 + |B| + 1)^{1/2})$, and $-\omega_1(A,B)E\{d = 0\}f_0 + E\{d = 0\}f_1 \in H_-((b_1^2 + b_2^2 + 1)^{1/2}(A,B)) =$

$H_-(|\omega_1 - \omega_2|(A,B))$. (The last equality follows from the fact that $\forall(\lambda,\mu) \in \sigma(A,B) \cap \{d=0\} : |\omega_1 - \omega_2|^2 = 4(b_1^2 + b_2^2) \geq \delta^2 > 0$.)

Now we consider separately $E(\Delta_i)y(t)$ in H_i for each of $i = 1,2,3,4$. To save room, here $\varphi(\lambda,\mu,t)$ stands for $\varphi(A,B,t)$; for instance, $e^{\omega_1 t}$ stands for $e^{\omega_1 t}(A,B)$.

1. $\Delta_1 \subseteq \{(\lambda,\mu) \in C^2 | \mu \neq 0\}$. Let $\exists h \in H_1 \stackrel{\text{def}}{=} E(\Delta_1)H : E(\Delta_1)y(t) \to h$ in H_1 as $t \to +\infty$. Then $h = 0$, by virtue of Corollary 2 to Theorem 10.3, i.e., $\lim_{t\to+\infty} \| E(\Delta_1)y(t) \| = 0$. But $\forall t \in R_+ : E(\Delta_1)y(t) = e^{\omega_1 t}(E(\Delta_1)f_0 + (1 - e^{(\omega_2-\omega_1)t})g_1)$ where $g_1 \stackrel{\text{def}}{=} (\omega_1 - \omega_2)^{-1}(-\omega_1 E(\Delta_1)f_0 + E(\Delta_1)f_1) \in H_1$. Taking into account that $e^{\omega_1 t}$ is a unitary operator in H_1, we obtain that $(E(\Delta_1)f_0 + g_1 - e^{(\omega_2-\omega_1)t}g_1)$ converges to 0 in H_1 as $t \to +\infty$. But $e^{(\omega_2-\omega_1)t}g_1$ converges to 0 in H_1 as $t \to +\infty$ since $\forall(\lambda,\mu) \in \Delta_1 : \operatorname{Re}(\omega_2 - \omega_1) < 0$ and $g_1 \in H_1$ (cf. the proof of Lemma 10.1). Thus, $E(\Delta_1)f_0 + g_1 = 0$. Conversely, if $E(\Delta_1)f_0 + g_1 = 0$ then $E(\Delta_1)y(t) = -e^{\omega_2 t}g_1 \to 0$ in H_1 as $t \to +\infty$.

It remains to notice that $-\omega_2 E(\Delta_1)f_0 + E(\Delta_1)f_1 = (\omega_1 - \omega_2)(E(\Delta_1)f_0 + g_1)$. Thus, $y_1(t)$ has a limit in H_1 as $t \to +\infty$ if and only if $-\omega_2 E(\Delta_1)f_0 + E(\Delta_1)f_1 = 0$.

2. Just as before, one can show that $E(\Delta_2)y(t) = e^{\omega_2 t}E(\Delta_2)f_0 + (e^{\omega_1 t} - e^{\omega_2 t})(\omega_1 - \omega_2)^{-1}(-\omega_2 E(\Delta_2)f_0 + E(\Delta_2)f_1)$ $(t \in R_+)$ where $E(\Delta_2)f_0 \in H$, $E(\Delta_2)f_1 \in H_-(|A|+1)$, and $-\omega_2 E(\Delta_2)f_0 + E(\Delta_2)f_1 \in H_-(|\omega_1 - \omega_2|(A,B))$. Notice that $\forall(\lambda,\mu) \in \sigma(A,B) \cap \Delta_2 : \mu = 0$, $|\lambda| = |\omega_1 - \omega_2| \geq \delta > 0$, $\omega_1 = 0$, $\omega_2 = -\lambda$.

Now $g_2 \stackrel{\text{def}}{=} (\omega_1 - \omega_2)^{-1}(-\omega_2 E(\Delta_2)f_0 + E(\Delta_2)f_1) \in H_2$. Just as previously, $e^{\omega_2 t}E(\Delta_2)f_0 \to 0$, $e^{\omega_2 t}g_2 \to 0$ in H_2 as $t \to +\infty$. But $e^{\omega_1 t} = 1$, $\forall(\lambda,\mu) \in \sigma(A,B) \cap \Delta_2$, $t \in R_+$. Thus, $E(\Delta_2)y(t)$ converges to $(\omega_1 - \omega_2)^{-1}(-\omega_2 E(\Delta_2)f_0 + E(\Delta_2)f_1) = E(\Delta_2)f_0 + A^{-1}E(\Delta_2)f_1$ in H_2 as $t \to +\infty$.

3. $\Delta_3 \subseteq \{(\lambda,\mu) \in C^2 | \mu \neq 0\}$. Let $\exists h \in H_3 : E(\Delta_3)y(t) \to h$ in H_3 as $t \to +\infty$. Similarly to the point 1 of this proof, one can show that for $g_3 \stackrel{\text{def}}{=} (\omega_1 - \omega_2)^{-1}(-\omega_1 E(\Delta_3)f_0 + E(\Delta_3)f_1) \in H_3 : E(\Delta_3)f_0 + g_3 - e^{(\omega_2-\omega_1)t}g_3$ converges to 0 in H_3 as $t \to +\infty$. In particular, this implies that $e^{(\omega_2-\omega_1)t}g_3$ has a limit in H_3 as $t \to +\infty$. By Lemma 10.2, this is possible only if $g_3 = 0$ (we make use of the fact that $\forall(\lambda,\mu) \in \sigma(A,B) \cap \Delta_3 : \operatorname{Re}\omega_2 = \operatorname{Re}\omega_1 = 0$, $|\omega_1 - \omega_2| \geq \delta > 0$). But then $E(\Delta_3)f_0 \to 0$ as $t \to +\infty$, i.e., $E(\Delta_3)f_0 = 0$. Now $E(\Delta_3)f_0 = 0$ and $(\omega_1 - \omega_2)g_3 = -\omega_1 E(\Delta_3)f_0 + E(\Delta_3)f_1 = 0$. Thus, $E(\Delta_3)f_1 = 0$ as well. Obviously, if $E(\Delta_3)f_0 = E(\Delta_3)f_1 = 0$ then $y_3(t) = 0$, $\forall t \in R_+$.

4. For any $(\lambda,\mu) \in \sigma(A,B) \cap \Delta_4 : \mu = 0$, $\lambda_1 = 0$, $|\lambda_2| = |\lambda| = |\omega_1 - \omega_2| \geq \delta > 0$. Let $g_4 \stackrel{\text{def}}{=} A^{-1}E(\Delta_4)f_1$; then $g_4 \in H_4$, and $E(\Delta_4)y(t) = E(\Delta_4)f_0 + (1 - e^{-i\lambda_2 t})g_4 = E(\Delta_4)f_0 + g_4 - e^{-i\lambda_2 t}g_4$.

Let $\exists h \in H_4 : E(\Delta_4)y(t) \to h$ in H_4 as $t \to +\infty$. This is the case if and only if $e^{-i\lambda_2 t}g_4$ has a limit in H_4 as $t \to +\infty$. By Lemma 10.2, this is possible only if $g_4 = 0$. Conversely, if $g_4 = 0$ then $E(\Delta_4)y(t) \equiv E(\Delta_4)f_0$ converges to $E(\Delta_4)f_0$ in H_4 as $t \to +\infty$. Finally, $g_4 = 0$ if and only if $E(\Delta_4)f_1 = 0$.

The proof of Theorem 10.4 is completed.

4 Stabilization of weak solutions and asymptotic stability of an equation

Definition 10.4. Let A and B be c.n.o. in H. Equation (1) is said to be stabilizable on R_+ if each weak solution of (1) on R_+ has a limit in H as $t \to +\infty$.

Obviously, if equation (1) is stabilizable on R_+ then it is stable on R_+.

At the beginning of Section 2, we noted that for the first order equation $y'(t) + Ay(t) = 0$ with a self-adjoint operator A in H, the converse implication is valid too: if this equation is stable on R_+ then it is stabilizable on R_+.

For second order equation (1) with c.s.o. A and B in H, the converse implication is not valid. Equation (1) which is stable on R_+ may not be stabilizable.

Theorem 10.5. *Let A and B be c.n.o. in H. Equation* (1) *is stabilizable on R_+ if and only if there hold* (10.3) *and one of the following equivalent conditions:*

1) $E(\{(\lambda, \mu) \in C^2 | d = 0\} \setminus \{(\lambda, \mu) | \lambda_1 > 0,\ \mu = 0\}) = 0$;
2) $E(\Delta_1 \cup \Delta_3 \cup \Delta_4) = 0$;
3) $E(\{(\lambda, \mu) \in C^2 | d = 0,\ \lambda_1 > 0,\ \mu_2 \neq 0\} \cup \{(\lambda, \mu) \in C^2 | d = 0,\ \lambda_1 = 0,\ \mu_2 = 0\}) = 0$.

See the notations and the remark before Theorem 10.4.

Theorem 10.5 follows immediately from Theorems 10.1 and 10.4.

Definition 10.5. Let A and B be c.n.o. in H. Equation (1) is said to be asymptotically stable on R_+ if each weak solution of (1) on R_+ converges to 0 in H as $t \to +\infty$.

Theorem 10.6. *Let A and B be c.n.o. in H. Equation* (1) *is asymptotically stable on R_+ if and only if* (10.3) *holds and* $E\{(\lambda, \mu) \in C^2 | d = 0\} = 0$.

See the notations and the remark before Theorem 10.4.

Theorem 10.6 follows immediately from Theorems 10.1 and 10.4.

5 Exponential stability and uniform exponential stability of an equation

Definition 10.6. Let A and B be c.n.o. in H. Equation (1) is said to be exponentially stable on R_+ if there exists $0 < \gamma_1 < +\infty$ such that for each weak solution $y(t)$ of (1) on R_+:

$$\exists C_y < +\infty : \| y(t) \| \leq C_y e^{-\gamma_1 t},\ \forall t \in R_+.$$

Theorem 10.7. *Let A and B be c.n.o. in H. Equation* (1) *is exponentially stable on R_+ if and only if one of the following equivalent conditions holds:*

$$1)\quad \exists 0 < \gamma < +\infty \forall (\lambda, \mu) \in \sigma(A, B) : \max_{i=1,2} \operatorname{Re} \omega_i \leq -\gamma; \tag{10.11}$$

$$2)\quad \exists 0 < \gamma < +\infty : \sigma(A, B) \subseteq \{(\lambda, \mu) \in C^2 | d \geq \gamma\}; \tag{10.12}$$

$$3)\quad \exists 0 < \gamma < +\infty : \sigma(A, B) \subseteq \{(\lambda, \mu) \in C^2 | \lambda_1 > 2\gamma,$$

$$\mu_1 > \frac{1}{(\lambda_1 - 2\gamma)^2}(\mu_2^2 - \lambda_1\lambda_2 \cdot \mu_2 + \gamma(\lambda_1 - \gamma)(\lambda_2^2 + (\lambda_1 - 2\gamma)^2))\}. \tag{10.13}$$

The proof is similar to the proof of Theorem 9.3.

Corollary 1. *For equation* (1) *with c.n.o. A and B to be exponentially stable on R_+, it is necessary that* $\inf \sigma(\operatorname{Re} A) > 0$.

Corollary 2. *Let A and B be c.n.o. in H, and* $\operatorname{Im} B = 0$. *For equation* (1) *to be exponentially stable on R_+, it is necessary that* $\inf \sigma(\operatorname{Re} B) > 0$ *and* $\operatorname{Re} A$ *be subordinate to* $\operatorname{Re} B$.

Thus, it is impossible that incomplete second order equation (1) with c.n.o. A and B in H, i.e., the equation $y''(t) + Ay'(t) = 0$ with a normal operator A or $y''(t) + By(t) = 0$ with a normal operator B, be exponentially stable on R_+.

Corollary 3. *Let A and B be c.n.o. in H, and $\sigma(P)$ be the spectrum of the operator pencil $P(z)$ associated with equation* (1) *(see Section* 6 *of Chapter* 3*). Equation* (1) *is exponentially stable on R_+ if and only if*

$$\sup_{z \in \sigma(P)} \operatorname{Re} z < 0.$$

This follows from Theorem 10.7 and Lemma 3.8.

This situation differs from the situation for stability of equation (1) on R_+: it is impossible to express a necessary and sufficient condition for stability of (1) on R_+ in terms of $\sigma(P)$.

Definition 10.7. Let A and B be c.n.o. in H, and $G \subseteq H \times \Phi'$ be a Banach space. Equation (1) is said to be uniformly exponentially stable on R_+ in the space of initial data G if there exist $0 < C < +\infty$ and $0 < \gamma_1 < +\infty$ such that for each $(f_0, f_1) \in G$: there exists a unique weak solution $y(t)$ of (1) on R_+ with i.d. (f_0, f_1), and

$$\| y(t) \| \leq C \| (f_0, f_1) \|_G \, e^{-\gamma_1 t}, \quad \forall t \in R_+.$$

Theorem 10.8. *Let A and B be c.n.o. in H, and $G = H \times H_{-1} = H \times H_{-}((|A|^2 + |B| + 1)^{1/2})$. Equation (1) is uniformly exponentially stable on R_+ in the space of initial data G if and only if (3.2) (i.e., (3.4)) and (10.12) hold, i.e., one of the following equivalent conditions 1), 2), 3) holds:*

1) (10.11) *and* (10.7) *hold;*
2) $\exists 0 < C < +\infty,\ 0 < \gamma < +\infty : \sigma(A,B) \subseteq \{(\lambda,\mu) \in \mathbb{C}^2 | d \geq \gamma,\ |\operatorname{Im}\lambda|^2 \leq C^2(d^2 + b_1^2 + b_2^2)\}$;
3) (10.13) *and* (10.9) *hold.*

This follows from Theorems 3.1 and 10.7 in the same manner as Theorem 10.2 follows from Theorems 3.1 and 10.1; in the proof of the sufficiency, $\psi_i(A,B,t)$ are replaced by $e^{\gamma_1 t}\psi_i(A,B,t)$.

6 Stabilization of $\frac{y(t)}{t}$ for weak solutions of an equation

In this section, we consider a weaker condition on equation (1) than condition of stability.

Denote: $\Delta_0 = \{(\lambda,\mu) \in \mathbb{C}^2 | \operatorname{Re}\omega_1 = 0,\ \omega_1 = \omega_2\} = \{(\lambda,\mu) \in \mathbb{C}^2 | d^2 + b_1^2 + b_2^2 = 0\} = \{(\lambda,\mu) | \lambda_1 = 0,\ \mu_1 = -\frac{\lambda_2^2}{4},\ \mu_2 = 0\}$.

Theorem 10.9. *Let A and B be c.n.o. in H. For every weak solution $y(t)$ of (1) on R_+ the function $t^{-1}y(t)$ $(t > 0)$ has a limit in H as $t \to +\infty$ if and only if one of the following equivalent conditions holds:*

1) (10.1);
2) $\sigma(A,B) \subseteq \{(\lambda,\mu) \in \mathbb{C}^2 | d \geq 0\}$;
3) (10.4).

If (10.1) holds, and $y(t)$ is a weak solution of equation (1) on R_+ with i.d. (f_0, f_1), then $t^{-1}y(t) \to \frac{1}{2}AE(\Delta_0)f_0 + E(\Delta_0)f_1$ in H as $t \to +\infty$.

Proof. The necessity of (10.1) can be proved just as in the proof of Theorem 10.1. In order to prove the sufficiency, denote:

$$\Delta_5 = \{(\lambda,\mu) \in \mathbb{C}^2 | d \geq 0,\ d^2 + b_1^2 + b_2^2 > 1\},$$
$$\Delta_6 = \{(\lambda,\mu) \in \mathbb{C}^2 | d \geq 0,\ d^2 + b_1^2 + b_2^2 \leq 1\}.$$

Obviously $\Delta_0 \subset \Delta_6$. For each of $i = 5, 6$, denote: $H_i = E(\Delta_i)H$, and let A_i and B_i be the restrictions of A and B to H_i. Now A_i and B_i are c.n.o. in H_i, $\sigma(A_i, B_i) \subseteq \sigma(A,B) \cap \overline{\Delta_i}$.

Let $y(t)$ be a weak solution of (1) on R_+ with initial data (f_0, f_1). By virtue of Theorem 7.5 and Remark 1 to it, we obtain:

$$f_0 \in H,\ f_1 \in H_{-}((((|\lambda|^2 + |\mu| + 1)^{1/2}(|d| + 1)^{1/2})(A,B)),$$

and

$$y(t) = e^{\omega_1 t}(A,B)f_0 + \psi_1(A,B,t)g \ (t \in R_+)$$

where

$$g = -\omega_1(A,B)f_0 + f_1 \in H_-(((d^2+b_1^2+b_2^2+1)^{1/2}(|d|+1)^{1/2})(A,B)).$$

For each of $i = 5,6$, denote: $y_i(t) = E(\Delta_i)y(t)$. For any $t > 0 : t^{-1}y(t) = t^{-1}y_5(t) + t^{-1}y_6(t)$.

1. $y_5(t)$ is a weak solution of the equation

$$y_5''(t) + A_5 y_5'(t) + B_5 y_5(t) = 0$$

on R_+. Taking into account that $\sigma(A_5,B_5) \subseteq \overline{\Delta_5}$, this equation is stable on R_+ (see Theorem 10.1). Thus, $\exists C_5 < +\infty \forall t > 0 : \| y_5(t) \| \leq C_5$; consequently, $t^{-1}y_5(t) \to 0$ in H_5 (i.e., in H) as $t \to +\infty$.

2. $y_6(t) = e^{\omega_1 t}(A,B)E(\Delta_6)f_0 + \psi_1(A,B,t)E(\Delta_6)g \ (t \in R_+)$. Taking into account that $\forall(\lambda,\mu) \in \Delta_6 : d^2+b_1^2+b_2^2 \leq 1$, we obtain that $E(\Delta_6)g \in E(\Delta_6)H$. Denote: $f_6 = E(\Delta_6)f_0$, $g_6 = E(\Delta_6)g$; then $f_6 \in H_6$, $g_6 \in H_6$, and $y_6(t) = e^{\omega_1 t}(A,B)f_6 + \psi_1(A,B,t)g_6(t \in R_+)$.

For any $t \geq 1$:

$$t^{-1}y_6(t) = (t^{-1}e^{\omega_1 t})(A,B)f_6 + (t^{-1}\psi_1(\lambda,\mu,t))(A,B)g_6.$$

Recall that $\max\limits_{z \in C^1, |z| \leq 1} |\frac{e^z-1}{z}| \leq 2$. For any $(\lambda,\mu) \in \Delta_6$, $t \geq 1 : |t^{-1}e^{\omega_1 t}| \leq 1$;

$$|t^{-1}\psi_1(\lambda,\mu,t)| = e^{-dt}\left|\frac{e^{(\omega_2-\omega_1)t}-1}{(\omega_2-\omega_1)t}\right| \leq 2e^{-dt} \leq 2.$$

For any $(\lambda,\mu) \in \Delta_6 : \lim\limits_{t\to+\infty}(t^{-1}e^{\omega_1 t}) = 0$; if $d^2+b_1^2+b_2^2 > 0$, i.e., $d > 0$ or $|\omega_1-\omega_2|^2 = 4(b_1^2+b_2^2) > 0$, then $\lim\limits_{t\to+\infty}(t^{-1}\psi_1(\lambda,\mu,t)) = 0$; if $d^2+b_1^2+b_2^2 = 0$, i.e., $d = 0$ and $|\omega_1-\omega_2| = 0$, then $\lim\limits_{t\to+\infty}(t^{-1}\psi_1(\lambda,\mu,t)) = 1$. Just as in the proof of Lemma 10.1, we obtain that

$$t^{-1}y_6(t) \to E\{(\lambda,\mu) \in \Delta_6 | d^2+b_1^2+b_2^2 = 0\}g_6 = E(\Delta_0)g \text{ in } H \text{ as } t \to +\infty.$$

Since $g = \omega_1(A,B)f_0 + f_1$, and $\forall(\lambda,\mu) \in \Delta_0 : \omega_1 = \omega_2 = -\frac{\lambda}{2}$, we obtain: $t^{-1}y(t) \to \frac{1}{2}AE(\Delta_0)f_0 + E(\Delta_0)f_1$ in H as $t \to +\infty$.

7 The case of self-adjoint A and B

Let's clarify the results of this chapter for the special case that A and B are self-adjoint operators.

Theorem 10.10. *Let A and B be c.s.o. in H.*

1. *Equation* (1) *is stable on* R_+ *if and only if one of the following equivalent conditions* 1), 2), 3) *holds:*

   1) $\sup_{(\lambda,\mu)\in\sigma(A,B),i=1,2} \operatorname{Re}\omega_i(\lambda,\mu) \leq 0$, *and* $(0,0) \notin \sigma(A,B)$;

   2) $\exists 0 < \delta < +\infty : \sigma(A,B) \subseteq \{(\lambda,\mu) \in R^2 | \lambda \geq 0, \mu \geq 0\} \setminus \{(\lambda,\mu) | \lambda^2 + \mu^2 \leq \delta^2\}$ (10.14)

   (see Fig. 10.1);

   3) $\inf \sigma(A) \geq 0$, $\inf \sigma(B) \geq 0$, *and* $(0,0) \notin \sigma(A,B)$.

2. *Equation* (1) *is uniformly stable on* R_+ *in the space of initial data* $G = H \times H$ *if and only if* (10.14) *holds.*

3. *The previous statement, with* $H \times H$ *replaced by* $H \times H_{-1} = H \times H_{-}((|A|^2 + |B| + 1)^{1/2})$.

4. *Stabilization of the Cesaro means of all weak solutions of equation* (1) *on* R_+ *is the case if and only if* (10.14) *holds.*

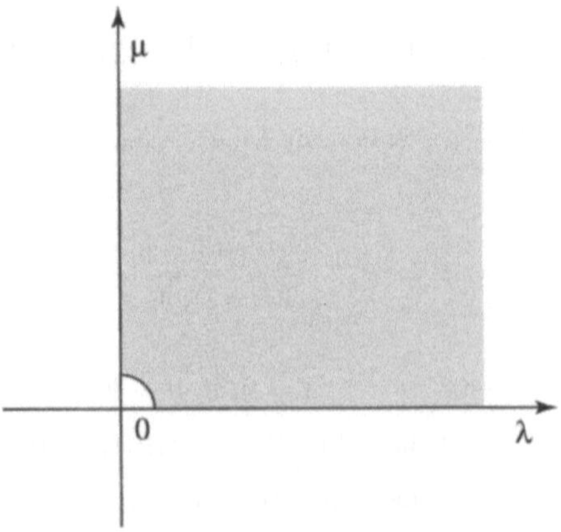

Fig. 10.1

Theorem 10.11. *Let A and B be c.s.o. in H such that* (10.14) *holds. Let $y(t)$ be a weak solution of equation* (1) *on* R_+ *with initial data* (f_0, f_1). *The Cesaro mean for $y(t)$ has a limit in H as $t \to +\infty$, and this limit is equal to* $E_B\{0\}f_0 + A^{-1}E_B\{0\}f_1$.

Corollary 1. *Let A and B be c.s.o. in H such that (10.14) holds. For the Cesaro means of all weak solutions of equation (1) on R_+ to converge to 0 in H as $t \to +\infty$, it is necessary and sufficient that* $\operatorname{Ker} B = \{0\}$.

Theorem 10.12. *Let A and B be c.s.o. in H such that (10.14) holds, $y(t)$ be a weak solution of equation (1) on R_+ with i.d. (f_0, f_1). The function $y(t)$ has a limit in H as $t \to +\infty$ if and only if $E_A\{0\}f_0 = E_A\{0\}f_1 = 0$. If this is the case then* $\lim_{t\to+\infty} y(t) = E_B\{0\}f_0 + A^{-1}E_B\{0\}f_1$.

Theorem 10.13. *Let A and B be c.s.o. in H. Equation (1) is stabilizable on R_+ if and only if (10.14) holds and* $\operatorname{Ker} A = \{0\}$.

Theorem 10.14. *Let A and B be c.s.o. in H. Equation (1) is asymptotically stable on R_+ if and only if (10.14) holds and* $\operatorname{Ker} A = \operatorname{Ker} B = \{0\}$.

Theorem 10.15. *Let A and B be c.s.o. in H.*

1. *Equation (1) is exponentially stable on R_+ if and only if one of the following equivalent conditions holds:*

   1) $\sup_{(\lambda,\mu)\in\sigma(A,B), i=1,2} \operatorname{Re}\omega_i(\lambda,\mu) < 0$;

   2) $\exists 0 < \gamma < +\infty : \sigma(A,B) \subseteq \{(\lambda,\mu) \in R^2 | \lambda \geq 2\gamma,\ \mu \geq \gamma\lambda - \gamma^2\}$ (10.15) (see Fig. 10.2);

   3) $\inf \sigma(A) > 0$, $\inf \sigma(B) > 0$, *and A is subordinate to B (i.e., $D(B) \subseteq D(A)$).*
2. *Equation (1) is uniformly exponentially stable on R_+ in the space of initial data $G = H \times H$ if and only if (10.15) holds.*
3. *The previous statement, with $H \times H$ replaced by $H \times H_{-1} = H \times H_{-}(((|A|^2 + |B| + 1)^{1/2})$.*

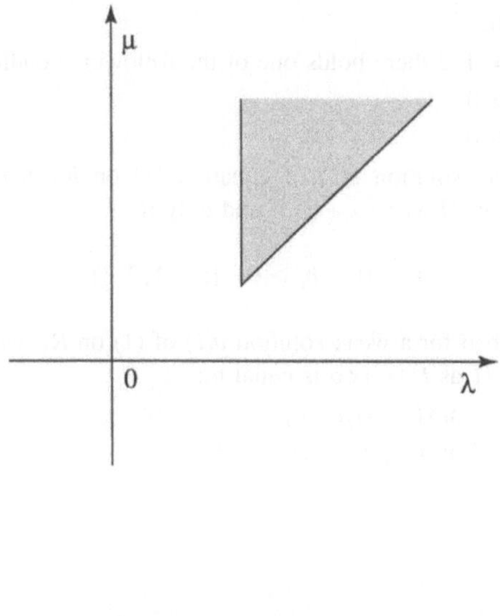

Fig. 10.2

Theorem 10.16. *Let A and B be c.s.o. in H. For every weak solution $y(t)$ of equation (1) on R_+ the function $t^{-1}y(t)$ $(t > 0)$ has a limit in H as $t \to +\infty$ if and only if one of the following equivalent conditions holds:*

1) $\sup_{(\lambda,\mu)\in\sigma(A,B),i=1,2} \operatorname{Re}\omega_i(\lambda,\mu) \le 0$;
2) $\sigma(A,B) \subseteq \{(\lambda,\mu) | \lambda \ge 0,\ \mu \ge 0\}$;
3) $\inf \sigma(A) \ge 0$, $\inf \sigma(B) \ge 0$.

If this is the case then for a weak solution $y(t)$ of equation (1) on R_+ with i.d. (f_0, f_1) : $\lim_{t\to+\infty} t^{-1}y(t) = E\{(0,0)\}f_1$.

Example 10.1. Let A and B be the c.s.o. from Example 1.1. Equation (1) is stable on R_+ if and only if:

1) $a_i \ge 0$, $b_i \ge 0$ $(i = 1,2,3)$;
2) $\max(a_3, b_3) > 0$.

Equation (1) is stabilizable on R_+ if and only if:

1) $a_i \ge 0$, $b_i \ge 0$ $(i = 1,2,3)$;
2) $\max(a_3, b_3) > 0$;
3) $\max(a_1, a_2, a_3) > 0$.

Notice that if 1) and 2) hold but 3) does not (i.e., $a_1 = a_2 = a_3 = 0$) then no weak solution of (1) on R_+ has a limit in H as $t \to +\infty$ (except for the trivial weak solution $y(t) \equiv 0$, $t \in R_+$).

Equation (1) is asymptotically stable on R_+ if and only if:

1) $a_i \ge 0$, $b_i \ge 0$ $(i = 1,2,3)$;
2) $\max(a_3, b_3) > 0$;
3) $\max(a_1, a_2, a_3) > 0$, $\max(b_1, b_2, b_3) > 0$.

Equation (1) is exponentially stable on R_+ if and only if:

1) $a_3 > 0$, $b_3 > 0$;
2) for each of $i = 1,2$ there holds one of the following conditions:
   a) $a_i > 0$, $b_i > 0$;
   b) $a_i = 0$, $b_i \ge 0$.

For every weak solution $y(t)$ of equation (1) on R_+ the function $t^{-1}y(t)$ $(t > 0)$ has a limit in H as $t \to +\infty$ if and only if:

$$a_i \ge 0, \quad b_i \ge 0 \quad (i = 1,2,3).$$

If this is the case then for a weak solution $y(t)$ of (1) on R_+ with i.d. (f_0, f_1) the limit of $t^{-1}y(t)$ in H as $t \to +\infty$ is equal to:

0 in the case of $\max(a_1, a_2, a_3, b_1, b_2, b_3) > 0$;

f_1 in the case of $\max(a_1, a_2, a_3, b_1, b_2, b_3) = 0$.

Part IV
Boundary-Value Problems on a Half-Line

Introduction to Part IV

We now return to the study of well-posedness of boundary-value problems. In Chapter 3, we investigated well-posedness of the Cauchy problem for equation (1) on the half-line $R_+ = [0, +\infty)$ as well as on a finite segment $[0, T]$. In Chapter 4, we studied well-posedness of the Dirichlet and Neumann problems for equation (1) on a finite segment $[0, T]$.

In the present part, we consider well-posedness of the Dirichlet and Neumann problems for equation (1) on the half-line R_+.

Definition 11.1. Let A and B be c.n.o. in H, $f \in H$. A solution $y(t)$ of equation (1) on R_+ such that $y(0) = f$ $(y'(0) = f)$ is said to be a solution of the Dirichlet (Neumann) problem for (1) on R_+ with boundary data f.

Definition 11.2. Let A and B be c.n.o. in H, $f \in H$ $(f \in \Phi')$. A weak solution $y(t)$ of equation (1) on R_+ such that $y(0) = f$ $((y(t), g)'|_{t=0} = (f, g),\ \forall g \in \Phi)$ is said to be a weak solution of the Dirichlet (Neumann) problem for (1) on R_+ with boundary data f.

In Chapter 11, we consider the Dirichlet problem for equation (1) on R_+. The situation for this problem on $[0, +\infty)$ differs essentially from the situation for this problem on a finite segment. Let us explain this.

Let A and B be arbitrary c.n.o. in H. A boundary condition $y(0) = f_0$ does not single out a unique solution (weak solution) of equation (1) on R_+. Indeed, let $f_0 = 0$. For any $k \geq 1$ and any $f_1 \in E_k H : y(t) = \psi_1(A, B, t) f_1 (t \in R_+)$ is a solution of (1) on R_+ such that $y(0) = 0$.

We obtain that in the space $C(R_+, H)$ there is more than one solution of equation (1) on R_+ such that $y(0) = 0$.

For the condition $y(0) = 0$ to single out a unique solution (weak solution) $y(t)$ of equation (1) on R_+, we seek this solution (weak solution) in a linear subspace of $C(R_+, H)$ which is narrower than the whole $C(R_+, H)$. In this narrower space, a solution (weak solution) $y(t)$ of (1) on R_+ such that $y(0) = 0$ may be unique.

Definition 11.3. Let A and B be c.n.o. in H. A linear subspace of $C(R_+, H)$ is said to be a class of uniqueness (weak uniqueness) for the Dirichlet problem for equation (1) on R_+ if there exists a unique solution (weak solution) $y(t)$ of (1) on R_+ from this subspace such that $y(0) = 0$.

Let A and B be arbitrary c.n.o. in H. The linear subspace of $C(R_+,H)$ which consists of the only function $y(t)=0$ ($t\in R_+$) is a class of (weak) uniqueness for the Dirichlet problem for equation (1) on R_+. The whole $C(R_+,H)$ is not a class of (weak) uniqueness for the Dirichlet problem for equation (1) on R_+.

In general, one and the same linear subspace of $C(R_+,H)$ may be a class of (weak) uniqueness for the Dirichlet problem for a certain equation (1) on R_+ but not be a class of (weak) uniqueness for another equation.

We consider different linear subspaces of $C(R_+,H)$. Each of these subspaces is defined by a certain condition on behaviour of functions at infinity.

Definition 11.4. Let $\gamma\in R^1$. Denote by Y_γ the space of all $y(t)\in C(R_+,H)$ such that $e^{-\gamma t}\parallel y(t)\parallel\to 0$ as $t\to+\infty$ (i.e., $\exists C_y(t)\in C(R_+)$ such that $C_y(t)\to 0$ as $t\to+\infty$ and $\parallel y(t)\parallel\leq C_y(t)e^{\gamma t}$, $\forall t\in R_+$).

For instance, if $\gamma=0$ then Y_γ is the space of all functions $y(t)\in C(R_+,H)$ such that $y(t)$ converges to 0 in H as $t\to+\infty$.

For each of these linear spaces Y_γ, we answer the following question: to indicate *all* pairs of c.n.o. A and B in H such that this space is a class of (weak) uniqueness for the Dirichlet problem for equation (1) on R_+.

This subspace of $C(R_+,H)$ corresponds to a certain condition on behaviour of functions at infinity. In fact, we indicate *all* equations (1) with c.n.o. A and B in H such that there exists a unique (weak) solution $y(t)$ of (1) on R_+ with $y(0)=0$ which satisfies this condition on behaviour at infinity.

Starting from these results, we consider (weak) well-posedness of the Dirichlet problem for equation (1) on R_+.

Definition 11.5. Let A and B be c.n.o. in H, U be a linear subspace of $C(R_+,H)$, and G be a linear subspace of H. We shall say that the Dirichlet problem in the class of functions U for equation (1) on R_+ is (weakly) well-posed in the space of boundary data G if for each $f\in G$ there exists a unique (weak) solution of equation (1) on R_+ from U such that $y(0)=f$.

Obviously, Definition 11.5 holds if and only if simultaneously:

1) U is a class of (weak) uniqueness for the Dirichlet problem for equation (1) on R_+;
2) for each $f\in G$ there exists a (weak) solution of equation (1) on R_+ from U with $y(0)=f$.

For each of the indicated linear subspaces of $C(R_+,H)$, we give a condition on c.n.o. A and B in H which is necessary and sufficient for the Dirichlet problem in this space of functions for equation (1) on R_+ to be weakly well-posed in G=H. Similarly, we consider well-posedness of the Dirichlet problem in G= $D(B)\cap D(A^2)$.

The Neumann problem for equation (1) on R_+ is treated analogously in Chapter 12. One may consider the Dirichlet and Neumann problems for equation (1) on $(-\infty,0]$ quite similarly (cf. Section 6 of Chapter 4).

Chapter 11
The Dirichlet problem on a half-line

1 Classes of (weak) uniqueness

Let A and B be arbitrary c.n.o. in H. By Lemma 6.1, to each weak solution $y(t)$ of equation (1) on R_+ there correspond $f_0 = y(0) \in H$ and an $\mathbb{E}$-sequence in H $\{f_1^{(k)}\}_{k=1}^{\infty}$ such that $\forall k \geq 1 : f_1^{(k)} = (E_k y)'(0)$, and

$$y(t) = \lim_{k \to +\infty} (\psi_0(A,B,t)E_k f_0 + \psi_1(A,B,t)f_1^{(k)}) \quad (t \in R_+).$$

Lemma 11.1. *Let $\gamma \in R^1$, A and B be c.n.o. in H such that $\sigma(A,B) \subseteq \{(\lambda,\mu) \in C^2 | \operatorname{Re}\omega_1 \geq \gamma\}$. If $y(t)$ is a weak solution of equation* (1) *on R_+ and $y(t) \in Y_\gamma$, then $\forall k \geq 1 : f_1^{(k)} = \omega_2(A,B)E_k f_0$, and $y(t) = e^{\omega_2 t}(A,B)f_0$ $(t \in R_+)$.*

Proof. It suffices to prove that $\forall m \geq 1 : f_1^{(m)} = \omega_2(A,B)E_m f_0$ and $E_m y(t) = e^{\omega_2 t}(A,B)E_m f_0$ $(t \in R_+)$; then

$$\forall t \in R_+ : \; y(t) = \lim_{m \to \infty} e^{\omega_2 t}(A,B)E_m f_0 = e^{\omega_2 t}(A,B)f_0.$$

Fix an arbitrary $k \geq 1$. Denote:

$$U_k = \square_k \cap \{(\lambda,\mu) \in C^2 | \, |\omega_1 - \omega_2| \geq \frac{1}{k}\},$$
$$V_k = \square_k \cap \{(\lambda,\mu) \in C^2 | \omega_1 = \omega_2\}.$$

By virtue of Lemma 6.1,

$$\begin{aligned} E(U_k)y(t) &= \psi_0(A,B,t)E(U_k)f_0 + \psi_1(A,B,t)E(U_k)f_1^{(k)} \\ &= e^{\omega_1 t}(A,B)h_1 + e^{\omega_2 t}(A,B)h_2 \; (t \in R_+), \end{aligned}$$

where

$$\begin{aligned} h_1 &= (\omega_1 - \omega_2)^{-1}(A,B)[-\omega_2(A,B)E(U_k)f_0 + E(U_k)f_1^{(k)}] \in E_k H, \\ h_2 &= (\omega_1 - \omega_2)^{-1}(A,B)[\omega_1(A,B)E(U_k)f_0 - E(U_k)f_1^{(k)}] \in E_k H. \end{aligned}$$

Consequently, $h_1 + e^{(\omega_2 - \omega_1)t}(A,B)h_2 = e^{-\omega_1 t}(A,B)E(U_k)y(t)$ $(t \in R_+)$.

Since $\sigma(A,B) \subseteq \{(\lambda,\mu) \in C^2 | \mathrm{Re}\,\omega_1 \geq \gamma\}$, we have for any $t \in R_+$: $\| e^{-\omega_1 t}(A,B)E(U_k)y(t) \| \leq e^{-\gamma t} \| E(U_k)y(t) \| \leq e^{-\gamma t} \| y(t) \|$. But $y(t) \in Y_\gamma$ means that $e^{-\gamma t} \| y(t) \| \to 0$ as $t \to +\infty$. Thus, $h_1 + e^{(\omega_2-\omega_1)t}(A,B)h_2 \to 0$ as $t \to +\infty$.

Accordingly, $(h_1 + e^{(\omega_2-\omega_1)t}(A,B)h_2)^{\tilde{}} \to 0$ as $t \to +\infty$ (recall that $\tilde{}$ denotes the Cesaro mean for a function, see Section 2 of Chapter 10). But for any $(\lambda,\mu) \in \sigma(A,B) \cap U_k$: $\mathrm{Re}(\omega_2 - \omega_1) \leq 0$ and $|\omega_1 - \omega_2| \geq \frac{1}{k}$. Therefore $(e^{(\omega_2-\omega_1)t}(A,B)h_2)^{\tilde{}} = \frac{e^{(\omega_2-\omega_1)t}-1}{(\omega_2-\omega_1)t}(A,B)h_2 \to 0$ as $t \to +\infty$. Thus, $(h_1)^{\tilde{}} = h_1 \to 0$ as $t \to +\infty$, i.e., $h_1 = 0$.

This means that $-\omega_2(A,B)E(U_k)f_0 + E(U_k)f_1^{(k)} = 0$, i.e., $E(U_k)f_1^{(k)} = \omega_2(A,B)E(U_k)f_0$. Now

$$h_2 = (\omega_1 - \omega_2)^{-1}(A,B)(\omega_1(A,B)E(U_k)f_0 - \omega_2(A,B)E(U_k)f_0) = E(U_k)f_0.$$

Finally, $E(U_k)y(t) = e^{\omega_1 t}(A,B)h_1 + e^{\omega_2 t}(A,B)h_2 = e^{\omega_2 t}(A,B)E(U_k)f_0$ $(t \in R_+)$.

Further,

$$E(V_k)y(t) = e^{\omega_1 t}(A,B)[E(V_k)f_0 + t(-\omega_2(A,B)E(V_k)f_0 + E(V_k)f_1^{(k)})] \ (t \in R_+),$$

whence

$$E(V_k)f_0 + t(-\omega_2(A,B)E(V_k)f_0 + E(V_k)f_1^{(k)}) = e^{-\omega_1 t}(A,B)E(V_k)y(t) \ (t \in R_+).$$

As previously, we obtain:

$$E(V_k)f_0 + t(-\omega_2(A,B)E(V_k)f_0 + E(V_k)f_1^{(k)}) \to 0 \text{ as } t \to +\infty.$$

This is the case if and only if $E(V_k)f_0 = 0$ and $-\omega_2(A,B)E(V_k)f_0 + E(V_k)f_1^{(k)} = 0$. Hence $E(V_k)f_1^{(k)} = \omega_2(A,B)E(V_k)f_0$ and $E(V_k)y(t) = 0 = e^{\omega_2 t}(A,B)E(V_k)f_0$ $(t \in R_+)$.

Summarizing, $\forall k \geq 1$:

$$\begin{aligned}
E(U_k)f_1^{(k)} &= \omega_2(A,B)E(U_k)f_0, \\
E(V_k)f_1^{(k)} &= \omega_2(A,B)E(V_k)f_0, \\
E(U_k)y(t) &= e^{\omega_2 t}(A,B)E(U_k)f_0 \ (t \in R_+), \\
E(V_k)y(t) &= e^{\omega_2 t}(A,B)E(V_k)f_0 \ (t \in R_+).
\end{aligned}$$

Hence $\forall m \geq 1 : f_1^{(m)} = \omega_2(A,B)E_m f_0$ and $E_m y(t) = e^{\omega_2 t}(A,B)E_m f_0$ $(t \in R_+)$. The proof is over.

Theorem 11.1. *Let $\gamma \in R^1$, A and B be c.n.o. in H. Y_γ is a class of uniqueness (weak uniqueness) for the Dirichlet problem for equation* (1) *on R_+ if and only if one of the following equivalent conditions holds:*

1) $\forall(\lambda,\mu) \in \sigma(A,B) : \max\limits_{i=1,2} \operatorname{Re}\omega_i \geq \gamma;$ (11.1)

2) $\sigma(A,B) \subseteq C^2 \setminus \{(\lambda,\mu)|\lambda_1 > -2\gamma,\ \mu_1 > (\lambda_1+2\gamma)^{-2}(\mu_2^2 - \lambda_1\lambda_2 \cdot \mu_2 - \gamma(\lambda_1+\gamma)(\lambda_2^2 + (\lambda_1+2\gamma)^2))\}.$ (11.2)

Proof. The equivalence of conditions 1) and 2) follows from the proof of Lemma 3.2. The sufficiency of condition 1) follows immediately from Lemma 11.1. Indeed, if $y(t) \in Y_\gamma$ is a weak solution of (1) on R_+ and $f_0 = y(0) = 0$, then, by Lemma 11.1, $y(t) = e^{\omega_2 t}(A,B)f_0 = 0$ $(t \in R_+)$.

It remains to prove the necessity of condition 1). In other words, we have to prove that if $\exists(\lambda_0,\mu_0) \in \sigma(A,B) : \max\limits_{i=1,2} \operatorname{Re}\omega_i(\lambda_0,\mu_0) < \gamma$, then there exists a nontrivial solution $y(t)$ of equation (1) on R_+ such that $y(t) \in Y_\gamma$ and $y(0) = 0$.

Let's indicate such a solution. By the assumption,

$$\exists 0 < \delta < +\infty : \ \gamma_0 = \max_{i=1,2} \operatorname{Re}\omega_i(\lambda_0,\mu_0) + \delta < \gamma.$$

Set

$$\begin{aligned}\Delta =& \{(\lambda,\mu) \in C^2 |\ |\lambda-\lambda_0|^2 + |\mu-\mu_0|^2 < 1\}\\ &\cap \{(\lambda,\mu) \in C^2 |\ |\operatorname{Re}\omega_1(\lambda,\mu) - \operatorname{Re}\omega_1(\lambda_0,\mu_0)| < \delta\}.\end{aligned}$$

Δ is an open bounded set in C^2. Since $\Delta \ni (\lambda_0,\mu_0)$ and $(\lambda_0,\mu_0) \in \sigma(A,B)$, we conclude that $E(\Delta) \neq 0$. Fix an arbitrary $f_1 \in E(\Delta)H$ such that $\| f_1 \| = 1$; this is possible since $E(\Delta) \neq 0$. Then $y(t) = \psi_1(A,B,t)f_1$ is a solution of (1) on R_+ such that $y(0) = 0$, $y'(0) = f_1$. Here $y'(0) \neq 0$, and so $y(t)$ is a nontrivial solution. For any $(\lambda,\mu) \in \Delta$, $t \in R_+$:

$$|\psi_1(\lambda,\mu,t)| = |te^{\omega_1 t}| \cdot \left|\frac{e^{(\omega_2-\omega_1)t} - 1}{(\omega_2-\omega_1)t}\right| \leq 2te^{\gamma_0 t};$$

hence $\| \psi_1(A,B,t)f_1 \| \leq 2te^{\gamma_0 t}$ $(t \in R_+)$ and so $y(t) \in Y_\gamma$. The proof is over.

Corollary 1. *Let $\gamma \in R^1$, A and B be commuting self-adjoint operators in H. Y_γ is a class of uniqueness (weak uniqueness) for the Dirichlet problem for equation* (1) *on R_+ if and only if $\sigma(A,B) \subseteq R^2 \setminus \{(\lambda,\mu)|\lambda > -2\gamma,\ \mu > -\gamma\lambda - \gamma^2\}$* (see Fig. 11.1).

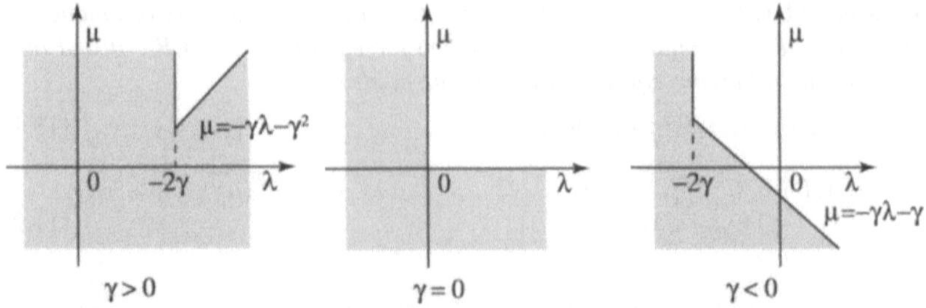

Fig. 11.1

2 Existence of (weak) solutions

Theorem 11.2. *Let $\gamma \in R^1$, A and B be c.n.o. in H. The following conditions are equivalent:*

1) *for each $f \in H$ there exists a weak solution of equation* (1) *on R_+ $y(t) \in Y_\gamma$ such that $y(0) = f$;*
2) *for each $f \in D(A^2) \cap D(B)$ there exists a solution of equation* (1) *on R_+ $y(t) \in Y_\gamma$ such that $y(0) = f$;*
3) $E\{(\lambda, \mu) \in C^2 | \min\limits_{i=1,2} \operatorname{Re} \omega_i \geq \gamma\} = 0;$ (11.3)
4) $\forall (\lambda, \mu) \in \sigma(A, B) : \min\limits_{i=1,2} \operatorname{Re} \omega_i \leq \gamma,$
and moreover, $E\{(\lambda, \mu) \in C^2 | \min\limits_{i=1,2} \operatorname{Re} \omega_i = \gamma\} = 0;$
5) $\sigma(A, B) \subseteq C^2 \setminus \{(\lambda, \mu) | \lambda_1 < -2\gamma,\ \mu_1 > (\lambda_1 + 2\gamma)^{-2}(\mu_2^2 - \lambda_1\lambda_2 \cdot \mu_2 - \gamma(\lambda_1 + \gamma)(\lambda_2^2 + (\lambda_1 + 2\gamma)^2))\}$, *and moreover,* $E(\{(\lambda, \mu) \in C^2 | \lambda_1 < -2\gamma,\ \mu_1 = (\lambda_1 + 2\gamma)^{-2}(\mu_2^2 - \lambda_1\lambda_2 \cdot \mu_2 - \gamma(\lambda_1 + \gamma)(\lambda_2^2 + (\lambda_1 + 2\gamma)^2))\} \cup \{(\lambda, \mu) \in C^2 | \lambda_1 = -2\gamma,\ \mu_2 = -\gamma\lambda_2,\ \mu_1 \geq -\frac{1}{4}\lambda_2^2 + \gamma^2\}) = 0.$

Proof. Similarly to the proof of Lemma 1.2, we obtain from the proof of Lemma 3.2 that $\forall \gamma \in R^1 : \{(\lambda, \mu) \in C^2 | \min\limits_{i=1,2} \operatorname{Re} \omega_i > \gamma\} = \{(\lambda, \mu) \in C^2 | \lambda_1 < -2\gamma,\ \mu_1 > (\lambda_1 + 2\gamma)^{-2}(\mu_2^2 - \lambda_1\lambda_2 \cdot \mu_2 - \gamma(\lambda_1 + \gamma)(\lambda_2^2 + (\lambda_1 + 2\gamma)^2))\}$. Further, $\forall \gamma_1 \in R^1 : \{(\lambda, \mu) \in C^2 | \min\limits_{i=1,2} \operatorname{Re} \omega_i = \gamma_1\} = (\bigcap\limits_{\gamma < \gamma_1} \{(\lambda, \mu) \in C^2 | \min\limits_{i=1,2} \operatorname{Re} \omega_i > \gamma\}) \setminus \{(\lambda, \mu) \in C^2 | \min\limits_{i=1,2} \operatorname{Re} \omega_i > \gamma_1\}$. This completes the proof of the equivalence of conditions 4) and 5).

Let's prove the necessity of condition 4) for 1); the necessity for 2) can be proved quite similarly.

Recall that $\forall (\lambda, \mu) \in C^2 : \operatorname{Re} \omega_2 = \min\limits_{i=1,2} \operatorname{Re} \omega_i$. Denote: $P_1 = \{(\lambda, \mu) \in C^2 | \operatorname{Re} \omega_2 > \gamma\}$, $P_2 = \{(\lambda, \mu) \in C^2 | \operatorname{Re} \omega_2 = \gamma\}$, $P_3 = \{(\lambda, \mu) \in C^2 | \operatorname{Re} \omega_2 < \gamma\}$. For each of $i = 1, 2, 3$, let $H_i = E(P_i)H$, A_i and B_i be the restrictions of A and

B to H_i. Then A_i and B_i are c.n.o. in H_i, $\sigma(A_i, B_i) \subseteq \sigma(A, B) \cap \overline{P_i}$. If $y(t)$ is a weak solution of (1) in H on R_+ with $y(0) = f$, then for each of $i = 1, 2, 3$: $y_i(t) = E(P_i)y(t)$ is a weak solution of the equation $y_i''(t) + A_i y_i'(t) + B_i y_i(t) = 0$ in H_i on R_+ with $y_i(0) = E(P_i)f$. Thus, if 1) holds for equation (1) in H, then it holds for the equation $y_i''(t) + A_i y_i'(t) + B_i y_i(t) = 0$ in H_i for each of $i = 1, 2, 3$.

Let 1) be valid.

Assume that $E(P_1) \neq 0$, i.e., $\exists(\lambda_0, \mu_0) \in \sigma(A, B) : \operatorname{Re}\omega_2(\lambda_0, \mu_0) > \gamma$; then $\exists 0 < \delta < +\infty : \gamma_0 = \operatorname{Re}\omega_2(\lambda_0, \mu_0) - \delta > \gamma$. Set $\Delta_1 = \{(\lambda, \mu) \in C^2 |\, |\lambda - \lambda_0|^2 + |\mu - \mu_0|^2 < 1\} \cap \{(\lambda, \mu) \in C^2 |\, |\operatorname{Re}\omega_2(\lambda, \mu) - \operatorname{Re}\omega_2(\lambda_0, \mu_0)| < \delta\}$. Δ_1 is an open bounded set in C^2. Since $\Delta_1 \ni (\lambda_0, \mu_0)$ and $(\lambda_0, \mu_0) \in \sigma(A, B)$, we conclude that $E(\Delta_1) \neq 0$. Fix an arbitrary $f_1 \in E(\Delta_1)H \subseteq E(P_1)H$ such that $\| f_1 \| = 1$. Condition 1) for the equation $y_1'' + A_1 y_1' + B_1 y_1 = 0$ holds; thus, there exists a weak solution of this equation in H_1 on R_+ $y_1(t) \in Y_\gamma$ such that $y_1(0) = f_1$. Further, $\sigma(A_1, B_1) \subseteq \{(\lambda, \mu) \in C^2 | \operatorname{Re}\omega_2 \geq \gamma\} \subseteq \{(\lambda, \mu) \in C^2 | \max_{i=1,2} \operatorname{Re}\omega_i \geq \gamma\}$. Applying Lemma 11.1 to A_1, B_1 in H_1, we have: $y_1(t) = e^{\omega_2 t}(A_1, B_1)f_1$ $(t \in R_+)$. For any $(\lambda, \mu) \in \Delta_1 : |e^{\omega_2 t}| = e^{\operatorname{Re}\omega_2 t} \geq e^{\gamma_0 t}$ $(t \in R_+)$; hence $e^{-\gamma t} \| y_1(t) \| \geq e^{(\gamma_0 - \gamma)t} \to +\infty$ as $t \to +\infty$, which contradicts $y_1(t) \in Y_\gamma$.

Thus, $\sigma(A, B) \subseteq \{(\lambda, \mu) \in C^2 | \operatorname{Re}\omega_2 \leq \gamma\}$. Assume that $E(P_2) \neq 0$. Then $\exists(\lambda_0, \mu_0) \in \sigma(A, B) \cap P_2$ such that for $\Delta_2 = \{(\lambda, \mu) \in C^2 |\, |\lambda - \lambda_0|^2 + |\mu - \mu_0|^2 < 1\} \cap P_2 : E(\Delta_2) \neq 0$. Fix an arbitrary $f_2 \in E(\Delta_2)H \subseteq E(P_2)H$ such that $\| f_2 \| = 1$. Just as previously, we obtain that $y_2(t) = e^{\omega_2 t}(A_2, B_2)f_2 \in Y_\gamma$. On the other hand, $\forall(\lambda, \mu) \in \Delta_2 : |e^{\omega_2 t}| = e^{\gamma t}$ $(t \in R_+)$, and so $e^{-\gamma t} \| y_2(t) \| = 1$ $(t \in R_+)$. This contradiction shows that $E(P_2) = 0$.

The necessity is now proved. Let's prove the sufficiency of 4) for 1) and 2).

Let 4) be valid. For each $f \in H$ we set $y(t) = e^{\omega_2 t}(A, B)f$ $(t \in R_+)$. Obviously $y(0) = f$. Analogously to the proof of Theorem 3.1, we conclude that for any $f \in D(B) \cap D(A^2)$: $y(t)$ is a solution of (1) on R_+; for any $f \in H$: $y(t)$ is a weak solution of (1) on R_+. Further, $\sigma(A, B) \subseteq \{(\lambda, \mu) \in C^2 | \operatorname{Re}(\omega_2 - \gamma) \leq 0\}$; moreover, $E\{(\lambda, \mu) \in C^2 | \operatorname{Re}(\omega_2 - \gamma) = 0\} = 0$. Similarly to the proof of Lemma 10.1, we deduce that for any $f \in H$: $e^{-\gamma t} \| y(t) \| = \| e^{(\omega_2 - \gamma)t}(A, B)f \| \to 0$ as $t \to +\infty$.

The proof is over.

Corollary 1. *Let $\gamma \in R^1$, A and B be c.s.o. in H. Condition* 1) *(condition* 2)*) of Theorem* 11.2 *holds if and only if both of the following conditions hold:*

1) $\sigma(A, B) \subseteq R^2 \setminus \{(\lambda, \mu) | \lambda < -2\gamma,\ \mu > -\gamma\lambda - \gamma^2\}$; (11.4)

2) $E(\{(\lambda, \mu) | \lambda \leq -2\gamma,\ \mu = -\gamma\lambda - \gamma^2\} \cup \{(\lambda, \mu) | \lambda = -2\gamma,\ \mu > \gamma^2\}) = 0.$ (11.5)

See Fig. 11.2; the set with E-measure 0 is shown by a discontinuous line.

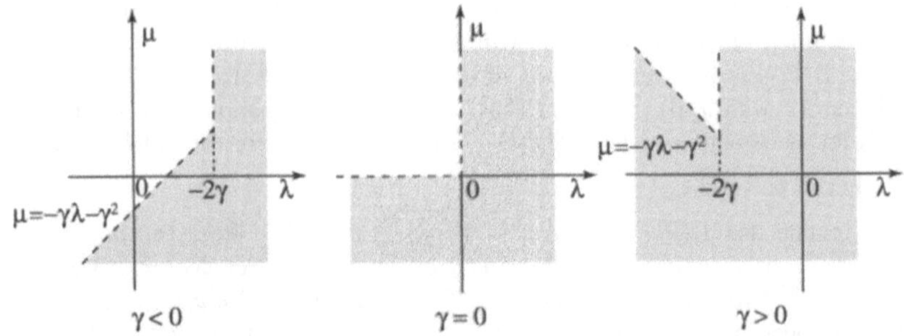

Fig. 11.2

3 A criterion for the (weak) well-posedness

Theorem 11.3. *Let $\gamma \in R^1$, A and B be c.n.o. in H. The Dirichlet problem in Y_γ for equation* (1) *on R_+ is weakly well-posed in H (well-posed in $D(B)\cap D(A^2)$) if and only if one of the following equivalent conditions holds:*

1) $E\{(\lambda,\mu)\in C^2|\min_{i=1,2}\operatorname{Re}\omega_i<\gamma,\ \max_{i=1,2}\operatorname{Re}\omega_i\geq\gamma\}=I$;

2) $\sigma(A,B)\subseteq\{(\lambda,\mu)\in C^2|\min_{i=1,2}\operatorname{Re}\omega_i\leq\gamma,\ \max_{i=1,2}\operatorname{Re}\omega_i\geq\gamma\}$, *and moreover,* $E\{(\lambda,\mu)\in C^2|\min_{i=1,2}\operatorname{Re}\omega_i=\gamma\}=0$;

3) $\sigma(A,B)\subseteq C^2\setminus\{(\lambda,\mu)\in C^2|\lambda_1\neq-2\gamma,\ \mu_1>(\lambda_1+2\gamma)^{-2}(\mu_2^2-\lambda_1\lambda_2\cdot\mu_2-\gamma(\lambda_1+\gamma)(\lambda_2^2+(\lambda_1+2\gamma)^2))\}$, *and moreover,* $E(\{(\lambda,\mu)\in C^2|\lambda_1<-2\gamma,\ \mu_1=(\lambda_1+2\gamma)^{-2}(\mu_2^2-\lambda_1\lambda_2\cdot\mu_2-\gamma(\lambda_1+\gamma)(\lambda_2^2+(\lambda_1+2\gamma)^2))\}\cup\{(\lambda,\mu)\in C^2|\lambda_1=-2\gamma,\ \mu_2=-\gamma\lambda_2,\ \mu_1\geq-\frac{1}{4}\lambda_2^2+\gamma^2\})=0$.

In this case, for every $f\in H$ the weak solution of the Dirichlet problem in Y_γ for equation (1) *on R_+ with boundary condition $y(0)=f$ has the form: $y(t)=e^{\omega_2 t}(A,B)f$ $(t\in R_+)$.*

The theorem follows immediately from Theorems 11.1 and 11.2 and Lemma 11.1, taking into account the remark after Definition 11.5 in Introduction to Part IV.

Corollary 1. *Let $\gamma\in R^1$, A and B be c.s.o. in H. The Dirichlet problem in Y_γ for equation* (1) *on R_+ is weakly well-posed in H (well-posed in $D(B)\cap D(A^2)$) if and only if $\sigma(A,B)\subseteq\{(\lambda,\mu)\in R^2|\mu\leq-\gamma\lambda-\gamma^2\}$, and moreover, $E\{(\lambda,\mu)|\lambda\leq-2\gamma,\ \mu=-\gamma\lambda-\gamma^2\}=0$.*

Let's analyze the last result.

For any $\lambda\in R^1$: $\frac{\lambda^2}{4}-(-\gamma\lambda-\gamma^2)=\frac{1}{4}(\lambda+2\gamma)^2$. Hence $\frac{\lambda^2}{4}\geq-\gamma\lambda-\gamma^2$, $\forall\lambda\in R^1$, with $\frac{\lambda^2}{4}=-\gamma\lambda-\gamma^2$ if and only if $\lambda=-2\gamma$.

In other words, the straight line $\{(\lambda,\mu)\in R^2|\mu=-\gamma\lambda-\gamma^2\}$ is tangent to the parabola $\mu=\frac{\lambda^2}{4}$ at the point $(-2\gamma,\gamma^2)$.

Now we may rewrite Corollary 1 in the following form:

To each $\gamma \in R^1$ there corresponds the straight line in R^2 tangent to the parabola $\mu = \frac{\lambda^2}{4}$ at the point $(-2\gamma, \gamma^2)$. We consider the half-plane in R^2 that is bounded by this straight line and situated below it (including the line). Moreover, we consider the ray on the tangent line having vertex at the point of tangency that lies to the left of this point (including the vertex). The Dirichlet problem in Y_γ for equation (1) on R_+ is weakly well-posed in H (well-posed in $D(B) \cap D(A^2)$) if and only if $\sigma(A, B)$ is situated in the indicated half-plane, and moreover, the indicated ray on the boundary of the half-plane has E-measure 0.

See Fig. 11.3; the set with E-measure 0 is shown by a discontinuous line.

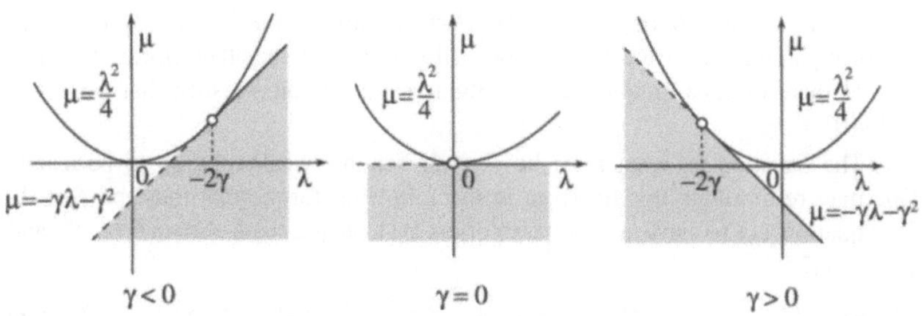

Fig. 11.3

Remark 1. Obviously, for $\gamma_1 > \gamma_2$: if Y_{γ_1} is a class of (weak) uniqueness for the Dirichlet problem for equation (1) on R_+, then Y_{γ_2} is such a class too. For $\gamma_1 < \gamma_2$: if Y_{γ_1} contains a (weak) solution of (1) on R_+ $y(t)$ with $y(0) = f$ for every $f \in D(B) \cap D(A^2)$ (for every $f \in H$), then Y_{γ_2} has the same property.

As for the (weak) well-posedness of the Dirichlet problem in Y_γ for equation (1) on R_+ in $G = D(B) \cap D(A^2)$ (in $G = H$), the situation is completely different. Equation (1) may have this property for a certain $\gamma_1 \in R^1$ but not have it for any other $\gamma_2 \in R^1$. This is the case, for instance, if A and B are c.s.o. in H and $\sigma(A, B)$ is the ray on the tangent line to $\mu = \frac{\lambda^2}{4}$ at the point $(-2\gamma_1, \gamma_1^2)$ having vertex at the point of tangency which lies to the right of this point (including it), and moreover, the vertex of the ray has E-measure 0.

Example 11.1. Let $\gamma \in R^1$, A and B be the c.s.o. from Example1.1. The Dirichlet problem in Y_γ for equation (1) on R_+ is weakly well-posed in $G = H$ (well-posed in $G = D(B) \cap D(A^2)$) if and only if simultaneously:

1) $b_3 \leq -\gamma a_3 - \gamma^2$;
2) for each of $i = 1, 2$ there holds one of the following conditions:
   a) $a_i > 0$, $\frac{b_i}{a_i} \leq -\gamma$;
   b) $a_i = 0$, $b_i \leq 0$;
   c) $a_i < 0$, $\frac{b_i}{a_i} \geq -\gamma$;

3) if $b_3 = -\gamma a_3 - \gamma^2$, $a_1 a_2 \neq 0$, and $\frac{b_1}{a_1} = \frac{b_2}{a_2} = -\gamma$, then $a_1 > 0$, $a_2 > 0$, and $a_3 \geq -2\gamma$;
if $b_3 = -\gamma a_3 - \gamma^2$, $a_1 = b_1 = 0$, $a_2 \neq 0$, and $\frac{b_2}{a_2} = -\gamma$, then $a_2 > 0$ and $a_3 \geq -2\gamma$;
if $b_3 = -\gamma a_3 - \gamma^2$, $a_2 = b_2 = 0$, $a_1 \neq 0$, and $\frac{b_1}{a_1} = -\gamma$, then $a_1 > 0$ and $a_3 \geq -2\gamma$;
if $b_3 = -\gamma a_3 - \gamma^2$ and $a_1 = a_2 = b_1 = b_2 = 0$, then $a_3 > -2\gamma$.

4 Boundary data of solutions

Let $\gamma \in R^1$, A and B be c.n.o. in H. Furthermore, let the Dirichlet problem in Y_γ for equation (1) on R_+ be weakly well-posed in H. In other words, for each $f_0 \in H$ there exists a unique weak solution $y(t)$ of (1) on R_+ such that $y(t) \in Y_\gamma$ and $y(0) = f_0$.

The question is: what must be $f_0 \in H$ for this weak solution to be usual? One may reformulate this question in the following form: what must be $f_0 \in H$ for equation (1) to have a unique solution $y(t)$ on R_+ such that $y(t) \in Y_\gamma$ and $y(0) = f_0$?

Theorem 11.4. *Let $\gamma \in R^1$, A and B be c.n.o. in H, and condition* 1) *of Theorem* 11.3 *be valid. There exists a unique solution $y(t)$ of equation* (1) *on R_+ such that $y(t) \in Y_\gamma$ and $y(0) = f_0$ if and only if $f_0 \in D(B) \cap D(\omega_2^2(A,B))$, i.e., $E\{(\lambda,\mu) \in C^2 | \, |\omega_1| \geq |\omega_2|\} f_0 \in D(B)$ and $E\{(\lambda,\mu) \in C^2 | \, |\omega_1| < |\omega_2|\} f_0 \in D(B) \cap D(A^2)$.*

Proof. Necessity. If such a solution $y(t)$ exists, then, by Definition 1.1, $f_0 = y(0) \in D(B)$. By Theorem 11.3, $y(t) = e^{\omega_2 t}(A,B) f_0$ $(t \in R_+)$. Now, it is easy to see that $y(t) \in C^2(R_+, H)$ implies $y''(t) = (\omega_2^2 e^{\omega_2 t})(A,B) f_0$ $(t \in R_+)$; cf. the proof of Theorem 3.2. Setting here $t = 0$, we obtain: $y''(0) = \omega_2^2(A,B) f_0$, and so $f_0 \in D(\omega_2^2(A,B))$.
Sufficiency. For any $(\lambda,\mu,t) \in \sigma(A,B) \times R_+ : |e^{\omega_2 t}| \leq e^{\gamma t}$; further, $|\omega_2| \leq \frac{1}{2}(|\omega_2|^2 + 1)$, $|\lambda \omega_2| = |(\omega_1 + \omega_2)\omega_2| = |\mu + \omega_2^2|$; summarizing, $|(e^{\omega_2 t})^{(i)}|$ $(i = 0,1,2)$, $|\lambda (e^{\omega_2 t})'|$, $|\mu e^{\omega_2 t}| \leq e^{\gamma t}(|\mu| + |\omega_2|^2 + 1)$. Since $f_0 \in D(B) \cap D(\omega_2^2(A,B))$, $y(t) = e^{\omega_2 t}(A,B) f_0$ is a solution of (1) on R_+. It remains to refer to Theorem 11.3.

Corollary 1. *Let $\gamma \in R^1$, A and B be c.s.o. in H, $\sigma(A,B) \subseteq \{(\lambda,\mu) \in R^2 | \mu \leq -\gamma\lambda - \gamma^2\}$, and $E\{(\lambda,\mu) | \lambda \leq -2\gamma,\ \mu = -\gamma\lambda - \gamma^2\} = 0$. Equation* (1) *has a unique solution $y(t)$ on R_+ such that $y(t) \in Y_\gamma$ and $y(0) = f_0$ if and only if $E_A(-\infty,0) f_0 \in D(B)$ and $E_A[0,+\infty) f_0 \in D(A^2) \cap D(B)$.*

Remark 1. If $\exists 0 < C < +\infty : \sigma(A,B) \subseteq R^2 \setminus \{(\lambda,\mu) | \lambda < 0,\ |\mu| < C\lambda^2\}$, then $D(B) \cap D(\omega_2^2(A,B)) = D(B) \cap D(A^2)$. In the general case, $D(B) \cap D(\omega_2^2(A,B))$ may be strictly wider than $D(B) \cap D(A^2)$. For instance, let $\gamma > 0$, $\sigma(A) = (-\infty, -2\gamma]$, $B = 0$; then $D(B) \cap D(\omega_2^2(A,B)) = H$, whereas $D(B) \cap D(A^2) = D(A^2) \subset H$.

Chapter 12
The Neumann problem on a half-line

1 Classes of uniqueness (weak uniqueness)

Definition 12.1. Let A and B be c.n.o. in H. A linear subspace of $C(R_+,H)$ is said to be a class of uniqueness (weak uniqueness) for the Neumann problem for equation (1) on R_+ if this problem with boundary data $f=0$ has a unique solution (weak solution) from this subspace.

Theorem 12.1. *Let $\gamma \in R^1$, A and B be c.n.o. in H. Y_γ is a class of uniqueness (weak uniqueness) for the Neumann problem for equation (1) on R_+ if and only if one of the following equivalent conditions holds:*

1) (11.1) *holds, and moreover, if $\gamma > 0$, then $E\{(\lambda,\mu) \in \sigma(A,B)|\omega_2 = 0\} = 0$;*
2) (11.2) *holds, and moreover, if $\gamma > 0$, then $E\{(\lambda,\mu) \in \sigma(A,B)|\lambda_1 \leq -\gamma,$ $\mu = 0\} = 0$.*

Proof. Notice that if $\gamma > 0$ and (11.1) holds, then $\forall(\lambda,\mu) \in \sigma(A,B) : \mathrm{Re}\,\omega_1 \geq \gamma > 0$, hence $\omega_1 \neq 0$, and so, $\omega_2 = 0$ if and only if $\lambda_1 \leq -\gamma$, $\mu = 0$. Thus, 1) and 2) are equivalent.

Let's prove the sufficiency of 1). Denote: $P_1 = \{(\lambda,\mu) \in \sigma(A,B)|\omega_2 \neq 0\}$, $P_2 = \{(\lambda,\mu) \in \sigma(A,B)|\omega_2 = 0\}$. Let $y(t) \in Y_\gamma$ be a weak solution of the Neumann problem for (1) on R_+ with b.d. $f_1 = 0$. By Lemma 11.1, $\forall k \geq 1$: $f_1^{(k)} = \omega_2(A,B)E_k f_0$, and $y(t) = e^{\omega_2 t}(A,B)f_0$ $(t \in R_+)$. By the assumption, $\forall k \geq 1 : f_1^{(k)} = 0$. Thus, $E(P_1)f_0 = 0$; hence $E(P_1)y(t) = e^{\omega_2 t}(A,B)E(P_1)f_0 = 0$ $(t \in R_+)$.

Now we must prove that $E(P_2)y(t) = 0$ $(t \in R_+)$. By Lemma 11.1, $E(P_2)y(t) = e^{\omega_2 t}(A,B)E(P_2)f_0 = E(P_2)f_0$ $(t \in R_+)$. If $\gamma \leq 0$, then $y(t) \in Y_\gamma$ implies that $\| E(P_2)f_0 \| = \| E(P_2)y(t) \| \leq \| y(t) \| \to 0$ as $t \to +\infty$; so, $E(P_2)f_0 = 0$. If $\gamma > 0$, then $E(P_2) = 0$ by virtue of 1).

Let's prove the necessity of 1). Let Y_γ be a class of (weak) uniqueness for the Neumann problem for equation (1) on R_+. One can prove the necessity of (11.1) for this similarly to the proof of Theorem 11.1, with $\psi_1(A,B,t)f_1$ replaced by $\psi_0(A,B,t)f_0$.

Let $\gamma > 0$, and assume that $E(P_2) \neq 0$. Fix an arbitrary $f_0 \in E(P_2)H$ such that $\| f_0 \| = 1$. For any $(\lambda,\mu) \in P_2 : \mu = 0$; so, $y(t) = f_0$ $(t \in R_+)$ is a solution of (1) on R_+. Moreover, $e^{-\gamma t} \| y(t) \| = e^{-\gamma t} \| f_0 \| \to 0$ as $t \to +\infty$; $y'(0) = 0$, but $y(0) \neq 0$. The existence of such $y(t)$ contradicts our assumption that Y_γ is a class of (weak) uniqueness for the Neumann problem for (1) on R_+. Thus, $E(P_2) = 0$.

Corollary 1. *Let $\gamma \in R^1$, A and B be c.s.o. in H. Y_γ is a class of uniqueness (weak uniqueness) for the Neumann problem for equation* (1) *on R_+ if and only if both of the following conditions hold:*

1) $\sigma(A,B) \subseteq R^2 \setminus \{(\lambda,\mu)|\lambda > -2\gamma,\ \mu > -\gamma\lambda - \gamma^2\}$ (see Fig. 11.1);
2) *if $\gamma > 0$, then, moreover,* $E\{(\lambda,\mu)|\lambda \le -\gamma,\ \mu = 0\} = 0$ (see Fig. 12.1).

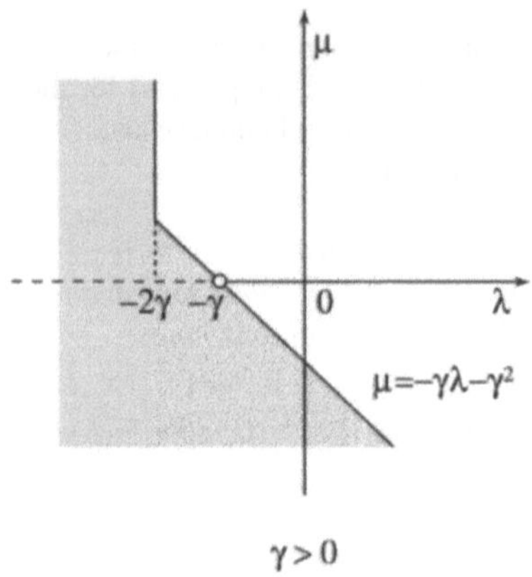

Fig. 12.1

2 Existence of solutions and weak solutions

Theorem 12.1. *Let $\gamma \in R^1$, A and B be c.n.o. in H. The following conditions* 1), 2) *are equivalent:*

1) *for each $f \in H_{-2}$ the Neumann problem for equation* (1) *on R_+ with b.d. f has a weak solution $y(t) \in Y_\gamma$;*
2) *a)* (11.3) *holds;*
   b) if $\gamma \ge 0$, then

$$\exists \varepsilon > 0 : E\{(\lambda,\mu) \in \sigma(A,B) | \operatorname{Re}\omega_1 \ge \gamma,\ \operatorname{Re}\omega_2 < \gamma,\ |\omega_2| < \varepsilon\} = 0;$$

   c) if $\gamma = 0$, then $(0,0) \notin \sigma(A,B)$.

Remark 1. For $\gamma > 0$ the following conditions are equivalent:

1) $\exists \varepsilon > 0 : E\{(\lambda,\mu) \in \sigma(A,B) | \operatorname{Re}\omega_1 \geq \gamma,\ \operatorname{Re}\omega_2 < \gamma,\ |\omega_2| < \varepsilon\} = 0$;
2) $\exists \varepsilon > 0 : E\{(\lambda,\mu) \in \sigma(A,B) | \operatorname{Re}\omega_1 \geq \gamma,\ \operatorname{Re}\omega_2 < \gamma,\ \min(|\omega_1|, |\omega_2|) < \varepsilon\} = 0$;
3) for $P \stackrel{\text{def}}{=} \{(\lambda,\mu) \in C^2 | \operatorname{Re}\omega_1 \geq \gamma,\ \operatorname{Re}\omega_2 < \gamma\} = C^2 \setminus (\{(\lambda,\mu)|\lambda_1 \neq -2\gamma,\ \mu_1 > (\lambda_1+2\gamma)^{-2}(\mu_2^2 - \lambda_1\lambda_2\cdot\mu_2 - \gamma(\lambda_1+\gamma)(\lambda_2^2 + (\lambda_1+2\gamma)^2))\} \cup \{(\lambda,\mu)|\lambda_1 < -2\gamma,\ \mu_1 = (\lambda_1+2\gamma)^{-2}(\mu_2^2 - \lambda_1\lambda_2\cdot\mu_2 - \gamma(\lambda_1+\gamma)(\lambda_2^2 + (\lambda_1+2\gamma)^2))\} \cup \{\lambda,\mu)|\lambda_1 = -2\gamma,\ \mu_2 = -\gamma\lambda_2,\ \mu_1 \geq -\frac{1}{4}\lambda_2^2 + \gamma^2\})$ there exists $\varepsilon > 0$ such that $E(P \cap \{(\lambda,\mu) \in C^2 | \min(|\mu|, |\lambda^{-1}\mu|) < \varepsilon\}) = 0$.

Proof of Theorem 12.2. Necessity. Let 1) be valid. Let's prove (11.3). At first, we prove that $E\{(\lambda,\mu) \in \sigma(A,B) | \operatorname{Re}\omega_2 \geq \gamma,\ \omega_2 \neq 0\} = 0$. The proof is quite similar to the proof of the necessity in Theorem 11.2, with Δ_i replaced by $\Delta_i \cap \{(\lambda,\mu) \in C^2 |\ |\omega_2(\lambda,\mu)| > \frac{1}{2}|\omega_2(\lambda_0,\mu_0)|\}$ and $y_i(t) = e^{\omega_2 t}(A_i,B_i)f_i$ replaced by $y_i(t) = (\omega_2^{-1}e^{\omega_2 t})(A_i,B_i)f_i$.

For $\gamma > 0 : \{(\lambda,\mu) \in \sigma(A,B) | \operatorname{Re}\omega_2 \geq \gamma\} = \{(\lambda,\mu) \in \sigma(A,B) | \operatorname{Re}\omega_2 \geq \gamma,\ \omega_2 \neq 0\}$.

For an arbitrary $\gamma \in R^1$, consider $\{(\lambda,\mu) \in \sigma(A,B) | \operatorname{Re}\omega_1 \geq \gamma,\ \omega_2 = 0\}$. Similarly to the proof of the necessity of Theorem 11.2, we obtain from Lemma 11.1 that any weak solution of equation (1) in $E\{(\lambda,\mu) \in \sigma(A,B) | \operatorname{Re}\omega_1 \geq \gamma,\ \omega_2 = 0\}H$ on R_+ $y(t) \in Y_\gamma$ has the form: $y(t) = f_0$ $(t \in R_+)$ where $f_0 \in H$. Thus, $y(t)$ is a solution of (1) on R_+ with $y'(t) = 0$. For 1) to be valid, it is necessary that $E\{(\lambda,\mu) \in \sigma(A,B) | \operatorname{Re}\omega_1 \geq \gamma,\ \omega_2 = 0\} = 0$.

For $\gamma \leq 0$:

$$\begin{aligned}&\{(\lambda,\mu) \in \sigma(A,B) | \operatorname{Re}\omega_2 \geq \gamma, \omega_2 = 0\}\\ &= \{(\lambda,\mu) \in \sigma(A,B) | \operatorname{Re}\omega_1 \geq \gamma, \omega_2 = 0\},\end{aligned}$$

and so $E\{(\lambda,\mu) \in \sigma(A,B) | \operatorname{Re}\omega_2 \geq \gamma,\ \omega_2 = 0\} = 0$.

Let $\gamma \geq 0$. We have shown that $E\{(\lambda,\mu) \in \sigma(A,B) | \operatorname{Re}\omega_1 \geq \gamma, \omega_2 = 0\} = 0$. Denote: $P = \{(\lambda,\mu) \in \sigma(A,B) | \operatorname{Re}\omega_1 \geq \gamma,\ \omega_2 \neq 0\}$. We now prove that $\exists \varepsilon > 0 : E(P \cap \{(\lambda,\mu)|\ |\omega_2| < \varepsilon\}) = 0$.

Consider an arbitrary $f \in E(P)H$. In view of 1), the Neumann problem for (1) on R_+ with b.d. f has a weak solution $y(t) \in Y_\gamma$. For this weak solution, $\forall k \geq 1 : f_1^{(k)} = E_k f$. By Lemma 11.1, $\forall k \geq 1 : E(\Box_k \cap \{(\lambda,\mu) \in P|\ |\omega_2| \geq \frac{1}{k}\})y(0) = \omega_2^{-1}(A,B)E(\Box_k \cap \{(\lambda,\mu) \in P|\ |\omega_2| \geq \frac{1}{k}\})f$. Thus, $y(0) = \omega_2^{-1}(A,B)f$.

We conclude that $\omega_2^{-1}(A,B)$ is defined on the whole $E(P)H$. Hence $\omega_2^{-1}(A,B)$ is a bounded linear operator on $E(P)H$, i.e.,

$$\exists \varepsilon > 0 :\ E(P \cap \{(\lambda,\mu)|\ |\omega_2| < \varepsilon\}) = 0.$$

Summarizing, $E\{(\lambda,\mu) \in \sigma(A,B) | \operatorname{Re}\omega_1 \geq \gamma,\ |\omega_2| < \varepsilon\} = 0$.

Let $\gamma = 0$. We have proved, in particular, that

$$\exists \varepsilon > 0: \ E\{(\lambda, \mu) \in \sigma(A,B) | \operatorname{Re} \omega_1 \geq 0, \max(|\omega_1|, |\omega_2|) < \varepsilon\} = 0.$$

In order to prove $(0,0) \notin \sigma(A,B)$, it remains to prove that

$$\exists \varepsilon_1 > 0: \ E\{(\lambda, \mu) \in \sigma(A,B) | \operatorname{Re} \omega_1 < 0, \ \max(|\omega_1|, |\omega_2|) < \varepsilon_1\} = 0.$$

Denote: $P = \{(\lambda, \mu) \in \sigma(A,B) | \max\limits_{i=1,2} \operatorname{Re} \omega_i < 0, \ \max\limits_{i=1,2} |\omega_i| < 1\}$. In view of 1), for any $f \in E(P)H$: the Neumann problem for (1) on R_+ with b.d. f has a weak solution $y_1(t) \in Y_0$. By Lemma 6.1, $y(t) = E(P)y_1(t)$ is a solution of (1) in $E(P)H$ on R_+ with $y'(0) = f$. Furthermore, $f_0 = y(0) \in E(P)H$, and $y(t) = e^{\omega_1 t}(A,B)f_0 + \psi_1(A,B,t)g$ $(t \in R_+)$, where $g = f - \omega_1(A,B)f_0$. Since $y_1(t) \in Y_0$, $\| y(t) \| = \| E(P)y_1(t) \| \to 0$ as $t \to +\infty$. Further, it was shown in the proof of Lemma 10.1 that $\| e^{\omega_1 t}(A,B)f_0 \| \to 0$ as $t \to +\infty$. Consequently, $\| \psi_1(A,B,t)g \| \to 0$ as $t \to +\infty$.

Thus, for any $f \in E(P)H$ there exists $f_0 \in E(P)H$ such that for $g = f - \omega_1(A,B)f_0 : \| \psi_1(A,B,t)g \| \to 0$ as $t \to +\infty$.

Assume that $(0,0) \in \sigma(A,B)$. For each $n \geq 1$, denote: $P_n = \{(\lambda, \mu) \in C^2 | \max\limits_{i=1,2} \operatorname{Re} \omega_i < 0, \ \max\limits_{i=1,2} |\omega_i| < \frac{1}{n^2}\}$. There exist sequences $\{(\lambda_n, \mu_n)\}_{n=1}^{\infty}$ and $\{\delta_n\}_{n=1}^{\infty}$ such that:

1) $\forall n \geq 1 : (\lambda_n, \mu_n) \in \sigma(A,B) \cap P_n$ and $\delta_n > 0$;
2) $\forall n \geq 1 : \Delta_n = \{(\lambda, \mu) \in C^2 | \, |\lambda - \lambda_n|^2 + |\mu - \mu_n|^2 < \delta_n^2\} \subseteq P_n$;
3) $\forall m > n \geq 1 : \Delta_m \cap \Delta_n = \emptyset$.

For each $n \geq 1 : (\lambda_n, \mu_n) \in \sigma(A,B)$, and so $E(\Delta_n) \neq 0$. We define $f^{(n)} \in E(\Delta_n)H$ so that $\| f^{(n)} \| = \frac{1}{n}$. Finally, we set $f = \sum\limits_{n=1}^{\infty} f^{(n)}$. Here $\| f \|^2 = \sum\limits_{n=1}^{\infty} \frac{1}{n^2} < +\infty$, and $f \in E(P)H$; moreover, $\forall n \geq 1 : \| E(P_n)f \| \geq \frac{1}{n}$.

Fix an arbitrary $f_0 \in E(P)H$. Clearly, $\forall n \geq 1 : \| E(P_n)\omega_1(A,B)f_0 \| \leq \frac{1}{n^2} \| f_0 \|$ and hence for $g = f - \omega_1(A,B)f_0 : \| E(P_n)g \| \geq \| E(P_n)f \| - \| E(P_n)\omega_1(A,B)f_0 \| \geq \frac{1}{n} - \frac{1}{n^2} \| f_0 \|$. Therefore $\exists n_0 \geq 1 \forall n \geq n_0 : \| E(P_n)g \| \geq \frac{1}{2n}$.

Notice that $\varepsilon_0 \stackrel{\text{def}}{=} \inf\limits_{|z| \leq 2} |\frac{e^z - 1}{z}| > 0$.

Take an arbitrary $n \geq n_0$ and set $t = n$. For any $(\lambda, \mu) \in P_n : |e^{\omega_1 t}| \geq e^{-1}$; $|\omega_1 - \omega_2| t \leq 2$, and so $|\frac{e^{(\omega_2 - \omega_1)t} - 1}{(\omega_2 - \omega_1)t}| \geq \varepsilon_0$; consequently, $|\psi_1(\lambda, \mu, t)| \geq \varepsilon_0 e^{-1} t = \varepsilon_0 e^{-1} n$. Hence $\| \psi_1(A,B,t)g \| \geq \| E(P_n)\psi_1(A,B,t)g \| = \| \psi_1(A,B,t)E(P_n)g \| \geq \varepsilon_0 e^{-1} n \cdot \frac{1}{2n} = \frac{1}{2}\varepsilon_0 e^{-1}$.

Thus, $f \in E(P)H$ is such that for any $f_0 \in E(P)H$ and $g = f - \omega_1(A,B)f_0$: $\| \psi_1(A,B,t)g \|$ does not converge to 0 as $t \to +\infty$. This contradicts our conclusion arrived at before the assumption that $(0,0) \in \sigma(A,B)$.

Consequently, $(0,0) \notin \sigma(A,B)$. We have completed the proof of the necessity.

Sufficiency. Let 2) be valid, and let $f \in H_{-2}$.

1) If $\gamma < 0$, set $y(t) = (\omega_2^{-1} e^{\omega_2 t})(A,B)f$ $(t \in R_+)$.
2) If $\gamma = 0$, then $\exists\delta > 0 : E\{(\lambda,\mu) \in \sigma(A,B) | \operatorname{Re}\omega_1 \geq 0, \operatorname{Re}\omega_2 < 0, |\omega_2| < \delta\} = 0$, $E\{(\lambda,\mu) \in \sigma(A,B) | \max(|\omega_1|,|\omega_2|) < \delta\} = 0$. Denote: $P_1 = \{(\lambda,\mu) \in \sigma(A,B) | \operatorname{Re}\omega_2 < 0, |\omega_2| \geq \delta\}$, $P_2 = \{(\lambda,\mu) \in \sigma(A,B) | \operatorname{Re}\omega_1 < 0, |\omega_2| < \delta, |\omega_1| \geq \delta\}$. Then $E(P_1 \cup P_2) = I$. Set
$$y(t) = (\omega_2^{-1} e^{\omega_2 t})(A,B)E(P_1)f + (\omega_1^{-1} e^{\omega_1 t})(A,B)E(P_2)f \ (t \in R_+).$$
3) If $\gamma > 0$, then $\exists 0 < \varepsilon < \frac{\gamma}{2} : E\{(\lambda,\mu) \in \sigma(A,B) | \operatorname{Re}\omega_1 \geq \gamma, \operatorname{Re}\omega_2 < \gamma, |\omega_2| < \varepsilon\} = 0$. Denote: $P_1 = \{(\lambda,\mu) \in \sigma(A,B) | \operatorname{Re}\omega_2 < \gamma, |\omega_2| \geq \varepsilon\}$, $P_2 = \{(\lambda,\mu) \in \sigma(A,B) | \operatorname{Re}\omega_1 < \gamma, |\omega_2| < \varepsilon, |\omega_1| \geq \varepsilon\}$, $P_3 = \{(\lambda,\mu) \in \sigma(A,B) | \max(|\omega_1|,|\omega_2|) < \varepsilon\}$. Then $E(P_1 \cup P_2 \cup P_3) = I$. Set $y(t) = (\omega_2^{-1} e^{\omega_2 t})(A,B)E(P_1)f + (\omega_1^{-1} e^{\omega_1 t})(A,B)E(P_2)f + \psi_1(A,B,t)E(P_3)f$ $(t \in R_+)$.

It is obvious that $y(t)$ is a weak solution of the Neumann problem for equation (1) on R_+ with b.d. f, and moreover, $y(t) \in Y_\gamma$.

Corollary 1. *Let A and B be c.s.o. in H.*

1. *Let $\gamma < 0$. Condition* 1) *of Theorem* 12.2 *holds if and only if conditions* (11.4) *and* (11.5) *are satisfied* (see Fig. 12.2a).
2. *Let $\gamma = 0$. Condition* 1) *of Theorem* 12.2 *holds if and only if $\exists\varepsilon > 0 : \sigma(A,B) \subseteq R^2 \setminus (\{(\lambda,\mu)|\lambda < 0, \mu > \varepsilon\lambda\} \cup \{(\lambda,\mu)|\lambda^2 + \mu^2 < \varepsilon^2\})$, and moreover, $E\{(\lambda,\mu)|\lambda = 0, \mu \geq \varepsilon\} = 0$* (see Fig. 12.2b).
3. *Let $\gamma > 0$. Condition* 1) *of Theorem* 12.2 *holds if and only if conditions* (11.4) *and* (11.5) *are satisfied, and moreover, $\exists\varepsilon > 0 : \sigma(A,B) \subseteq R^2 \setminus \{(\lambda,\mu)|\mu < -\gamma\lambda - \gamma^2, |\mu| < \varepsilon|\lambda|\}$ and $E\{(\lambda,\mu) \in R^2|\mu = -\gamma\lambda - \gamma^2, |\mu| < \varepsilon|\lambda|\} = 0$* (see Fig. 12.2c).

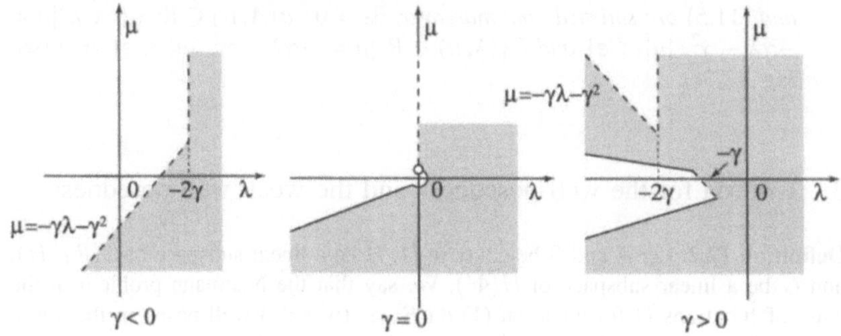

Fig. 12.2a, 12.2b, 12.2c

In order to prove Corollary 1, recall that for any $(\lambda,\mu)\in R^2$: if $\lambda\le 0$, then $|\omega_2|=\min(|\omega_1|,|\omega_2|)$; if $\lambda\ge 0$, then $|\omega_2|=\max(|\omega_1|,|\omega_2|)$. Now Corollary 1 follows immediately from Theorem 12.2, by virtue of Lemmas 1.1 and 1.2.

Theorem 12.3. *Let $\gamma\in R^1$, A and B be c.n.o. in H. The following conditions are equivalent:*

1) *for each $f\in H_1=D(A)\cap D(|B|^{1/2})$ there exists a solution of equation* (1) *on R_+ $y(t)\in Y_\gamma$ such that $y'(0)=f$;*
2) *a)* (11.3) *holds;*
   b) if $\gamma>0$, then $\exists\varepsilon>0: E\{(\lambda,\mu)\in\sigma(A,B)|\operatorname{Re}\omega_1\ge\gamma,\ \operatorname{Re}\omega_2<\gamma,\ |\mu|<\varepsilon\}=0$;
   if $\gamma=0$, then $\exists\varepsilon>0: E\{(\lambda,\mu)\in\sigma(A,B)|\operatorname{Re}\omega_1\ge\gamma,\ \operatorname{Re}\omega_2<\gamma,\ \max(|\omega_2|^2,|\mu|)<\varepsilon\}=0$, and $(0,0)\notin\sigma(A,B)$.

Remark 1. For $\gamma>0$ the following conditions are equivalent:

1) $\exists\varepsilon>0: E\{(\lambda,\mu)\in\sigma(A,B)|\operatorname{Re}\omega_1\ge\gamma,\ \operatorname{Re}\omega_2<\gamma,\ \max(|\omega_2|^2,|\mu|)<\varepsilon\}=0$;
2) $\exists\varepsilon>0: E\{(\lambda,\mu)\in\sigma(A,B)|\operatorname{Re}\omega_1\ge\gamma,\ \operatorname{Re}\omega_2<\gamma,\ |\mu|<\varepsilon\}=0$;
3) $\exists\varepsilon>0: E(P\cap\{(\lambda,\mu)\in C^2|\ |\mu|<\varepsilon\})=0$, with P from Remark 1 to Theorem 12.2.

In general, for $\gamma\ge 0$ condition 2) of Theorem 12.2 is stronger than condition 2) of Theorem 12.3. This is analogous to the situation for Theorems 4.4 and 4.5.

Corollary 1. *Let A and B be c.s.o. in H.*

1. *Let $\gamma<0$. Condition* 1) *of Theorem* 12.3 *holds if and only if conditions* (11.4) *and* (11.5) *are satisfied* (see Fig. 12.3a).
2. *Let $\gamma=0$. Condition* 1) *of Theorem* 12.3 *holds if and only if $\exists\varepsilon>0$: $\sigma(A,B)\subseteq R^2\setminus(\{(\lambda,\mu)|\lambda<0,\ \mu>-\varepsilon\}\cup\{(\lambda,\mu)|\lambda^2+\mu^2<\varepsilon^2\})$, and moreover, $E\{(\lambda,\mu)|\lambda=0,\ \mu\ge\varepsilon\}=0$* (see Fig. 12.3b).
3. *Let $\gamma>0$. Condition* 1) *of Theorem* 12.3 *holds if and only if conditions* (11.4) *and* (11.5) *are satisfied, and moreover, $\exists\varepsilon>0:\sigma(A,B)\subseteq R^2\setminus\{(\lambda,\mu)|\mu<-\gamma\lambda-\gamma^2,\ |\mu|<\varepsilon\}$ and $E\{(\lambda,\mu)\in R^2|\mu=-\gamma\lambda-\gamma^2,\ |\mu|<\varepsilon\}=0$* (see Fig. 12.3c).

3 Criteria for the well-posedness and the weak well-posedness

Definition 12.2. Let A and B be c.n.o. in H, U be a linear subspace of $C(R_+,H)$, and G be a linear subspace of $H(\Phi')$. We say that the Neumann problem in the class of functions U for equation (1) on R_+ is (weakly) well-posed in the space of boundary data G if for each $f\in G$ the Neumann problem for equation (1) on R_+ with boundary data f has a unique (weak) solution from U.

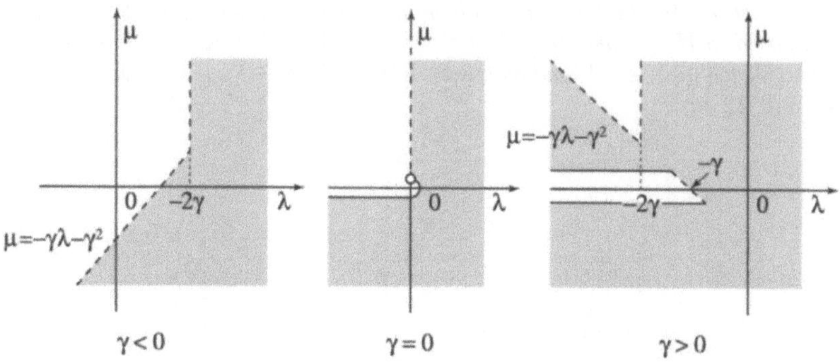

Fig. 12.3a, 12.3b, 12.3c

Theorem 12.4. *Let $\gamma \in R^1$, A and B be c.n.o. in H. The Neumann problem in Y_γ for equation* (1) *on R_+ is weakly well-posed in H_{-2} if and only if one of the following equivalent conditions holds:*

1) $\exists\varepsilon > 0 : E\{(\lambda,\mu) \in C^2 | \operatorname{Re}\omega_1 \geq \gamma,\ \operatorname{Re}\omega_2 < \gamma,\ |\omega_2| \geq \varepsilon\} = I$;
2) *a) condition* 2) *of Theorem* 11.3 *holds;*
   b) if $\gamma > 0$, then $\exists\varepsilon > 0 : \sigma(A,B) \subseteq \{(\lambda,\mu) \in C^2 | \min(|\omega_1|, |\omega_2|) \geq \varepsilon\}$;
   if $\gamma = 0$, then $\exists\varepsilon > 0 : E\{(\lambda,\mu) \in \sigma(A,B) |\ |\omega_2| < \varepsilon\} = 0$;
3) *a) condition* 3) *of Theorem* 11.3 *holds;*
   b) if $\gamma > 0$, then $\exists\varepsilon > 0 : \sigma(A,B) \subseteq \{(\lambda,\mu) \in C^2 | \min(|\mu|, |\lambda^{-1}\mu|) \geq \varepsilon\}$;
   if $\gamma = 0$, then $\exists\varepsilon > 0 : \sigma(A,B) \cap P \subseteq \{(\lambda,\mu) \in C^2 |\ |\omega_2| \geq \varepsilon\}$ (for P from Remark 1 *to Theorem* 12.2*).*

In this case, the Neumann problem in Y_γ for equation (1) *on R_+ is weakly well-posed in $G = H_{-2}$ not only over $D(A^2) \cap D(B)$ but over the whole $D(A) \cap D(B)$ (which is the maximal Φ by Definition* 1.4*). For every $f \in H_{-2}$ the weak solution of the Neumann problem for equation* (1) *on R_+ with b.d. f $y(t) \in Y_\gamma$ has the form: $y(t) = (\omega_2^{-1} e^{\omega_2 t})(A,B)f$ $(t \in R_+)$.*

Cf. Section 4 of Chapter 4 and Section 3 of Chapter 11.

Remark 1. If 1) holds, then $E\{(\lambda,\mu) \in C^2 |\ |\omega_2| < \varepsilon\} = 0$, and so $E\{(\lambda,\mu) \in C^2 | \max(|\omega_1|, |\omega_2|) < \varepsilon\} = 0$; hence $(0,0) \notin \sigma(A,B)$.

Corollary 1. *Let A and B be c.s.o. in H.*

1. *Let $\gamma < 0$. The Neumann problem in Y_γ for equation* (1) *on R_+ is weakly well-posed in H_{-2} if and only if $\sigma(A,B) \subseteq \{(\lambda,\mu) \in R^2 | \mu \leq -\gamma\lambda - \gamma^2\}$, and moreover, $E\{(\lambda,\mu) | \lambda \leq -2\gamma,\ \mu = -\gamma\lambda - \gamma^2\} = 0$* (see Fig. 12.4a).

2. *Let $\gamma = 0$. The Neumann problem in Y_γ for equation* (1) *on R_+ is weakly well-posed in H_{-2} if and only if $\exists\varepsilon > 0 : \sigma(A,B) \subseteq \{(\lambda,\mu) \in R^2 | \lambda \leq \varepsilon,\ \mu \leq \varepsilon\lambda - \varepsilon^2\} \cup \{(\lambda,\mu) \in R^2 | \lambda > \varepsilon,\ \mu \leq 0\}$* (see Fig. 12.4b).

3. *Let $\gamma > 0$. The Neumann problem in Y_γ for equation* (1) *on R_+ is weakly well-posed in H_{-2} if and only if $\exists \varepsilon > 0 : \sigma(A, B) \subseteq \{(\lambda, \mu) \in R^2 | \mu \leq -\gamma\lambda - \gamma^2\} \setminus \{(\lambda, \mu) | \, |\mu| < \varepsilon|\lambda|\}$, and moreover, $E\{(\lambda, \mu) | \lambda \leq -2\gamma, \mu = -\gamma\lambda - \gamma^2\} = 0$* (see Fig. 12.4c).

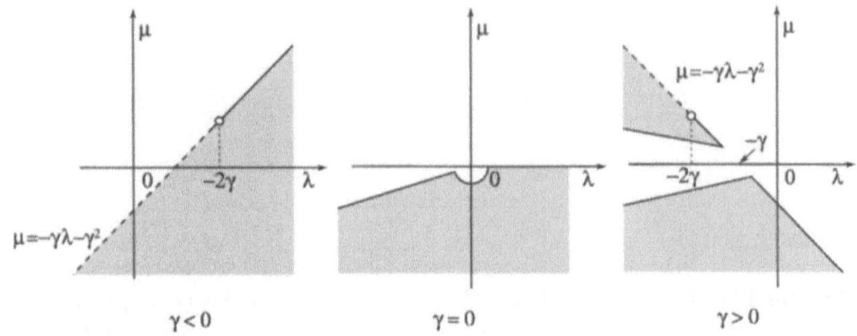

Fig. 12.4a, 12.4b, 12.4c

Example 12.1. Let A and B be the c.s.o. from Example 1.1.

For $\gamma < 0$, the Neumann problem in Y_γ for equation (1) on R_+ is weakly well-posed in H_{-2} if and only if the Dirichlet problem in Y_γ for (1) on R_+ is weakly well-posed in H (see Example 11.1).

For $\gamma = 0$, the Neumann problem in Y_γ for equation (1) on R_+ is weakly well-posed in H_{-2} if and only if simultaneously:

1) $b_1 \leq 0$, $b_2 \leq 0$, and $b_3 \leq 0$;
2) if $b_3 = 0$, then $a_3 > 0$;
3) for each of $i = 1, 2$: if $b_i = 0$, then $a_i \geq 0$.

Let $\gamma > 0$. The Neumann problem in Y_γ for equation (1) on R_+ is weakly well-posed in H_{-2} if and only if $b_3 \leq -\gamma a_3 - \gamma^2$ and one of the following cases holds.

I $b_3 > 0$ and simultaneously:

1) for each of $i = 1, 2$ there holds one of the following cases:
   a) $a_i = b_i = 0$;
   b) $a_i < 0$, $b_i > 0$, and $\frac{b_i}{a_i} \geq -\gamma$;
2) if $b_3 = -\gamma a_3 - \gamma^2$ and for each of $i = 1, 2$:
   $a_i = b_i = 0$ or: $a_i < 0$ and $\frac{b_i}{a_i} = -\gamma$,
   then $a_1 = a_2 = b_1 = b_2 = 0$ and $a_3 > -2\gamma$.

II $b_3 < 0$, and for each of $i = 1, 2$ there holds one of the following cases:
   a) $a_i = b_i = 0$;
   b) $b_i < 0$, $\frac{a_i}{b_i} \geq -\frac{1}{\gamma}$.

Theorem 12.5. *Let $\gamma \in R^1$, A and B be c.n.o. in H. The Neumann problem in Y_γ for equation* (1) *on R_+ is well-posed in $H_1 = D(A) \cap D(|B|^{1/2})$ if and only if one of the following equivalent conditions holds:*

1) $\exists \varepsilon > 0 : E\{(\lambda, \mu) \in C^2 | \operatorname{Re}\omega_1 \geq \gamma, \operatorname{Re}\omega_2 < \gamma, \max(|\mu|, |\omega_2|^2) \geq \varepsilon\} = I;$
2) *a) condition* 2) *of Theorem* 11.3 *holds;*
   b) if $\gamma > 0$, then $\exists \varepsilon > 0 : \sigma(A, B) \subseteq \{(\lambda, \mu) \in C^2 |\, |\mu| \geq \varepsilon\}$;
   if $\gamma = 0$, then $\exists \varepsilon > 0 : E\{(\lambda, \mu) \in \sigma(A, B) | \max(|\mu|, |\omega_2|^2) < \varepsilon\} = 0$;
3) *a) condition* 3) *of Theorem* 11.3 *holds;*
   b) if $\gamma > 0$, then $\exists \varepsilon > 0 : \sigma(A, B) \subseteq \{(\lambda, \mu) \in C^2 |\, |\mu| \geq \varepsilon\}$;
   if $\gamma = 0$, then $\exists \varepsilon > 0 : \sigma(A, B) \cap P \subseteq \{(\lambda, \mu) \in C^2 | \max(|\mu|, |\omega_2|^2) \geq \varepsilon\}$ (for P from Remark 1 *to Theorem* 12.2*).*

In this case, for every $f \in H_1$ the solution of the Neumann problem for equation (1) *on R_+ with b.d. f $y(t) \in Y_\gamma$ has the form:*

$$y(t) = (\omega_2^{-1} e^{\omega_2 t})(A, B) f \quad (t \in R_+).$$

Corollary 1. *Let A and B be c.s.o. in H.*

1. *Let $\gamma < 0$. The Neumann problem in Y_γ for equation* (1) *on R_+ is well-posed in H_1 if and only if $\sigma(A, B) \subseteq \{(\lambda, \mu) \in R^2 | \mu \leq -\gamma\lambda - \gamma^2\}$, and moreover,* $E\{(\lambda, \mu) | \lambda \leq -2\gamma, \mu = -\gamma\lambda - \gamma^2\} = 0$ (see Fig. 12.5a).

2. *Let $\gamma = 0$. The Neumann problem in Y_γ for equation* (1) *on R_+ is well-posed in H_1 if and only if $\exists \varepsilon > 0 : \sigma(A, B) \subseteq \{(\lambda, \mu) \in R^2 | \lambda < \varepsilon, \mu \leq -\varepsilon\} \cup \{(\lambda, \mu) \in R^2 | \lambda \geq \varepsilon, \mu \leq 0\}$* (see Fig. 12.5b).

3. *Let $\gamma > 0$. The Neumann problem in Y_γ for equation* (1) *on R_+ is well-posed in H_1 if and only if $\exists \varepsilon > 0 : \sigma(A, B) \subseteq \{(\lambda, \mu) \in R^2 | \mu \leq -\gamma\lambda - \gamma^2\} \setminus \{(\lambda, \mu) |\, |\mu| < \varepsilon\}$, and moreover, $E\{(\lambda, \mu) | \lambda \leq -2\gamma, \mu = -\gamma\lambda - \gamma^2\} = 0$* (see Fig. 12.5c).

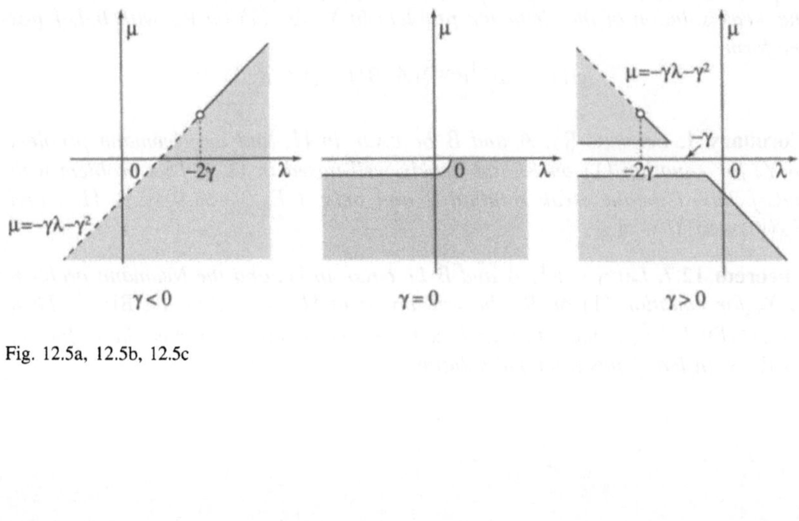

Fig. 12.5a, 12.5b, 12.5c

Example 12.2. Let A and B be the c.s.o. from Example 1.1.

For $\gamma < 0$, the Neumann problem in Y_γ for equation (1) on R_+ is well-posed in H_1 if and only if the Dirichlet problem in Y_γ for (1) on R_+ is well-posed in $D(A^2) \cap D(B)$ (see Example 11.1).

For $\gamma = 0$, the Neumann problem in Y_γ for equation (1) on R_+ is well-posed in H_1 if and only if simultaneously:

1) $b_1 \leq 0$, $b_2 \leq 0$, and $b_3 \leq 0$;
2) if $b_3 = 0$, then $a_3 > 0$;
3) if $b_3 = 0$, then for each of $i = 1, 2$: if $b_i = 0$, then $a_i \geq 0$.

Let $\gamma > 0$. The Neumann problem in Y_γ for equation (1) on R_+ is well-posed in H_1 if and only if $b_3 \leq -\gamma a_3 - \gamma^2$ and one of the following cases holds:

I $b_3 > 0$ and simultaneously:

1) for each of $i = 1, 2$ there holds one of the following cases:
   a) $a_i = b_i = 0$;
   b) $a_i < 0$, $b_i \geq 0$, and $\frac{b_i}{a_i} \geq -\gamma$;
2) if $b_3 = -\gamma a_3 - \gamma^2$ and for each of $i = 1, 2$:
   $a_i = b_i = 0$ or: $a_i < 0$ and $\frac{b_i}{a_i} = -\gamma$,
   then $a_1 = a_2 = b_1 = b_2 = 0$ and $a_3 > -2\gamma$.

II $b_3 < 0$, and for each of $i = 1, 2$ there holds one of the following cases:
   a) $b_i = 0$, $a_i \leq 0$;
   b) $b_i < 0$, $\frac{a_i}{b_i} \geq -\frac{1}{\gamma}$.

4 Boundary data of solutions and weak solutions

Theorem 12.6. *Let $\gamma \in R^1$, A and B be c.n.o. in H, and condition* 1) *of Theorem* 12.4 *hold. The Neumann problem in Y_γ for equation* (1) *on R_+ with b.d. f has a unique weak solution if and only if $f \in H_-((|\omega_2| + 1)(A, B))$, i.e., $E\{(\lambda, \mu) \in C^2 |\ |\omega_1| \geq |\omega_2|\} f \in H_{-2}$ and $E\{(\lambda, \mu) \in C^2 |\ |\omega_1| < |\omega_2|\} f \in H_{-1}$. In this case, the weak solution of the Neumann problem in Y_γ for* (1) *on R_+ with b.d. f has the form:*

$$y(t) = (\omega_2^{-1} e^{\omega_2 t})(A, B) f \quad (t \in R_+).$$

Corollary 1. *Let $\gamma \in R^1$, A and B be c.s.o. in H, and the Neumann problem in Y_γ for equation* (1) *on R_+ be weakly well-posed in H_{-2}. This problem with b.d. f has a unique weak solution if and only if $E_A((-\infty, 0)) f \in H_{-2}$ and $E_A([0, +\infty)) f \in H_{-1}$.*

Theorem 12.7. *Let $\gamma \in R^1$, A and B be c.n.o. in H, and the Neumann problem in Y_γ for equation* (1) *on R_+ be well-posed in $H_1 = D(A) \cap D(|B|^{1/2})$. Then $D(A) \cap D(|B|^{1/2})$ is the set of all f such that the Neumann problem in Y_γ for* (1) *on R_+ with b.d. f has a unique solution.*

Commentaries on the literature

In this monograph, we considered complete second order homogeneous linear differential equations with commuting normal operator coefficients in Hilbert spaces. For some results on nonhomogeneous equations see [136, 311], for fourth and higher order equations and applications to equations of the stratified fluid motion see [134–136, 311, 314, 315], for equations with anti-commuting self-adjoint operator coefficients see [313].

For other approaches to the study of nondegenerate complete second order linear differential equations

$$y''(t) + Ay'(t) + By(t) = 0,$$

where A and B are linear operators in Banach spaces, see monographs [156, 372], the corresponding chapters in monographs [22, 93, 131, 179, 317], memoirs [329, 330, 332], and papers [4, 11–14, 17–20, 23, 24, 38–40, 42–45, 50, 59, 63, 64, 68, 69–73, 77–79, 83, 84, 87–91, 94, 97–104, 107–110, 113–117, 119–122, 124, 125, 132–136, 146–148, 153–155, 165, 166, 169, 171–173, 175, 182, 186, 194, 201, 202, 204, 209, 216, 218, 223–230, 232–235, 237–239, 247, 248, 250–255, 259–265, 277, 280–285, 289, 293–296, 300, 301, 303, 307–309, 311–316, 322–324, 328, 331, 335, 345–347, 350, 354, 356–371, 373, 374, 376–380, 385–388]. First works on these equations were P. D. Lax [204], J.-L. Lions [209], B. S. Mityagin [239], P. E. Sobolevskii [322–324], S. Ya. Yakubov [363–366] and B. Weiss [356]; see also S. G. Krein [179].

Also included in Bibliography are some works on mechanics and mathematical physics where nondegenerate complete second order linear equations (1) and higher order equations arise; some works explaining operator pencils approach to solution of these problems of mechanics; some fundamental works on operator theory which may be useful for the study of equation (1); finally, a series of monographs, reviews and original papers giving a flavour of approaches developed and results obtained for first order and incomplete second order equations, without any pretense of completeness.

In this list, we restrict ourselves to nondegenerate complete second order and higher order linear differential equations in Banach spaces. In fact, we do not even mention boundless literature on nonlinear partial differential equations and their abstract analogues (in particular, on asymptotic behaviour of their solutions; see, for instance, the monographs [16, 140, 143, 144, 340] and references therein). As for linear equations, we do not touch on an extensive bibliography on classical partial differential equations; the advances in this area go back to Hadamard, Petrovskii, Sobolev, Leray, Shilov, Gårding, and further progress is connected with pseudodifferential operators and microlocal analysis; see, for example, [152].

Even in the area of abstract linear differential equations, we do not touch on works on equations with an operator on the higher derivative, degenerate equations, Sobolev type equations; see, for example, the monographs [37, 174, 381,

382] and the reviews [180, 352] and references therein. Operator pencils are represented only in the context of nondegenerate operator differential equations; for a comprehensive presentation and extended bibliography see recent monographs [221, 292].

Even with such limitations on the subject, this list makes no pretense of completeness; its only purpose is to represent variety of existing approaches to the study of the problem.

Bibliography

1. S. Agmon and L. Nirenberg, *Properties of solutions of ordinary differential equations in Banach space*, Comm. Pure Appl. Math. **16** (1963), no. 2, 121–239.
2. N. I. Ahiezer and I. M. Glazman, *The Theory of Linear Operators in Hilbert Spaces* (transl. from the Russian), vols. I, II, Ungar, New York, 1961.
3. N. Ahmed, *Semigroup Theory with Applications to Systems and Control*, John Wiley & Sons Inc., New York, 1991.
4. B. A. Aliev, *The Cauchy problem for differential-operator equations of hyperbolic type*, Dokl. Akad. Nauk Azerb. SSR **36** (1980), no. 3, 6–9.
5. I. P. Andreichikov and V. I. Yudovich, *On stability of viscoelastic beams*, Izv. Akad. Nauk SSSR, Solid Mechanics (1974), no. 2, 78–87.
6. P. Antosik, J. Mikusinski and R. Sikorski, *Theory of Distributions: the Sequential Approach*, Elsevier Science Publishers B. V., Amsterdam, 1973.
7. W. Arendt, *Vector-valued Laplace transforms and Cauchy problems*, Israel J. Math **59** (1987), 327–352.
8. W. Arendt and C. Batty, *Domination and ergodicity for positive semigroups*, Proc. Amer. Math. Soc. **114** (1992), no. 3, 743–747.
9. W. Arendt, F. Neubrander and U. Schlotterbeck, *Interpolation of semigroups and integrated semigroups*, Semigroup Forum **45** (1992), no. 1, 26–37.
10. W. Arendt and J. Prüss, *Vector-valued Tauberian theorems and asymptotic behavior of linear Volterra equations*, SIAM J. Math. Anal. **23** (1992), no. 2, 412–448.
11. A. Arosio, *Equations différentielles opérationelles linéaires du deuxième ordre: problème de Cauchy et comportement asymptotique lorsque $t \to \infty$*, Comp. Rend. Acad. Sci. Paris, Sér. A, **295** (1982), 83–86.
12. A. Arosio and S. Panizzi, *Global bounded weak solutions for an abstract nonlinear Timoshenko beam equation with four propagation speeds*, Funkcial. Ekvac. **36** (1993), no. 1, 109–121.
13. P. Aviles and J. Sandefur, *Nonlinear second order equations with applications to partial differential equations*, J. Differential Equations **58** (1985), 404–427.
14. J. Avrin, *Convergence properties of the strongly damped nonlinear Klein-Gordon equation*, J. Differential Equations **67** (1987), no. 2, 243–255.
15. T. Ya. Azizov and I. S. Iokhvidov, *Foundations of the Theory of Linear Operators in Spaces with Indefinite Metric* (in Russian), Nauka, Moscow, 1986.
16. A. V. Babin and M. I. Vishik, *Attractors of Evolution Equations*, North-Holland, Amsterdam, 1992.

17. M. K. Balaev, *Well-posedness of the Cauchy problem for arbitrary order differential-operator equations of parabolic type*, Izv. Akad. Nauk Azerb. SSR, Ser. Fiz. Tehn. Mat. Nauk (1976), no. 5, 25–31.

18. M. K. Balaev, *On evolution equations of arbitrary order of parabolic type*, Dokl. Akad. Nauk Azerb. SSR **41** (1988), 7–10.

19. M. K. Balaev, *On well-posedness of operator-differential equations of arbitrary order*, Dokl. Akad. Nauk SSSR **317** (1991), no. 3, 526–529.

20. M. K. Balaev and S. Ya. Yakubov, *Unbounded quadratic operator and differential pencils* (in Russian), in: Some problems in the theory of nonlinear analysis, No. 2, Elm, Baku, 1991, pp. 175–191.

21. J. Ball, *Continuity properties of nonlinear semigroups*, J. Funct. Anal. **17** (1974), no. 1, 91–103.

22. V. Barbu, *Nonlinear Semigroups and Differential Equations in Banach Spaces*, Nordhoff, Leyden, 1976.

23. L. Barkwell and P. Lancaster, *Overdamped and gyroscopic vibrating systems*, J. Appl. Mech. **59** (1992), 176–181.

24. L. Barkwell, P. Lancaster and A. S. Markus, *Gyroscopically stabilized systems: a class of quadratic eigenvalue problems with real spectrum*, Canad. J. Math. **44** (1992), 221–233.

25. H. Bart, *Periodic strongly continuous semigroups*, Ann. Mat. Pura Appl. **115** (1977), 311–318.

26. H. Bart and S. Goldberg, *Characterisations of almost periodic strongly continuous groups and semigroups*, Math. Ann. **236** (1978), no. 2, 105–116.

27. C. Batty, *Tauberian theorems for the Laplace-Stieltjes transform*, Trans. Amer. Math. Soc. **322** (1990), no. 2, 783–804.

28. R. Beals, *On the abstract Cauchy problem*, J. Funct. Anal. **10** (1972), no. 3, 281–299.

29. A. Belleni-Morante, *Applied Semigroups and Evolution Equations*, Oxford Univ. Press, Oxford, 1979.

30. C. D. Benchimol, *Feedback stabilizability in Hilbert spaces*, Appl. Math. Optim. **4** (1978), no. 3, 225–248.

31. Yu. M. Berezanskii, *Spaces with negative norms*, Uspekhi Mat. Nauk **18** (1963), no. 1, 63–96.

32. Yu. M. Berezanskii, *Expansion in Eigenfunctions of Self-Adjoint Operators* (transl. from the Russian), Amer. Math. Soc., Providence, R.I., 1968.

33. Yu. M. Berezanskii, *Self-Adjoint Operators in Spaces of Functions of Infinitely Many Variables* (transl. from the Russian), Amer. Math. Soc., Providence, R.I., 1986.

34. M. Sh. Birman and M. Z. Solomyak, *Spectral Theory of Self-Adjoint Operators in Hilbert Spaces* (transl. from the Russian), D. Reidel, Dordrecht, 1987.

35. M. Sh. Birman, A. M. Vershik and M. Z. Solomyak, *A product of commuting spectral measures may not be countably-additive*, Funkts. Anal. Prilozh. **13** (1979), no. 1, 61–62.

36. V. V. Bolotin, *The Dynamic Stability of Elastic Systems* (transl. from the Russian), Holden Day Inc., San Francisco, 1964.

37. R. Carrol and R. E. Showalter, *Singular and Degenerate Cauchy problems*, Academic Press, New York, 1976.

38. J. Chazarain, *Problèmes de Cauchy au sens des distributions vectorielles et applications*, Comp. Rendus Acad. Sci. Paris, Ser. A, **266** (1968), 10–13.

39. J. Chazarain, *Problèmes de Cauchy abstraits et applications à quelques problèmes mixtes*, J. Functional Analysis **7** (1971), 386–446.

40. G. Chen and D. L. Russel, *A mathematical model for linear elastic systems with structural damping*, Quart. Appl. Math. **39** (1981/2), no. 4, 433–454.

41. G. Chen and J.-X. Zhou, *Vibration and Damping in Distributed Systems*, Vol. 1, CRC Press, Boca Raton, FL, 1993.

42. S. Chen and R. Triggiani, *Proof of two conjectures of G. Chen and D. L. Russel on structural damping for elastic systems: the case* $\alpha = \frac{1}{2}$, in: Lect. Notes Math. **1354**, Springer, New York, 1988, pp. 234–256.

43. S. Chen and R. Triggiani, *Proof of two conjectures on structural damping for elastic systems: the case* $\frac{1}{2} \leq \alpha \leq 1$, Pacific J. Math. **136** (1989), no. 1, 15–56.

44. S. Chen and R. Triggiani, *Gevrey class semigroups arising from elastic systems with gentle dissipation: the case* $0 < \alpha < \frac{1}{2}$, Proc. Amer. Math. Soc. **110** (1990), 401–415.

45. S. Chen and R. Triggiani, *Characterization of domains of fractional powers of certain operators arising in elastic systems with applications*, J. Differential Equations **88** (1990), 279–293.

46. N. G. Chetaev, *Stability of Motion: Works in Analytical Mechanics* (in Russian), Izd. Akad. Nauk SSSR, Moscow, 1962.

47. I. Cioranescu, *La caractérisation spectrale d'opérateurs, générateurs des semigroupes distributions d'ordre fini de croissance*, J. Math. Anal. Appl. **41** (1973), no. 3, 539–544.

48. I. Cioranescu, *On periodic distribution groups*, in: Top. Mod. Oper. Theory (Proc. Timisoara, 1980), Birkhäuser, Basel, 1981, pp. 53–69.

49. I. Cioranescu, *On the second order Cauchy problem associated with a linear operator*, J. Math. Anal. Appl. **154** (1991), no. 1, 238–242.

50. Ph. Clement and J. Prüss, *On second order differential equations in Hilbert space*, Boll. Univ. Mat. Ital. 3-B **7** (1989), 623–638.

51. R. Cook, *Infinite Matrices and Spaces of Sequences*.

52. G. Da Prato, *Semigruppi di crescenza n*, Ann. Scuola Norm. Sup. Pisa Sci. Fis. Mat. **20** (1966), no. 4, 753–782.

53. G. Da Prato, *Weak solutions for linear abstract differential equations in Banach spaces*, Advances Math. **5** (1970), no. 2, 181–245.

54. G. Da Prato, *Applications Croissantes et Equations d'Evolution dans les Espaces de Banach*, Academic Press, New York, 1976.

55. G. Da Prato and E. Giusti, *Una caratterizzazione dei generatori di funzioni coseno astratte*, Boll. Un. Mat. Italiana 3 **22** (1967), 357–362.

56. G. Da Prato and P. Grisvard, *Sommes d'opérateurs linéaires et équations différentielles opérationnelles*, J. Math. Pures Appl. **54** (1975), no. 3, 305–387.

57. Yu. L. Daletskii and M. G. Krein, *Stability of Solutions of Differential Equations in Banach Space* (transl. from the Russian), Amer. Math. Soc., Providence, R.I., 1974.

58. R. Datko, *Uniform asymptotic stability of evolutionary processes in a Banach space*, SIAM J. Math. Anal. **3** (1972), no. 3, 428–445.

59. R. Datko and Y. C. You, *Some second-order vibrating systems cannot tolerate small time delays in their damping*, J. Optim. Theory Appl. **70** (1991), no. 3, 521–537.

60. E. Davies, *One-Parameter Semigroups*, Academic Press, New York, 1980.

61. E. Davies and M. M. Pang, *The Cauchy problem and a generalization of the Hille-Yosida theorem*, Proc. London Math. Soc. **55** (1987), 181–208.

62. A. A. Dezin, *Partial Differential Equations: an Introduction to a General Theory of Linear Boundary Value Problems* (transl. from the Russian), Springer-Verlag, Berlin-Heidelberg-New York, 1987.

63. G. Di Blasio, K. Kunisch and E. Sinestrari, *L^2-regularity for parabolic partial integro-differential equations with delay in the highest order derivatives*, J. Math. Anal. Appl. **102** (1984), 38–57.

64. G. Di Blasio, K. Kunisch and E. Sinestrari, *Mathematical models for the elastic beam with structural damping*, Appl. Anal. **48** (1993), no. 1–4, 133-156.

65. *Differential Equations in Banach Spaces* (Proc. Bologna, 1985), ed. by A. Favini and E. Obrecht, Lecture Notes in Math. **1223**, Springer, New York, 1986.

66. *Differential Equations in Banach Spaces* (Proc. Bologna, 1991), ed. by G. Dore, A. Favini, E. Obrecht and A. Venni, Lecture Notes in Pure and Appl. Math. **148**, Marcel Dekker, New York, 1993.

67. J. Donaldson, *The abstract Cauchy problem*, J. Differential Equations **25** (1977), 400–409.

68. J. Donaldson and R. Hersh, *A perturbation series for Cauchy's problem for higher-order abstract parabolic equations*, Proc. Nat. Acad. Sci. USA **67** (1970), 41–42.

69. Yu. A. Dubinskii, *The Cauchy problem for operator-differential equations*, Dokl. Akad. Nauk SSSR **181** (1968), 1046–1049.

70. Yu. A. Dubinskii, *On one class of higher order differential-operator equations*, Dokl. Akad. Nauk SSSR **187** (1969), 982–985.

71. Yu. A. Dubinskii, *Boundary value problems for certain classes of higher order differential-operator equations*, Dokl. Akad. Nauk SSSR **196** (1971), 32–35.

72. Yu. A. Dubinskii, *On certain differential-operator equations of general type*, Dokl. Akad. Nauk SSSR **201** (1971), 1033–1036.

73. Yu. A. Dubinskii, *On certain arbitrary order differential-operator equations*, Mat. Sbornik **90** (1973), 3–22.

74. Yu. A. Dubinskii, *The algebra of pseudodifferential operators with analytic symbols and its applications to mathematical physics*, Uspekhi Mat. Nauk **37** (1982), no. 5, 96–137.

75. N. Dunford and J. T. Schwartz, *Linear Operators*, vols. I, II, Interscience Inc., New York, 1988 (new edn).

76. S. D. Eidelman, *Parabolic Systems* (in Russian), Nauka, Moscow, 1964.

77. K.-J. Engel, *Growth estimates for semigroups generated by* 2×2 *operator-matrices*, Resultate Math. **20** (1991), no.1–2, 444–453.

78. K.-J. Engel, *On singular perturbations of second order Cauchy problems*, Pacific J. Math. **152** (1992), 79–91.

79. K.-J. Engel, *On dissipative wave equations in Hilbert space*, J. Math. Anal. Appl. **184** (1994), no. 2, 302–316.

80. *Evolution Equations, Control Theory, and Biomathematics* (Proc. Han sur Lesse, 1991), ed. by Ph. Clement and G. Lumer, Lecture Notes in Pure and Appl. Math. **155**, Marcel Dekker, New York, 1994.

81. P. Fatou, *Séries trigonométriques et séries de Taylor*, Acta Math. **30** (1906), 335–400.

82. H. O. Fattorini, *Ordinary differential equations in linear topological spaces*, I, J. Differential Equations **5** (1969), 72–105.

83. H. O. Fattorini, *Ordinary differential equations in linear topological spaces*, II, J. Differential Equations **6** (1969), 50–70.

84. H. O. Fattorini, *Extension and behaviour at infinity of solutions of certain linear operational differential equations*, Pacific J. Math. **33** (1970), 583–615.

85. H. O. Fattorini, *The undetermined Cauchy problem in Banach spaces*, Math. Ann. **200** (1973), 103–112.

86. H. O. Fattorini, *Two point boundary value problems for operational differential equations*, Ann. Scuola Norm. Sup. Pisa (1974), no. 1, 63–79.

87. H. O. Fattorini, *Some remarks on convolution equations for vector valued distributions*, Pacific J. Math. **66** (1976), 347–371.

88. H. O. Fattorini, *Vector valued distributions having a smooth convolution inverse*, Pacific J. Math. **90** (1980), 347–372.

89. H. O. Fattorini, *Some remarks on second order abstract Cauchy problems*, Funkcialaj Ekvacioj **24** (1981), 331–344.

90. H. O. Fattorini, *Convergence and approximation theorems for vector valued distributions*, Pacific J. Math. **105** (1983), 77–114.

91. H. O. Fattorini, *Singular perturbation and boundary layer for an abstract Cauchy problem*, J. Math. Anal. Appl. **97** (1983), 529–571.

92. H. O. Fattorini, *The Cauchy Problem*, Encyclopedia of Math. and its Appl. **18**, Addison-Wesley, London-Amsterdam, 1983.

93. H. O. Fattorini, *Second Order Linear Differential Equations in Banach Spaces*, North-Holland Mathematics Studies **108**, North-Holland, Amsterdam-New York-Oxford, 1985.

94. H. O. Fattorini, *The hyperbolic singular perturbation problem: an operator theoretic approach*, J. Differential Equations **70** (1987), 1–41.

95. H. O. Fattorini and A. Radnitz, *The Cauchy problem with incomplete initial data in Banach spaces*, Mich. Math. J. **18** (1971), 291–330.

96. A. Favini, *Laplace transform method for a class of degenerate evolution problems*, Rend. Mat. **12** (1979), no. 3–4, 511–536.

97. A. Favini, *Parabolicity of second order differential equations in Hilbert space*, Semigroup Forum **42** (1991), no. 3, 303–312.

98. A. Favini, *Second order parabolic equations in Banach space*, in: Differential Equations with Applications in Biology, Physics and Engineering (Proc. Leibnitz, 1989), Marcel Dekker, New York, 1991, pp. 129–137.

99. A. Favini, I. Lasiecka and H. Tanabe, *Abstract differential equations and nonlinear dispersive systems*, Dif. Int. Equations **6** (1993), no. 5, 995–1008.

100. A. Favini and E. Obrecht, *Conditions for parabolicity of second order abstract differential equations*, Dif. Int. Equations **4** (1991), no. 5, 1005–1022.

101. A. Favini and H. Tanabe, *Linear parabolic differential equations of higher order in time*, in: Differential Equations in Banach Spaces (Proc. Bologna, 1991), Lecture Notes in Pure and Appl. Math. **148**, Marcel Dekker, New York, 1993, pp. 85–92.

102. A. Favini and H. Tanabe, *On regularity of solutions to n-order differential equations of parabolic type in Banach spaces*, Osaka J. Math. **31** (1994), no. 2, 225–246.

103. A. Favini and A. Venni, *On the Cauchy problem for a second order semilinear parabolic equation with factored linear part*, Semigroup Forum **45** (1992), no. 2, 135–147.

104. A. Favini and A. Yagi, *On second order implicit differential equations in Banach spaces*, in: Evolution Equations, Control Theory, and Biomathematics

(Proc. Han sur Lesse, 1991), Lecture Notes in Pure and Appl. Math. **155**, Marcel Dekker, New York, 1994, pp. 205–213.

105. W. Feller, *The parabolic differential equations and the associated semigroups of transformations*, Ann. Math. **55** (1952), 468–519.

106. W. Feller, *An Introduction to Probability Theory and Its Applications*, vol. II, John Wiley & Sons Inc., New York-London, 1971.

107. I. P. Fishman, *On the Cauchy problem for differential equations in Hilbert space*, Ukrain. Math. J. **36** (1984), 122–125.

108. I. P. Fishman, *On representation of the general solution of a differential-operator equation*, Ukrain. Math. J. **36** (1984), 804–808.

109. I. P. Fishman, *On boundary values of solutions of differential-operator equations*, Ukrain. Math. J. **37** (1985), 388–393.

110. W. E. Fitzgibbon, *Strongly damped quasilinear evolution equations*, J. Math. Anal. Appl. **79** (1981), 536–550.

111. G. Foias, *Remarque sur les semigroupes distributions d'opérateurs normaux*, Portug. Math. **19** (1960), no. 3–4, 227–242.

112. A. Friedman, *Partial Differential Equations*, Holt, New York, 1969.

113. S. A. Gabov, G. Yu. Malysheva and A. G. Sveshnikov, *On one equation of mixed type related to the oscillations of the compressible stratified fluid*, Differents. Uravneniya **19** (1983), no. 7, 1171–1180.

114. S. A. Gabov, G. Yu. Malysheva, A. G. Sveshnikov and A. K. Shatov, *On some equations arising in dynamics of the rotating, stratified and compressible fluid*, Zhurn. Vych. Mat. Matem. Fiz. **24** (1984), no. 12, 1850–1863.

115. S. A. Gabov and K. S. Mamedov, *On one operator pencil arising in dynamics of the compressible stratified fluid*, Dokl. Akad. Nauk SSSR **280** (1985), no. 1, 23–26.

116. S. A. Gabov, B. B. Orazov and A. G. Sveshnikov, *On one evolutionary equation of fourth order arising in hydroacustics of the stratified fluid*, Differents. Uravneniya **22** (1986), no. 1, 19–25.

117. S. A. Gabov and A. G. Sveshnikov, *Problems of Dynamics of Stratified Fluids* (in Russian), Nauka, Moscow, 1986.

118. F. R. Gantmakher, *Theory of Matrices* (in Russian), Nauka, Moscow, 1988.

119. M. G. Gasymov, *To the theory of evolutionary equations of regular type*, Dokl. Akad. Nauk SSSR **200** (1971), 13–16.

120. M. G. Gasymov, *To the theory of higher order evolutionary equations*, Dokl. Akad. Nauk SSSR **206** (1972), 780–783.

121. M. G. Gasymov, *On solvability of boundary-value problems for a class of operator-differential equations*, Dokl. Akad. Nauk SSSR **235** (1977), no. 3, 505–508.

122. M. G. Gasymov and S. S. Mirzoev, *On solvability of boundary-value problems for operator-differential equations of elliptic type of second order*, Differents. Uravneniya **28** (1992), no. 4, 651–661.

123. M. I. Gekhtman and A. A. Kalyuzhny, *On the orthogonal polynomials in several variables*, Integr. Equat. Oper. Theory **19** (1994).

124. L. M. Gershtein, *On the solvability of a comlete second order differential equation in a Banach space*, Ukrain. Math. J. **45** (1993), no. 10, 1449–1454.

125. L. M. Gerstein and P. E. Sobolevskii, *«Commutant method» in the theory of operators and the theory of differential equations with operator coefficients*, Integr. Equat. Oper. Theory **16** (1993), no. 3, 400–429.

126. I. Gohberg and S. Goldberg, *Basic Operator Theory*, Birkhäuser, Basel, 1981.

127. I. Gohberg, S. Goldberg and M. A. Kaashoek, *Classes of Linear Operators*, Vol. 1, Birkhäuser, Basel, 1990.

128. I. C. Gohberg and M. G. Krein, *Introduction to the Theory of Linear Nonselfadjoint Operators in Hilbert Space*, Amer. Math. Soc., Providence, R.I., 1969.

129. J. A. Goldstein, *Semigroups and second-order differential equations*, J. Functional Analysis **4** (1969), 50–70.

130. J. A. Goldstein, *Variable domain second order evolution equations*, Appl. Anal. **5** (1976), 283–291.

131. J. A. Goldstein, *Semigroups of Linear Operators and Applications*, Oxford University Press, New York, 1985.

132. J. A. Goldstein, R. de Laubenfels and J. T. Sandefur, *Regularized semigroups, iterated Cauchy problems and equipartition of energy*, Monatsch. Math. **115** (1993), no. 1–2, 47–66.

133. J. A. Goldstein and J. Sandefur (Jr.), *Asymptotic equipartition of energy for differential equations in Hilbert space*, Trans. Amer. Math. Soc. **219** (1976), 397–406.

134. M. L. Gorbachuk and I. V. Fedak, *The Cauchy problem for a differential-operator equation connected with oscillations of stratified fluids*, Dokl. Akad. Nauk SSSR **297** (1987), 14–17.

135. M. L. Gorbachuk and A. Ya. Shklyar, *On solvability of the Cauchy problem for a fourth order differential-operator equation*, Uspekhi Mat. Nauk **42** (1987), no. 4, 161.

136. M. L. Gorbachuk and A. Ya. Shklyar, *Equations connected with the stratified fluid motion*, in: Nonlinear and Turbulent Processes in Physics (Proceedings of the III International Workshop, Kiev 1987), Naukova Dumka, Kiev, vol. II (1988), 32–35.

137. V. I. Gorbachuk and M. L. Gorbachuk, *Boundary Value Problems for Operator Differential Equations* (transl. from the Russian), Kluwer Academic Publishers, Dordrecht, 1990.

138. V. I. Gorbachuk and A. V. Knyazyuk, *Boundary values of solutions of differential-operator equations*, Uspekhi Mat. Nauk **44** (1989), no. 3, 55–91.

139. A. K. Gushchin and V. P. Mikhailov, *On uniform quasiasymptotics of solutions of the second mixed problem for a hyperbolic equation*, Mat. Sbornik **131** (1986), 419–437.

140. J. K. Hale, *Asymptotic Behavior of Dissipative Systems*, Math. Surveys and Monographs **25**, Amer. Math. Soc., Providence, R.I., 1988.

141. P. R. Halmos, *Measure Theory*, Van Nostrand, Princeton, N.J., 1962.

142. P. R. Halmos, *A Hilbert Space Problem Book*, Springer, New York, 1982.

143. A. Haraux, *Nonlinear Evolution Equations — Global Behavior of Solutions*, Lecture Notes in Math. **841**, Springer, Berlin-New York, 1981.

144. A. Haraux, *Systèmes Dynamiques Dissipatifs et Applications*, RMA, Vol. 17, Masson, Paris, 1991.

145. G. Hardy, *Divergent Series*.

146. K. A. Heimes, *Green's functions for linear second order systems*, SIAM J. Math. Anal. **9** (1978), no. 1, 207–214.

147. R. Hersh, *Direct solutions of general one-dimensional linear parabolic equation via an abstract Plansherel formula*, Proc. Nat. Acad. Sci. USA **63** (1969), 648–654.

148. R. Hersh, *Explicit solution of a class of higher order abstract Cauchy problems*, J. Differential Equations **8** (1970), 570–579.

149. E. Hille, *Une généralisation du problème de Cauchy*, Ann. Inst. Fourier **4** (1953), 31–48.

150. E. Hille and R. Phillips, *Functional Analysis and Semigroups*, Amer. Math. Soc., Providence, R.I., 1957.

151. P. Holmes and J. Marsden, *Bifurcation to divergence and flutter in flow induced oscillations: An infinite dimensional analysis*, Automatica **14** (1978), 367–384.

152. L. Hörmander, *The Analysis of Linear Partial Differential Operators*, Vols. 1, 2, 3, 4, Springer, Berlin-New York, 1983–1985.

153. Huang Fa Lun, *On the holomorphic property of the semigroup associated with linear elastic systems with structural damping*, Acta Math. Sci. (English Ed.) **5** (1985), no. 3, 271–277.

154. Huang Fa Lun, *On the mathematical model for linear elastic systems with structural damping*, SIAM J. Contr. Optim. **26** (1988), no. 3, 714–724.

155. Huang Fa Lun, *Some problems for linear elastic systems with damping*, Acta Math. Sci. (English Ed.) **10** (1990), no. 3, 319–326.

156. V. K. Ivanov, I. V. Mel'nikova and A. I. Filinkov, *Differential-Operator Equations and Ill-Posed Problems* (in Russian), Nauka, Moscow, 1993.

157. P. E. T. Jorgensen and R. T. Moore, *Operator Commutation Relations*, Reidel, Dordrecht, 1984.

158. P. Jonas and H. Langer, *Compact perturbations of definitizable operators*, J. Operator Theory **2** (1979), 63–77.

159. L. M. Kachanov, *Mechanics of Plastic Media* (in Russian), Gostekhizdat, Moscow-Leningrad, 1948.

160. S. Kantorovitz, *The Hille-Yosida space of an arbitrary operator*, J. Math. Anal. Appl. **136** (1988), 107–111.

161. T. Kato, *Perturbation Theory for Linear Operators*, Springer, Berlin-New York, 1965.

162. M. V. Keldysh, *On eigenvalues and eigenfunctions of some classes of non-selfadjoint equations*, Dokl. Akad. Nauk SSSR **77** (1951), no. 1, 11–14.

163. M. V. Keldysh, *On completeness of eigenfunctions of some classes of non-selfadjoint linear operators*, Uspekhi Mat. Nauk **26** (1971), no. 4, 15–41.

164. A. Ya. Khelemskii, *Gomology in Banach and Topological Algebras* (in Russian), Moscow State Univ., Moscow, 1986.

165. P. A. Kirichuk, *Well-posedness of the Cauchy problem for a complete third-order equation in a Banach space*, Dokl. Akad. Nauk Ukrainy (1992), no. 1, 10–16.

166. J. Kisynski, *On second order Cauchy's problem in a Banach space*, Bull. Acad. Polon. Sci. **18** (1970), 371–374.

167. J. Kisynsky, *Semigroups of operators and some of their applications to partial differential equations*, in: Contr. Theory and Top. Funct. Anal., Vol. 3, Vienna, 1976, 305–405.

168. A. V. Knyazyuk, *The Dirichlet problem for second order differential equations with an operator coefficient*, Ukrain. Math. J. **37** (1985), no. 2, 256–260.

169. B. I. Knyukh, *On representation and boundary values of solutions of a second order homogeneous differential-operator equation*, Ukrain. Math. J. **38** (1986), 101–104.

170. K. Kobayasi and I. Miyadera, *On the strong convergence of the Cesaro means of contractions in Banach spaces*, Proc. Jap. Acad. A **56** (1980), no. 6, 45–49.

171. A. N. Kochubei, *Fundamental solutions of differential-operator equations*, Differents. Uravnen. **13** (1977), 1588–1597.

172. A. N. Kochubei, *Differential-operator equations in spaces of ultradistributions*, Differents. Uravnen. **16** (1980), 405–413.

173. A. N. Kochubei, *To the theory of generalized solutions of differential-operator equations*, Ukrain. Math. J. **37** (1985), 703–707.

174. N. D. Kopachevskii, S. G. Krein and Ngo Zui Kan, *Operator Methods in Linear Hydrodynamics: Evolutionary and Spectral Problems* (in Russian), Nauka, Moscow, 1989.

175. N. D. Kopachevskii, L. D. Orlova and Yu. S. Pashkova, *Differential-operator and integro-differential equations in the problem of small oscillations of*

hydrodynamical systems, Uchen. Zap. Simferopol Univ. (1995), no. 2 (41), 96–108.

176. A. G. Kostyuchenko and M. B. Orazov, *The problem on oscillations of an elastic semi-cylinder and self-adjoint quadratic pencils connected with it*, Trudy Sem. I. G. Petrovskogo **6** (1981), 97–146.
177. M. G. Krein and H. Langer, *To the theory of quadratic pencils of self-adjoint operators*, Dokl. Akad. Nauk SSSR **154** (1964), no. 6, 1258–1261.
178. M. G. Krein and H. Langer, *On some mathematical principles of the linear theory of damped vibrations of a continuum* (in Russian), in: Applications of Theory of Functions in Mechanics of Solid Medium (Proc. Int. Symp.), Vol. 2, Nauka, Moscow, 1965, pp. 283–322.
179. S. G. Krein, *Linear Differential Equations in Banach Space* (transl. from the Russian), Amer. Math. Soc., Providence, R.I., 1971.
180. S. G. Krein and M. I. Khazan, *Differential equations in Banach space*, Itogi Nauki i Tekhniki. Matematicheskii Analiz **21** (1983), 130–264.
181. S. G. Krein, G. I. Laptev and G. A. Tsvetkova, *On well-posedness by Hadamard of the Cauchy problem for an evolutionary equation*, Dokl. Akad. Nauk SSSR **192** (1970), no. 5, 980–983.
182. R. Labbas, *Ellipticity conditions and solution of a complete abstract second-order differential equation*, Boll. Un. Mat. Ital. A(7) **8** (1994), no. 3, 413–424.
183. G. E. Ladas and V. Lakshmikantham, *Differential Equations in Abstract Spaces*, Academic Press, New York, 1972.
184. J. E. Lagnese, *On equations of evolution and parabolic equations of higher order in t*, J. Math. Anal. Appl. **32** (1970), no. 1, 15–37.
185. P. Lancaster, *Theory of Matrices*, Academic Press, New York, 1969.
186. P. Lancaster, A. S. Markus and V. I. Matsaev, *Definitizable operators and quasihyperbolic operator polynomials*, J. Functional Analysis **131** (1995), no. 1, 1–28.
187. P. Lancaster and A. A. Shkalikov, *Damped vibrations of beams and related spectral problems*, Canad. Appl. Math. Quart. **2** (1994), no. 1, 45–90.
188. P. Lancaster, A. A. Shkalikov and Q. Ye, *Strongly definitizable linear pencils in Hilbert space*, Integr. Equat. Oper. Theory **17** (1993), 338–360.
189. P. Lancaster and Q. Ye, *Definitizable hermitian matrix pencils*, Aequationes Math. **46** (1993), 44–55.
190. H. Langer, *Spectraltheorie Linearer Operatoren in J-Räumen und einige Anwendungen auf die Schar $L(\lambda) = \lambda^2 I + \lambda B + C$*, Habilitationsschrift, Tech. Univ. Dresden, 1964.
191. H. Langer, *Über stark gedämpfte Scharen in Hilbertraum*, J. Math. Mech. **17** (1968), 685–705.
192. H. Langer, *Über eine Klasse polynomialer Scharen selbstadjungierter Operatoren in Hilbertraum*, J. Funct. Anal. **12** (1973), 13–29.

193. H. Langer, *Factorization of operator pencils*, Acta Sci. Math. **13** (1976), 83–96.

194. E. A. Larionov, *Spectral properties of quasihyperbolic pencils*, Sibirsk. Mat. Zh. **32** (1991), no. 5, 182–186.

195. I. Lasiecka, *Unified theory for abstract parabolic boundary problems — a semigroup approach*, Appl. Math. Optim. **6** (1980), no. 4, 287–333.

196. I. Lasiecka, *Existence and uniqueness of the solutions to second order abstract equations with nonlinear and nonmonotone boundary conditions*, Nonlinear Anal. **23** (1994), no. 6, 797–823.

197. I. Lasiecka, J.-L. Lions and R. Triggiani, *Nonhomogeneous boundary-value problems for second-order hyperbolic operators*, J. Math. Pures Appl. **65** (1986), 149–172.

198. I. Lasiecka and R. Triggiani, *Feedback semigroups and cosine operators for boundary feedback parabolic and hyperbolic equations*, J. Differential Equations **47** (1983), 246–271.

199. I. Lasiecka and R. Triggiani, *Finite rank, relatively bounded perturbations of C_0-semigroups, Part I: Well-posedness and boundary feedback of hyperbolic dynamics*, Ann. Scuola Norm. Sup. Pisa (4) **12** (1985), no. 4, 641–668.

200. I. Lasiecka and R. Triggiani, *Finite rank, relatively bounded perturbations of C_0-semigroups, Part II: Spectrum and Riesz basis assignment with applications to feedback systems*, Ann. Mat. Pura Appl. IV, CXLIII (1987), 47–100.

201. I. Lasiecka and R. Triggiani, *A lifting theorem for the time regularity of solutions to abstract equations with unbounded operators and applications to hyperbolic equations*, Proc. Amer. Math. Soc. **104** (1988), no. 3, 745–755.

202. R. de Laubenfels, *Incomplete iterated Cauchy problems*, J. Math. Anal. Appl. **168** (1992), no. 2, 552–579.

203. R. de Laubenfels, *C-semigroups and the Cauchy problem*, J. Functional Analysis **111** (1993), 44–61.

204. P. D. Lax, *Operator theoretic treatment of hyperbolic equations*, Bull. Amer. Math. Soc. **58** (1952), no. 2, 182.

205. P. D. Lax and R. S. Phillips, *Local boundary conditions of dissipative systems of linear partial differential operators*, Comm. Pure Appl. Math. **13** (1960), 427–455.

206. P. D. Lax and R. S. Phillips, *Scattering Theory*, Academic Press, New York, 1967.

207. J. R. Leigh, *Functional Analysis and Linear Control Theory*, Academic Press, London-New York, 1980.

208. A. V. Likhtarnikov, *Criteria of absolute stability of nonlinear operator equations*, Izv. Akad. Nauk SSSR, Mathematics **41** (1977), no. 5, 1064–1083.

209. J.-L. Lions, *Une remarque sur les applications du théorème de Hille-Yosida*, J. Math. Soc. Japan **9** (1957), 62–70.

210. J.-L. Lions, *Les semi-groupes distributions*, Portug. Math. **19** (1960), 3–4, 141–161.

211. J.-L. Lions, *Equations Différentielles Opérationelles et Problèmes aux Limites*, Springer, Berlin-New York, 1961.

212. J.-L. Lions, *Optimal Control of Systems Governed by Partial Differential Equations*, Springer, Berlin-New York, 1971.

213. J.-L. Lions, *Some aspects of the theory of linear evolution equations*, in: Boundary Value Probl. Linear Evolut. Partial Differential Equations (Proc. NATO Adv. Study Inst., Liege, 1976), Dordrecht-Boston, 1977, pp. 175–238.

214. J.-L. Lions, *Controlabilité Exacte des Systèmes Distribués*, Masson, Paris, 1980.

215. J.-L. Lions and E. Magenes, *Nonhomogeneous Boundary Value Problems and Applications*, Vols. 1, 2, Springer, Berlin-New York, 1972.

216. W. G. Litvinov, *A problem on non-steady flow of a nonlinear viscous fluid in a deformable pipe*, Funct. Anal. and Topology (1995), no. 4.

217. G. Lumer and R. S. Phillips, *Dissipative operators in a Banach space*, Pacific J. Math. **11** (1961), 679–698.

218. A. Lunardi, *Neumann boundary stabilization of structurally damped time periodic wave and plate equations*, in: Differential Equations with Applications in Biology, Physics, and Egineering (Proc. Leibnitz, 1989), Marcel Dekker, New York, 1991, pp. 241–257.

219. D. Lutz, *Strongly continuous operator cosine functions*, in: Functional Analysis (Proc. Dubrovnik, 1981), Lecture Notes in Math. **948**, Springer, Berlin-New York, 1982, pp. 73–97.

220. Yu. I. Lyubich, *Classical and local Laplace transform in the abstract Cauchy problem*, Uspekhi Mat. Nauk **21** (1966), no. 3, 3–51.

221. A. S. Markus, *Introduction to the Spectral Theory of Linear Operator Pencils* (transl. from the Russian), Amer. Math. Soc., Providence, R.I., 1988.

222. V. N. Maslennikova, *Solution in an explicit form of the Cauchy problem for one system of partial differential equations*, Izv. Akad. Nauk SSSR, Mathematics **22** (1958), no. 1, 135–160.

223. P. Massat, *Limiting behavior for strongly damped nonlinear wave equations*, J. Differential Equations **48** (1983), 334–349.

224. M. P. Matos and D. C. Pereira, *On a hyperbolic equation with strong damping*, Funkcial. Ekvac. **34** (1991), no. 2, 303–311.

225. V. G. Maz'ya and B. A. Plamenevskii, *On asymtotics of solutions of differential equations with operator coefficients*, Dokl. Akad. Nauk SSSR **196** (1971), 512–515.

226. I. V. Mel'nikova, *Family of M, N operator-functions and second order equations in Banach space*, Izv. Vuzov. Matem. (1985), no. 2, 45–52.

227. I. V. Mel'nikova, *A Miyadera-Feller-Phillips theorem for a complete second order equation in Banach space*, Izv. Vuzov. Matem. (1985), no. 4, 34–40.

228. I. V. Mel'nikova and A. I. Filinkov, *A connection between well-posedness of the Cauchy problem for an equation and a system in Banach space*, Dokl. Akad. Nauk SSSR, **300** (1988), 280–284.

229. I. V. Mel'nikova and A. I. Filinkov, *On classification by boundary-value problems of complete second order equations in Banach spaces*, Izv. Vuzov. Matem. (1990), no. 6, 39–45.

230. I. V. Mel'nikova and A. I. Filinkov, *An integrated family of M and N functions*, Dokl. Akad. Nauk **326** (1992), no. 1, 35–39.

231. I. V. Mel'nikova and A. I. Filinkov, *Integrated semigroups and C-semigroups. Well-posedness and regularization of differential-operator problems*, Uspekhi Mat. Nauk **49** (1994), no. 6, 111–150.

232. A. I. Miloslavskii, *To the justification of the spectral approach in nonconservative problems of the theory of elastic stability*, Function. Anal. Prilozh. **17** (1983), no. 3, 83–84.

233. A. I. Miloslavskii, *On stability of some classes of evolutionary equations*, Sibirsk. Mat. Zh. **26** (1985), no. 5, 118–132.

234. A. I. Miloslavskii, *On destabilizing effect of small damping on abstract nonconservative systems*, Uspekhi Mat. Nauk **41** (1986), no. 1, 189–190.

235. A. I. Miloslavskii, *On operator differential equations related to oscillations of a viscoelastic medium*, Dokl. Akad. Nauk SSSR **309** (1989), no. 3.

236. A. I. Miloslavskii, *On the spectrum of instability of an operator pencil*, Mat. Zametki **49** (1991), no. 4, 88–94.

237. S. S. Mirzoev, *On multiple completeness of root vectors of polynomial operator pencils corresponding to boundary-value problems on the semiaxis*, Function. Anal. Prilozh. **17** (1983), no. 2, 84–85.

238. S. S. Mirzoev, *Correct solvability of boundary-value problems for elliptic equations in a Hilbert space*, Dokl. Akad. Nauk Azerb. SSR **46** (1990), no. 1, 17–20.

239. B. S. Mityagin, *Differential equations with a small parameter in Banach space*, Izv. Akad. Nauk Azerb. SSR, Ser. Fiz. Tehn. Mat. Nauk (1961), no. 1, 23–38.

240. I. Miyadera, S. Oharu and N. Okazawa, *Generation theorems of semigroups of linear operators*, Publs. Res. Inst. Math. Sci. **8** (1973), no. 3, 509–555.

241. S. Mizohata, *The Theory of Partial Differential Equations*, Cambridge Univ. Press, Cambridge, 1973.

242. A. A. Movchan, *On one problem of stability of a pipe conveying fluid*, Prikl. Matem. Mekh. (1965), no. 4, 760–762.

243. R. Nagel (ed.), *One-parameter Semigroups of Positive Operators*, Lecture Notes in Math. **1184**, Springer, Berlin-New York, 1986.

244. R. Nagel, *On the stability of positive semigroups generated by operator matrices*, in: Complex Geometry and Analysis (Pisa, 1988), Lecture Notes in Math. **1422**, Springer, Berlin-New York, 1990, 77–83.

245. M. A. Naimark, *Normed Algebras* (transl. from the Russian), Nordhoff, Leyden, 1972.

246. S. Nemat-Nasser, S. N. Prasad and G. Herrmann, *Destabilizing effect of velocity-dependent forces in non-conservative continuous systems*, American Institute of Aeronautics and Astronautics Journal **4** (1966), 1276–1280.

247. F. Neubrander, *On well-posedness of abstract Cauchy problem*, Semigroup Forum **10** (1984), 375–385.

248. F. Neubrander, *Well-posedness of higher order abstract Cauchy problems*, Trans. Amer. Math. Soc. **295** (1986), 257–290.

249. F. Neubrander, *Integrated semigroups anf their application to the abstract Cauchy problem*, Pac. J. Math. **135** (1988), 111–157.

250. F. Neubrander, *Integrated semigroups and their applications to complete second order Cauchy problems*, Semigroup Forum **38** (1989), 233–251.

251. E. Obrecht, *Sul problema di Cauchy per le equazioni paraboliche astratte di ordine n*, Rend. Sem. Math. Univ. Padova **53** (1975), 231–256.

252. E. Obrecht, *Evolution operators for higher order abstract parabolic equations*, Czech. Math. J. **36** (111) (1986), 210–222.

253. E. Obrecht, *Second-order abstract Cauchy problems*, in: Semesterbericht Functional-analysis Tübingen, summer semester 1986, pp. 191–206.

254. E. Obrecht, *The Cauchy problem for time-dependent abstract parabolic equations of higher order*, J. Math. Anal. Appl. **125** (1987), no. 2, 508–530.

255. E. Obrecht, *Phase space for an n-th order differential equation in Banach space*, in: Semigroup Theory and Evolution Equations (Proc. Delft, 1989), Lecture Notes in Pure and Appl. Math. **134**, Marcel Dekker, New York, 1991, pp. 391–399.

256. S. Oharu, *Semigroups of linear operators in a Banach space*, Publs. Res. Inst. Math. Sci. **7** (1971), no. 2, 205–260.

257. N. Okazawa and N. Takenaka, *Wellposedness of abstract Cauchy problems for second order differential equations*, Israel J. Math. **69** (1990), no. 3, 257–288.

258. N. Okazawa and N. Tanaka, *Local C-semigroups and local integrated semigroups*, Proc. London Math. Soc. (3) **61** (1990), no. 1, 63–90.

259. M. B. Orazov, *On completeness of elementary solutions for some operator differential equations on a semiaxis and a segment*, Dokl. Akad. Nauk SSSR **245** (1979), no. 4, 788–792.

260. M. B. Orazov, *On completeness of elementary solutions in a problem on vibrations of an elastic rectangle plate*, Function. Anal. Prilozh. **14** (1980), no. 1, 77–78.

261. V. P. Orlov, *On the stability of the zero solution of a one-dimensional mathematical model of viscoelasticity*, Dif. Integr. Equations **4** (1991), no. 1, 89–101.

262. V. P. Orlov and P. E. Sobolevskii, *On mathematical models of viscoelasticity with memory*, Dif. Integr. Equations **4** (1991), no. 1, 103–115.

263. G. D. Orudzhev, *Polynomial approximation of solutions of operator-differential equations*, Ukrain. Math. J. **45** (1993), no. 3, 440–442.

264. G. D. Orudzhev, *On solvability of boundary-value problems for an abstract polyharmonic equation*, Dokl. Akad. Nauk **334** (1994), no. 3, 281–283.

265. B. G. Pachpatte, *On abstract second order differential equations*, Demonstratio Math. **23** (1990), no. 2, 357–366.

266. M. P. Paidoussis, *Dynamics of tubular cantilevers conveying fluid*, J. Mechanical Engineering Science **12** (1970), 85–103.

267. M. P. Paidoussis and N. T. Issid, *Dynamic stability of pipes conveying fluid*, J. Sound and Vibration **33** (1974), no. 3, 267–294.

268. A. Pazy, *Semigroups of Linear Operators and Applications to Partial Differential Equations*, Springer-Verlag, Berlin-Heidelberg-New York, 1986.

269. R. S. Phillips, *A note on the abstract Cauchy problem*, Proc. Nat. Acad. Sci. USA **40** (1954), no. 4, 244–248.

270. R. S. Phillips, *Dissipative operators and hyperbolic systems of partial differential equations*, Trans. Amer. Math. Soc. **90** (1959), 193–254.

271. R. S. Phillips, *On the integration of the diffusion equation with boundaries*, Trans. Amer. Math. Soc. **98** (1961), 62–84.

272. V. N. Pivovarchik, *A boundary-value problem connected with the oscillation of elastic beams with internal and viscous damping*, Vestn. Moscow State Univ. Ser 1 (1987), no. 3, 68–71.

273. V. N. Pivovarchik, *On vibration of a semi-infinite beam with internal and external damping*, Prikladn. Matem. Mekh. **52** (1988), no. 5, 829–836.

274. V. N. Pivovarchik, *On the spectrum of quadratic operator pencils in the right half-plane*, Mat. Zametki **45** (1989), no. 6, 101–103.

275. V. N. Pivovarchik, *On total algebraic multiplicity of the spectrum in the right half-plane for one class of quadratic operator pencils*, Algebra i Analiz **3** (1991), no. 2, 223–230.

276. V. N. Pivovarchik, *A necessary condition of gyroscopic stabilization in one problem of mechanics*, Mat. Zametki **53** (1993), no. 6, 89–96.

277. B. A. Plamenevskii, *On existence and asymptotics of solutions of differential equations with unbounded operator coefficients in Banach space*, Izv. Akad. Nauk SSSR, Mathematics **36** (1972), no. 6, 1348–1401.

278. A. G. Postnikov, *The Tauberian theory and its applications*, Trudy Mat. Inst. Akad. Nauk SSSR **144** (1979), 1–148.

279. I. I. Privalov, *Boundary Properties of Single-Valued Analytic Functions* (in Russian), Gostehizdat, Moscow-Leningrad, 1950.

280. Ya. V. Radyno and N. I. Yurchuk, *The Cauchy problem for certain abstract even order hyperbolic equations*, Differents. Uravnen. **12** (1976), 331–342.

281. G. V. Radzievskii, *On uniqueness of solutions of boundary-value problems on a finite interval and a half-line for operator differential equations*, Ukrain. Math. J. **46** (1994), no. 3, 279–292.

282. M. B. Ragimov, *Operator pencils and almost periodicity of bounded solutions of second order differential equations*, Dokl. Akad. Nauk SSSR **314** (1990), no. 5, 1079–1082.

283. V. G. Raskin, *On solvability of arbitrary order differential-operator equations*, Differents. Uravnen. **12** (1976), 362–364.

284. V. G. Raskin and P. E. Sobolevskii, *The Cauchy problem for second order differential equations in Banach spaces*, Sibir. Mat. Zh. **8** (1967), no. 1, 70–90.

285. V. G. Raskin and P. E. Sobolevskii, *Expansions of solutions of second order differential equations in Banach spaces in series of exponential solutions*, Differents. Uravnen. **5** (1969), 543–545.

286. M. Reed, *Abstract Non-Linear Wave Equations*, Lecture Notes in Math. **507**, Springer, Berlin-New York, 1975.

287. M. Reed and B. Simon, *Methods of Modern Mathematical Physics, Vol. 1. Functional Analysis*, Academic Press, New York-London, 1972.

288. M. Reed and B. Simon, *Methods of Modern Mathematical Physics*, Vols. 2, 3, 4, Academic Press, New York-London, 1975–1978.

289. M. Reissig, *An abstract Cauchy-Covalevsky problem of higher order*, Math. Nachr. **150** (1991), 25–39.

290. F. Riesz, *Über die Randwerte einer analytischen Function*, Math. Z. **18** (1922), 87–95.

291. F. Riesz and B. Szekefelvy-Nagy, *Functional Analysis* (transl. from the French), Ungar, New York, 1955.

292. L. Rodman, *An Introduction to Operator Polynomials*, Birkhäuser, Basel, 1989.

293. V. K. Romanko, *Boundary-value problems for a class of differential operators*, Differents. Uravnen. **10** (1974), 117–131.

294. V. K. Romanko, *Boundary-value problems for certain differential-operator equations*, Dokl. Akad. Nauk SSSR **227** (1976), 812–815.

295. V. K. Romanko, *Unique solvability of boundary-value problems for certain differential-operator equations*, Differents. Uravnen. **13** (1977), 324–335.

296. V. K. Romanko, *Solvability of boundary-value problems for higher order differential-operator equations*, Differents. Uravnen. **14** (1978), 1081–1092.

297. D. L. Russel, *Controllability and stabilizability theory for linear partial differential equations: recent progress and open questions*, SIAM Rev. **20** (1978), 639–740.

298. D. L. Russel, *A general framework for the study of indirect damping mechanisms in elastic systems*, J. Math. Anal. Appl. **173** (1993), no. 2, 339–358.

299. Yu. S. Samoilenko, *Spectral Theory of Families of Self-Adjoint Operators* (transl. from the Russian), Kluwer Academic Publishers, Dordrecht, 1991.

300. J. T. Sandefur, *Higher order abstract Cauchy problems*, J. Math. Anal. Appl. **60** (1977), 728–742.

301. J. T. Sandefur, *Existence and uniqueness of solutions of second order nonlinear differential equations*, SIAM J. Math. Anal. **14** (1983), 477–487.

302. N. Sanekata, *Some remarks on the abstract Cauchy problem*, Publs. Res. Inst. Math. Sci. **11** (1975), no. 1, 51–65.

303. A. Y. Schoene, *Semigroups and a class of singular perturbation problems*, Indiana Univ. Math. J. **20** (1970), 247–263.

304. *Semigroups of Linear and Nonlinear Operators and Applications* (Proc. Curacao, 1992), ed. by G. R. Goldstein and J. A. Goldstein, Kluwer, Dordrecht, 1993.

305. S.-Y. Shaw, *Ergodic projections of continuous and discrete semigroups*, Proc. Amer. Math. Soc. **78** (1980), no. 1, 69–76.

306. E. I. Shemyakin, *Propagation of non-steady perturbations in a viscoelastic medium*, Dokl. Akad. Nauk SSSR **104** (1955), no. 1, 34–37.

307. A. A. Shkalikov, *Operator differential equations on the semiaxis and related spectral problems for self-adjoint operator pencils*, Dokl. Akad. Nauk SSSR **276** (1984), no. 2, 309–314.

308. A. A. Shkalikov, *Strongly damped operator pencils and solvability of associated operator differential equations*, Matem. Sbornik **135** (1988), no. 1, 96–118.

309. A. A. Shkalikov, *Elliptic equations in Hilbert space and associated spectral problems*, Trudy Sem. I. G. Petrovskogo **14** (1989), 140–224.

310. A. Ya. Shklyar, *Completeness of generalized eigen- and root vectors of certain classes of linear nonselfadjoint operators*, Dokl. Akad. Nauk Ukrain. SSR, Ser A (1987), no. 5, 30–33.

311. A. Ya. Shklyar, *The Cauchy Problem for Differential Equations with Commuting Normal Operator Coefficients* (in Russian), Thesis, Inst. Math. Acad. Sci. Ukrain. SSR (1990).

312. A. Ya. Shklyar, *Joint spectrum of commuting self-adjoint operators and criteria for well-posedness and stability for differential-operator equations*, Ukrain. Math. J. **43** (1991), 406–414.

313. A. Ya. Shklyar, *Well-posedness of the Cauchy problem for differential equations with anti-commuting self-adjoint operator coefficients*, in: Application

of Methods of Functional Analysis in Mathematical Physics (in Russian), Inst. Math. Acad. Sci. Ukraine (1991), 110–119.

314. A. Ya. Shklyar, *A criterion for unique solvability of the Cauchy problem for a fourth order differential-operator equation*, in: Boundary-Value Problems for Differential-Operator Equations (in Russian), Inst. Math. Acad. Sci. Ukraine (1991), 109–120.

315. A. Ya. Shklyar, *Well-posedness of the Cauchy problem for higher order trinomial differential-operator equations*, Ukrain. Math. J. **45** (1993), 704–714.

316. A. Ya. Shklyar, *Criteria for well-posedness and weak well-posedness of the Cauchy problem for complete second order linear differential equations in Hilbert spaces*, preprint 93.3, Inst. Math. Nat. Acad. Sci. Ukraine, Kiev, 1993.

317. R. Showalter, *Hilbert Space Methods for Partial Differential Equations*, Pitman, London, 1973.

318. Yu. T. Sil'chenko, *Differential equations in Banach space and parabolic equations in spaces of smooth functions*, Thesis, Voronezh State University, 1978.

319. Yu. T. Sil'chenko, *On an evolutionary equation with an operator generating a nonanalytic semigroup*, Differents. Uravneniya **15** (1979), no. 2, 363–366.

320. E. Skudrzyk, *Simple and Complex Vibratory Systems*, Pennsylvania State Univ. Press, 1968.

321. E. P. Sobolevskii and P. E. Sobolevskii, *The commutant method and the coercitivity of the Cauchy problem*, Dokl. Akad. Nauk SSSR **316** (1991), no. 4, 825–829.

322. P. E. Sobolevskii, *On second order differential equations in Banach space*, Dokl. Akad. Nauk SSSR **146** (1962), 774–777.

323. P. E. Sobolevskii, *On one type of second order differential equations in Banach space*, Uchen. Zap. Azerb. Gos. Univ. (1962), no. 3, 87–106.

324. P. E. Sobolevskii, *On second order equations with a small parameter on the higher order derivative*, Uspekhi Mat. Nauk **19** (1964), no. 6, 217–219.

325. P. E. Sobolevskii, *Coercitiveness inequalities for abstract parabolic equations*, Dokl. Akad. Nauk SSSR **157** (1964), no. 1, 52–55.

326. P. E. Sobolevskii, *Investigation of a mathematical model of thermoelasticity*, Dif. Int. Equations **3** (1990), no. 4, 643–652.

327. M. Sova, *Cosine operator functions*, Rozprawy Mat. **49** (1966), 1–46.

328. M. Sova, *Problème de Cauchy pour équations hyperboliques opérationelles à coefficients constants non-bornés*, Ann. Scuola Norm. Sup. Pisa **22** (1968), 67–100.

329. M. Sova, *Equations différentielles opérationelles linéaires du second ordre à coefficients constants*, Rozprawy Mat. **80** (1970), no. 7, 1–80.

330. M. Sova, *Linear differential equations in Banach spaces*, Rozprawy Mat. **85** (1975), no. 6, 1–82.

331. M. Sova, *On the Timoshenko type equations*, Cas. Pestov. Mat. **100** (1975), no. 3, 217–254.

332. M. Sova, *Linear differential equations in Hilbert spaces*, Rozprawy Mat. **91** (1981), no. 4, 1–63.

333. J. V. Strutt (Lord Rayleigh), *The Theory of Sound*, Vol. 2.

334. H. Tanabe, *Equation of Evolution*, Pitman, London, 1979.

335. H. Tanabe, *On fundamental solutions of linear parabolic equations of higher order in time and associated Volterra equations*, J. Differential Equations **73** (1988), 288–308.

336. J. L. Taylor, *A joint spectrum for several commuting operators*, J. Funct. Anal. **6** (1970), 172–191.

337. J. L. Taylor, *The analytic-functional calculus for several commuting operators*, Acta Math. **125** (1970), 1–38.

338. J. L. Taylor, *A general framework for a multi-operator functional calculus*, Adv. Math. **9** (1972), 183–252.

339. J. L. Taylor, *Foundations of several noncommuting variables*, Bull. Amer. Math. Soc. **79** (1973), 1–34.

340. R. Teman, *Infinite-dimensional Dynamical Systems in Mechanics and Physics*, Springer, New York, 1988.

341. W. Thompson (Lord Kelvin) and P. Tait, *Treatise on Natural Philosophy, Part 1*, Cambridge Univ. Press, Cambridge, 1921.

342. S. P. Timoshenko, *On the correction for shear of the differential equation for transverse vibrations of prismatic bars*, Phil. Mag. **41**, Ser. 6 (1921), 744–746.

343. S. P. Timoshenko, *The Theory of Elasticity*.

344. O. Toeplitz, *Über allgemeine lineare Mittelbildungen*, Prace Mat. Fiz. **22** (1911), 113–120.

345. R. Triggiani, *Improvement of stability properties of hyperbolic damped wave equations via boundary feedback*, in: Distributed Parameter Systems, Springer, New York, 1985, pp. 400–409.

346. R. Triggiani, *Finite rank, relatively bounded perturbations of semigroups generators, Part III: A sharp result on the lack of uniform stabilization*, Dif. Integr. Equations **3** (1990), 503–522.

347. R. Triggiani, *Regularity of some structurally damped problems with point control and with boundary control*, J. Math. Anal. Appl. **161** (1991), no. 2, 299–331.

348. Ya. S. Uflyand, *Propagation of waves in transverse vibrations of beams and plates*, Prikl. Matem. Mekh. **12** (1948), 287–300.

349. Yu. A. Ustinov and V. I. Yudovich, *On completeness of elementary solutions of the biharmonic equation in a half-strip*, Prikl. Matem. Mekh. **37** (1973), no. 4, 706–714.

350. Yu. N. Valitskii, *Well-posedness of a multi-point problem for an equation with operator coefficients*, Sibirsk. Mat. Zh. **29** (1988), no. 4, 547–555.

351. F.-H. Vasilescu, *Multi-dimensional Analytic Functional Calculus* (in Romanian), Edit. Acad., Bucuresti, 1979.

352. V. V. Vasil'ev, S. G. Krein and S. I. Piskarev, *Operator semigroups, cosine operator functions and linear differential equations*, Itogi Nauki i Tekhniki. Matematicheskii Analiz **28** (1990), 87–202.

353. S. S. Voit, *Propagation of initial condensations in a viscous gas*, Uch. Zap. Moscow State Univ., Mechanics, **5**, no. 172 (1954), 125–142.

354. G. F. Webb, *Existence and asymptotic behavior for a strongly damped nonlinear wave equation*, Canad. J. Math. **32** (1980), 631–643.

355. J. Weidmann, *Linear Operators in Hilbert Space*, Springer, Berlin-New York, 1980.

356. B. Weiss, *Abstract vibrating systems*, J. Math. Mech. **17** (1967), 241–255.

357. Xiao Ti Jun, *On the well-posedness of the Cauchy problem for a kind of complete second order linear differential equation* (in Chinese), Sichuan Univ. Nat. Sci. Ed. **25** (1988), 421–428.

358. Xiao Ti Jun and Liang Jin, *On complete second order linear differential equations in Banach spaces*, Pacific J. Math. **142** (1990), 175–196.

359. Xiao Ti Jun and Liang Jin, *A note on the propagators of second order linear differential equations in Hilbert spaces*, Proc. Amer. Math. Soc. **113** (1991), no. 3, 663–667.

360. Xiao Ti Jun and Liang Jin, *Analyticity of the propagators of second order linear differential equations in Banach spaces*, Semigroup Forum **44** (1992), no. 3, 356–363.

361. Xiao Ti Jun and Liang Jin, *Complete second order linear differential equations with almost periodic solutions*, J. Math. Anal. Appl. **163** (1992), no. 1, 136–146.

362. Xiao Ti Jun and Liang Jin, *Parabolicity of a class of higher order abstract differential equations*, Proc. Amer. Math. Soc. **120** (1994), no. 1, 173–181.

363. S. Ya. Yakubov, *On solvability of the Cauchy problem for evolutionary equations*, Dokl. Akad. Nauk SSSR **156** (1964), 1041–1044.

364. S. Ya. Yakubov, *On the Cauchy problem for second order differential equations in Banach space*, Dokl. Akad. Nauk SSSR **168** (1966), 759–762.

365. S. Ya. Yakubov, *Higher order differential equations with variable unbounded operators in Banach space*, Izv. Akad. Nauk Azerb. SSR, Ser. Fiz. Tehn. Mat. Nauk (1966), no 1, 20–27.

366. S. Ya. Yakubov, *Investigation of the Cauchy problem for differential equations of hyperbolic type in Banach space*, Dokl. Akad. Nauk Azerb. SSR **22** (1966), no. 4, 3–7.

367. S. Ya. Yakubov, *Uniform well-posedness of the Cauchy problem for evolutionary equations and applications*, Funkts. Anal. Prilozh. **4** (1970), no. 3, 86–94.

368. S. Ya Yakubov, *The uniformly well-posed Cauchy problem for abstract hyperbolic equations*, Izv. Vuzov. Matem. (1970), no. 12, 108–113.

369. S. Ya. Yakubov, *Solvability of the Cauchy problem for abstract quasilinear hyperbolic equations of second order and their applications*, Trudy Moscow Math. Soc. **23** (1970), 37–60.

370. S. Ya. Yakubov, *Solution of mixed problems for partial differential equations by an operator method*, Dokl. Akad. Nauk SSSR **196** (1971), 545–548.

371. S. Ya. Yakubov, *Well-posedness of a boundary-value problem for second order linear evolutionary equations*, Izv. Akad. Nauk Azerb. SSR, Ser. Fiz. Tehn. Mat. Nauk (1973), no. 2, 37–42.

372. S. Ya. Yakubov, *Linear Differential-Operator Equations and Their Applications* (in Russian), Elm, Baku, 1985.

373. S. Ya. Yakubov, *Multiple completeness for a system of operator pencils and elliptic boundary value problems*, Mat. Sbornik **181** (1990), no. 1, 95–113.

374. S. Ya. Yakubov, *Fredholm property of elliptic boundary value problems for partial differential and differential-operator equations*, Results Math. **24** (1993), no. 3–4, 372–388.

375. K. Yosida, *Functional Analysis*, Springer-Verlag, Berlin-Heidelberg-New York, 1965.

376. M. Yu. Yunusov, *Some boundary-value problems containing a small parameter*, Izv. Vuzov. Matem. (1992), no. 12, 67–79

377. N. I. Yurchuk, *On boundary-value problems for equations containing operators of the form* $\frac{d^{2m+1}}{dt^{2m+1}} + A$ *in the main part*, Differents. Uravnen. **10** (1974), 759–762.

378. N. I. Yurchuk, *On boundary-value problems for equations with operators of the form* $\frac{d^{2m}}{dt^{2m}} + A$, Differents. Uravnen. **10** (1974), 950–952.

379. N. I. Yurchuk, *A priori estimates for solutions of boundary-value problems for certain differential-operator equations*, Differents. Uravnen. **12** (1976), 729–739.

380. N. I. Yurchuk, *Solvability of boundary-value problems for certain differential-operator equations*, Differents. Uravnen. **13** (1977), 626–636.

381. S. Zaidman, *Differential Equations in Hilbert Spaces*, Pitman, London, 1979.

382. S. Zaidman, *Topics in Abstract Differential Equations*, Longman, Harlow, 1994.

383. E. E. Zaiac, *The Kelvin-Tait-Chetaev theorem and extensions*, J. Astronaut. Sci. **11** (1964), no. 2, 46–49.

384. V. N. Zefirov, V. V. Kolesov and A. I. Miloslavskii, *An investigation of fundamental frequences of a rectilinear pipe*, Izv. Akad. Nauk SSSR, Solid Mechanics (1985), no. 1, 179–188.

385. Zheng Quan, *A Hille-Yosida theorem for the higher-order abstract Cauchy problem*, Bull. London Math. Soc. **24** (1992), no. 6, 531–539.

386. Zheng Quan, *Applications of integrated semigroups to higher order abstract Cauchy problems*, Systems Sci. Math. Sci. **5** (1992), no. 4, 316–327.

387. Zheng Quan, *Higher order abstract Cauchy problems with N-closedness*, Math. Japon. **38** (1993), no. 3, 531–539.

388. Zheng Quan, *Strongly continuous M, N families of bounded operators*, Integr. Equat. Oper. Theory **19** (1994), no. 1, 105–119.

389. H. Ziegler, *Principles of Structural Stability*, Blaisdell, Waltham, Massachusetts, 1968.

Index

List of Symbols

Titles previously published in the series

OPERATOR THEORY: ADVANCES AND APPLICATIONS

BIRKHÄUSER VERLAG

Edited by **I. Gohberg,**

School of Mathematical Sciences, Tel-Aviv University, Ramat Aviv, Israel

80. **I. Gohberg, H. Langer** (Eds): Operator Theory and Boundary Eigenvalue Problems International Workshop in Vienna, July 27-30, 1993, 1995 (ISBN 3-7643-5259-0)

81. **H. Upmeier**: Toeplitz Operators and Index Theory in Several Complex Variables, 1995 (ISBN 3-7643-5280-5)

82. **T. Constantinescu**: Schur Parameters, Factorization and Dilation Problems, 1996 (ISBN 3-7643-5285-X)

83. **A.B. Antonevich**: Linear Functional Equations. Operator Approach, 1995 (ISBN 3-7643-2931-9)

84. **L.A. Sakhnovich**: Integral Equations with Difference Kernels on Finite Intervals, 1996 (ISBN 3-7643-5267-1)

85. **Y.M. Berezansky, G.F. Us, Z.G. Sheftel**: Functional Analysis, **Vol. I**, 1996 (ISBN 3-7643-5344-9)

86. **Y.M. Berezansky, G.F. Us, Z.G. Sheftel**: Functional Analysis, **Vol. II**, 1996 (ISBN 4-7643-5345-7)

87. **I. Gohberg / P. Lancaster / P.N. Shivakumar** (Eds): Recent Developments in Operator Theory and Its Applications, International Conference in Winnipeg, October 2–6, 1994, 1996 (ISBN 3-7643-5413-5)

88. **J. van Neerven** (Ed.): The Asymptotic Behaviour of Semigroups of Linear Operators 1996 (ISBN 3-7643-5455-0)

89. **Y. Egorov / V. Kondratiev**: On Spectral Theory of Elliptic Operators, 1996 (ISBN 3-7643-5390-2)

90. **A. Böttcher / I. Gohberg** (Eds): Singular Integral Operators and Related Topics. Joint German-Israeli Workshop, Tel Aviv, March 1-10, 1995, 1996 (ISBN 3-7643-5466-6)

91. **A.L. Skubachevskii**: Elliptic Functional Differential Equations and Applications, 1997 (ISBN 3-7643-5404-6)

MIX
Papier aus verantwortungsvollen Quellen
Paper from responsible sources
FSC® C105338

If you have any concerns about our products, you can contact us on
ProductSafety@springernature.com

In case Publisher is established outside the EU, the EU authorized representative is:
Springer Nature Customer Service Center GmbH
Europaplatz 3, 69115 Heidelberg, Germany

Printed by Libri Plureos GmbH
in Hamburg, Germany